Management Planning for Natur

A Theoretical Basis & Practical Guid

Mike Alexander

Management Planning for Nature Conservation

A Theoretical Basis & Practical Guide

Second Edition

Mike Alexander
CMSC, Great House Barn
New Street, Talgarth, Brecon
Powys, LD3 0AH
UK

ISBN 978-94-007-5115-6 ISBN 978-94-007-5116-3 (eBook)
DOI 10.1007/978-94-007-5116-3
Springer Dordrecht Heidelberg New York London

Library of Congress Control Number: 2012952136

Printed on acid-free paper

Springer is part of Springer Science+Business Media (www.springer.com)

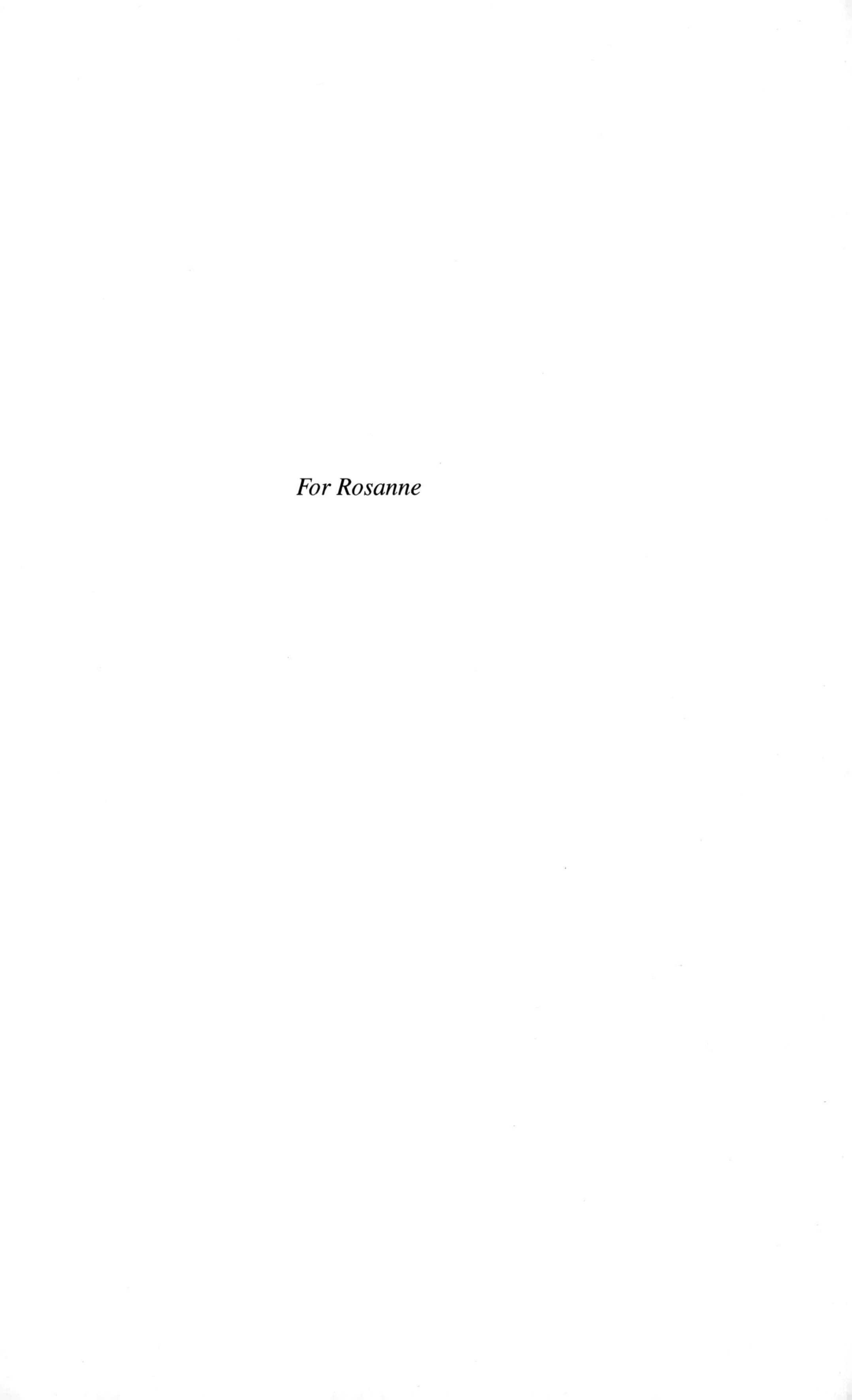

For Rosanne

Foreword

Several things conspire against the serious manager in the world of nature conservation. In the first place, for those inspired by the richness of wildlife and the diversity of their habitats, there can be a temptation to imagine that 'nature knows best' and can manage on its own. Yet, for most of us living and working in Western Europe, and in many other parts of the world, the landscapes that we cherish are the product of long and complex co-habitation between humankind, other animals and plants and the fabric of land that we all share. Much of the nature we wish to conserve then is already the result of some kind of management – or mismanagement, accidental encounter or serendipitous intervention – and needs continuing wise management to sustain it. For many of us, real wilderness is far away and, even for those landscapes that are closer at hand, we can forget that non-intervention, apparently 'leaving nature to itself', is actually another kind of management, or at least a deliberate choice that needs justification and planning, and one that comes as the most recent of probably a long history of interactions. Setting ourselves thoughtfully and responsibly within this continuing process is what this book is all about.

For some (and this is perhaps a particularly English trait), hesitancy about the need to manage is compounded by a resistance to professionalism in the practice of nature conservation as if too much seriousness would be inimical to that sense of wonder which is so important a motivation for many in their relationships with the natural world, and would make some crucial element in their commitment evaporate. Of course, the amateur (a word which means 'enthusiast' or 'devotee' in its origins) has been enormously important in the appreciation of wildlife, habitats and landscapes and in the drive to value and sustain them; and non-professionals go on making a huge contribution to their stewardship and their practical management, through providing volunteer labour, wardening and financial support. But this is no justification at all for amateurism in nature conservation and the plain and uncomplicated language this book uses, its clear structure, its organisation as a planning guide as much as a text book, are all meant to commend professionalism to amateurs as well as to those whose job it is to conserve.

Almost all of the habitats to which we now ascribe nature conservation value and which prompt our concern to sustain them have not been produced by wildlife management at all. They are the heritage of agriculture or other forms of activity that have harvested products from the land, of industrial exploitation or its aftermath, of landscape design, or are the by-product of development or the almost incidental results of long occupancy in which human cultures have made themselves at home. Nature conservation as we know it now, in the twenty-first century, is a young enthusiasm and profession: in fact, it is about as old as those of its most senior practitioners who are still at work today, like me. In Britain, it sprang like us from the commitment to a brave new world during and after the Second World War. Often then, in managing for nature conservation – to sustain the integrity and value of habitats and landscapes, and ensure the survival of the plants and animals that live there and depend on them – we are faced with the challenge of contriving interventions that must mimic activities of socio-economic milieux, of cultures, of sets of dependencies and expectations that are now lost, maybe gone for ever. The fact that the biodiversity of woodlands and meadows, heaths and mires, however attractive we rate it now, however much we think it has intrinsic value, was not the first concern of the generations who have bequeathed these landscapes to us, this is a crucial realisation in thinking what kind of management might be appropriate and necessary today and tomorrow.

Thinking clearly about management is the prime concern of the author of this book and the colleagues among whom he has developed these ideas. The book is meant to stimulate and challenge us to a clear understanding of what we are about in nature conservation. It poses some key questions, it develops some central concepts and principles and it provides a simple structure for management plans, with an instructive range of case studies. It contextualises the planning process within the legislative frame and reviews the ethical questions which we ought to ask about why we value other creatures and their habitats. It helpfully tells us how we might cut short our reading by taking different routes through the chapters. Above all, it makes common sense of what is for many a forbidding subject and what has become for some practitioners a realm of mumbo jumbo.

However, for the author and his colleagues, management planning for nature conservation is only partly a science. Science can tell us much about what species and habitats are like, how they function, what they depend upon and how they might respond should conditions change. For non-scientists (some scientists, too, regrettably), science has a degree of assurance in its analysis and predictions that can make it seem a comforting guarantor, if not of success, at least of our own authority as practitioners. But science in general has its limits in being able to understand the world in which we live, and ecology and those related sciences which aim to comprehend the world of nature have especially challenging material with which to work. In helping us shape our plans and projects in nature conservation, then, we need to appreciate the limits of science. It does not deal in certainties but in probabilities, and management as conceived in this book is necessarily about managing risk, risking failure, learning how to be a little more successful next time, and about adjusting our understanding about what success is. We can thus build our confidence, yet see that we do this by realising the limits of what we know. The notion that management planning for nature conservation is a learning process – iterative, adaptable, and developmental – is thus crucial to the book's aim.

Over the past 20 years or so, I have had the privilege of seeing that process at work as Mike Alexander and his colleagues have continued to learn how to manage the species, habitats and landscapes that are in their care. I have also seen in the author how a personal passion can inspire and craft a life of professional commitment. At a time when biodiversity is all too readily marketed as a commodity, and ecosystems seen as providing socio-economic, health and entertainment services, the chance to share such expertise and wisdom could help us find our own place in nature, where we ourselves belong.

Lancaster
July 2007

John Rodwell

Acknowledgements

Although the process of putting words onto paper took little more than a year, the thinking behind this book has evolved gradually from a lifetime's career in conservation. Inevitably, it was the people I met and worked with that helped these ideas grow and develop from the first thoughts of what conservation management should be to the realisation of what it could be. With some of them I have discussed and argued the detail over many years; for others it was perhaps just a chance remark that cleared some obstacle in the thinking process and allowed me to move forward. There are far too many people to mention, but I would like to thank them all.

More specifically, I would like to express my gratitude to the authors of the five case studies: Doug Oliver, Paul Culyer, Rosanne Alexander, Martin Vernik, Jurij Gulic, David Wheeler, James Perrins and David Mitchell. They all had special skills to bring, which have broadened and elaborated the main text.

The development of this planning system, and my confidence in the end product, is entirely dependent on thorough and rigorous testing. In this context, I must mention, in addition to their contributions to the case studies, David Wheeler and Doug Oliver for having the patience to write and rewrite plans endlessly until eventually we were satisfied with the results.

I would particularly like to thank Tom Hellawell, who has made such a substantial contribution to developing the practice of management planning. It was back in 1990 that Tom produced the first quantified and measurable conservation objectives; today we would not consider doing otherwise.

James Perrins has been a valued friend for almost as long as I can remember, so it was particularly rewarding that we could find a way of working together on the development of CMS. It was James' computer skills that turned the concept of a dynamic plan into reality, and it is his commitment and enthusiasm over many years that have allowed that innovation to continue.

John Rodwell has been central in guiding my understanding of plant communities, and this has played an essential part in developing the planning process, particularly the habitat objectives. Working with John is always a pleasure as well as being extremely informative. Adam Cole-King provided the note on the value of legislation and patiently discussed the issues which helped me to develop the sections on ethics and values. John Bacon, Mike McCabe and Mike Howe generously made time to give helpful comments on various sections of the text. I am also especially indebted to the following for their guidance, support and encouragement over the years: Denis Bellamy, Roger Bray, Richard Clarke, Sally Edmondson, Stephen Evans, Geraint George, Keith Jones, David Mitchell, Andrew Peterken, Tim Reed, Paul Rooney and Ian Tillotson.

I have taught a large number of courses at Plas Tan y Bwlch, the Snowdonia National Park field centre. Each course is different and each student brings a fresh perspective. This interaction with people from so many backgrounds has helped enormously in developing my ideas and understanding how to convey them. I would like to thank all the students who have participated in my courses, as well as the staff at Plas, and particularly Andrew Weir who has supplied such patient and dependable support.

In so many years working first for NCC and then CCW, it has been a privilege to spend time with friends and colleagues who have contributed so much and who truly care about the work they do. They have inspired me and been a constant source of moral support. Before that, one of my very first mentors was Chris Perrins, who took the time to discuss and answer my endless questions when I was still feeling lost and trying to find my way through the confusing contradictions of conservation.

I would like to thank Eric Cowell for all his encouragement and support. Eric is one of my oldest friends and mentors, and it has been such a pleasure to discuss my book with him. I must also mention Mike Howe: his commitment to the countryside, and his music, have been such an inspiration.

Pete Collishaw's unfailing interest in the progress of both editions of this work has inspired me to keep going (even when the sun was shining outside). His encouragement whenever my concentration wavered was so important.

I am grateful to Catherine Cotton for, in the first place, recognising the potential for this book and enabling me to turn it into a reality. I would also like to thank her for suggesting and enabling this new edition. Also, thanks to Ria Kanters for giving so much help and support.

And finally, very special thanks to my wife, Rosanne: she has made this possible. As with all things, I rely so much on her support. She has edited the various versions of this book so many times that I am certain that she knows most of it by heart. She has also had to tolerate, for many years, my impatience and bad temper on the frequent occasions when problems seemed insoluble.

Contents

Introduction

Yng nghesail y moelydd unig,
Cwm tecaf y cymoedd yw,-
Cynefin y carlwm a'r cadno
A hendref yr hebog a'i ryw:
Ni feddaf led troed ohono,
Na chymaint a dafad na chi;
Ond byddaf yn teimlo fin nos wrth fy nhan
Mai arglwydd y cwm ydwyf fi.

Eifion Wyn 1867–1926

This book represents a single instant on a journey towards understanding the science and art of conservation management and planning. An 'instant' because, although we may have some understanding, albeit imperfect, of how far we have travelled, we have no means of judging how far we have to go. It is a journey that began with the earliest pioneers who took those first tentative steps, and one that should continue for as long as humanity engages with nature.

For me, it also represents a personal journey. As a 13-year-old schoolboy I was so inspired and motivated by a visit to a nature reserve that, from that time on, I wanted nothing more than to become a reserve manager. Later in that same year I read Rachel Carson's *Silent Spring*. Although I could not understand everything that I read, it evoked deeper feelings that added to my initial sense of inspiration, and I realised that I had to do something: simply being an observer would not satisfy my emerging ambitions. Eventually, I became the manager of the nature reserve that was the source of my inspiration: Skomer, the most wonderful, wild, Atlantic Welsh island. It was my home for 10 years, and will always remain my spiritual haven. In Wales we talk of 'cynefin'. There is no English equivalent. It means 'the land or place where a person belongs', and is quite different to the concept of 'land that we own'. Skomer is my cynefin.

While on Skomer I spent a long time struggling with the idea that it should somehow be possible to understand and describe what it was that we were trying to achieve. Looking back, I realise that I did gain some understanding which, regrettably, I failed to articulate. The best that I could do before leaving was to list and describe all the work that I believed was necessary to manage the reserve.

When I eventually left Skomer, I became the manager of five spectacular National Nature Reserves in Meirionnydd, the wildest part of North Wales. These were five extraordinarily important places, and I was responsible for their management, but I understood so little of the habitats: upland and woodland, sand dune and salt marsh. What I really knew were cliffs and rocky shores, seals and sea birds. Simply organising the day-to-day management was overwhelming. The solution, I learned, to dealing with this seemingly impossible and chaotic situation was management planning. At first, I concentrated on the management activities, describing, programming and organising all the work that should be done. Then I returned to the questions: why are we here; why are we doing these things; what are we trying to achieve; how will we know when, or if, we achieve our objectives? I then revisited and reorganised the activities. Planning, or at least planning as I understood it, became: why are we here; what have we got; what is important; what do we want; what must we do; what should we monitor?

That was over 25 years ago, and since that time I have continued to learn. It has not been a lonely journey. I have been fortunate and privileged to work with many people, with different backgrounds and ideas, in many different parts of the world. This book, therefore, represents my account of a shared experience, one where I contributed a little but learned so much.

This book is written for students and practitioners. It deals with the development both of the conceptual and the practical aspects of management planning for nature conservation. It is about preparing plans, and it will guide you step by step through the initial stages, the description, objectives and action plan, leading to the implementation of a planning and management process. Although the focus is nature conservation, the other essential and integral components, access, recreation and stakeholder planning, are also included.

The planning process that I describe can be applied to any place that is managed entirely, or in part, for wildlife. It is equally relevant to nature reserves, where conservation is the primary land use, and country parks, where wildlife management may be a secondary interest. It can be applied to the management of species or habitats in any circumstance, regardless of any site designation. It is as relevant at a landscape scale as it is on a small agri-environment scheme.

I have quite deliberately used plain and uncomplicated language. This is because I believe that one of the most important functions of management planning is communication. The audience should not be restricted to professionals, and, as a consequence, the language that we use in planning should be plain and accessible to all. It would make little sense if I failed to apply that simple rule to this book. Occasionally, I have repeated similar ideas but in a different context and, on a few occasions, I have used the same text in more than one location. This is because the book also functions as a planning guide, and people must be able to dip in and out of the various chapters without having to constantly search for additional information held elsewhere.

Introduction to the Second Edition

The privilege and pleasure of having a book published is only equalled by an opportunity to prepare a new edition. Almost as soon as the first edition was published I began to notice the things that I could have expressed more clearly, and inevitably there were new ideas or simply a different perspective on older concepts, all of which led to a dissatisfaction with some of the original content. I guess that most of us live in the world of, 'I would have done that differently had I known better at the time'. This has been a wonderful opportunity for amendments and improvements, and for the inclusion of new material and ideas. There are some changes to every chapter.

Although only 5 years have passed since the first edition was published, there have been some significant changes in the world of nature conservation. We have suffered a global recession and, as a consequence, the resources available for conservation have diminished. But, of much greater concern is the increase in scepticism and, worse, the growth of a coherent though, in my opinion, often ill-informed opposition to nature conservation and all things 'environmental'. The 'better dead than red' sentiment of the 1960s is being replaced by 'greens are the new reds'.

Some sceptics imply that conservation has failed and question both our vision and our methods. Globally our endeavours to conserve wildlife have not been a complete failure, but it is an inescapable fact that we could and should have achieved so much more. Our impact has been limited by many factors. The most obvious is the ever increasing human population and the corresponding demands on our natural environment. Perhaps less obvious is that, despite all the international efforts and intentions, for example, the 1992 Rio declaration, governments throughout the world have failed to deliver even a fraction of the legislation and resources needed to protect our planet and ensure that all life, human and non-human, is sustainable.

Nature conservation is, and always will be, resource dependent. For many of the most important international conservation areas the primary purpose of land management is the protection of wildlife and natural landscapes. This is nearly always extremely expensive. Other conservation measures, including legislation and legal protection, are difficult and costly to enforce, and often rely on some form of compensatory payments. But these costs are eclipsed, in the minds of the commercial sectors, by the implied costs which are a consequence of the limits imposed on exploitation, development and economic growth. Unfortunately, so many people regard nature conservation as a luxury, something that we can only justify during periods of economic growth or in the more developed and more affluent parts of the world: a luxury that we certainly cannot afford during an economic recession. We, of course, might argue that nature conservation, the maintenance of biodiversity, the protection of the very fabric that makes life possible on this planet, is not a luxury but an essential necessity. These arguments fall on the deaf ears of politicians and developers, who live in their comfortable, but illusory, short-term time capsule, which apparently has no regard for tomorrow, let alone the long term.

So how should we, the people responsible for planning and managing our special places, respond? Our role must be to do our very best, regardless of how little support we receive. The first thing is to ensure that planning is given appropriate attention. I recently heard a chief executive of a wildlife organisation say that, because of financial difficulties, they were thinking about suspending planning activities. Planning is so much more important in times of financial constraint: it is always easier to manage something when resources are plentiful. When we have so little we must be very careful to ensure that we get the best possible value for our money, and we must also put ourselves in a position where we can demonstrate that this is the case. Good planning must be about seeking clarity of purpose, and then finding the most effective and efficient means of achieving that purpose. Both purpose and process will inevitably evolve, and this is why I have completely revised and extended Chapter 6, which introduces and explains adaptive management.

I mentioned a growth in scepticism. So often, I encounter statements that suggest we have failed to meet global biodiversity targets because the approaches to conservation – our methods – are inappropriate. Some people are extremely critical of nature conservation, or any form of environmental protection, which is reliant on the establishment of protected areas. Others criticise conservation which is focused on the protection of individual species.

We have known for a very long time that wildlife cannot survive as fragmented islands in a desolate, industrial or urban sea. Anyone who has had any experience of trying to manage land to conserve wildlife knows that they will often have to put in more effort outside the site than on a site. We understand the consequences of fragmentation and island bio-geography; we know that a holistic approach is essential and we recognise the neglect that areas outside the protected areas have endured. Occasionally, I encounter people who claim that they 'do not believe in special islands of conservation', and suggest that we must abandon sites and manage the entire countryside in the same way. These sentiments are nothing new, but they are now encountered more frequently. Unfortunately, they represent a naive and utopian view of what might be possible. This is a view that ignores decades of experience and sound evidence. Our first response to this criticism must be to point out that we have not failed and that we must not, under any circumstance, abandon the very special places: the designated and protected areas. Even if the concept of 'managing our entire countryside in the same way' is ever attainable, it will be very long time before it can be achieved. In the mean time, if we blindly adopt these ideas, lose our focus on protected areas and manage everything in the same way, the inevitable consequence is that, where there may be some gains in currently derelict areas, the best places will diminish, and we will leave a legacy of light green mediocrity to our future generations.

Support has grown for the ecosystem approach. Unfortunately the approach, as defined by the Convention on Biological Diversity, is so often misrepresented. The main issue is that people think that it is about managing 'ecosystems' (whatever they may be), and fail to realise that the 'ecosystem approach' is in fact a series of 12 principles which can be used to guide nature conservation management in any place and at any scale. The ecosystem approach is not about abandoning protected areas or setting aside species protection, but it does offer governments and others a means of enabling the integration of a wide range of different measures aimed at maintaining biodiversity and protecting our natural environment. I have added a new chapter (Chapter 7) which introduces the ecosystem approach and offers a very simple explanation of an ecosystem. We must continue to seek new and improved mechanisms for maintaining biodiversity: yesterday's solutions are not necessarily the same as tomorrow's. But we must be cautious: we must be sure that we have sufficient evidence to demonstrate, beyond any doubt, that new approaches are suitable alternatives before we abandon anything which has a proven track record.

Finally, I cannot claim that this is the best way to plan, but it is the best that I know. There is no point in claiming that adaptive management is a proven system. We have not been doing it for long enough. The response of habitats to management is slow, and I have no experience of adaptive management that has been in place for longer than 20 years. Only time will tell, but if the approach is less than satisfactory it can be changed or adapted.

How to Use This Book

The first chapter, 'The Need for Planning and Core Management Planning Principles', provides a brief justification for planning. It is important that I establish planning as the essential intellectual component of conservation management. I place most emphasis on dealing with the questions of why plans are not written and why so many fail.

The second chapter is the key to understanding this book. It begins with an outline of the structure of a management plan, the equivalent of a 'route map' to the planning process. It provides an essential overview for all readers and is particularly important if you are using this book as a guide to preparing a plan. It recommends the structure and contents for a plan, and provides a reference to the subsequent chapters which are relevant to each stage in the plan. The chapter concludes with sections on preparation and other important pre-planning issues and considerations.

Chapters 3, 4, 5, 6, 7, 8, and 9 set the scene. Although they are not strictly concerned with the process of preparing a plan, there is little purpose in attempting to understand planning or write a plan without considering these essential issues. Chapters 3 and 4 are about audience, communication and relationships. Nature conservation should not be regarded as a stand-alone activity, something that has no relevance to other people. It is most successful when stakeholders, and particularly local communities, are consulted, involved and gain a sense of ownership. Chapter 5 is about definitions. Many of the words that we use in conservation science and management have a multitude of different meanings, and it is important that I share a common language with you, the reader.

In Chapter 6, I introduce the most important planning concepts. In the context of this book, it is essential reading. It is where I describe planning as an iterative, developmental and cyclical process. If you are familiar with 'adaptive' planning and, as a consequence, think that you can give this chapter a miss, please don't.

Chapter 7 introduces the ecosystem approach to management. The most important point to bear in mind when reading this chapter is that the ecosystem approach is not a management system but a series of 12 principles that can be applied to planning and management. The term 'ecosystem approach' is unfortunately misleading, because many people believe, erroneously, that it is about managing ecosystems. I must stress that the application of the ecosystem approach is not restricted to ecosystems: it can be used to guide biodiversity management in any situation.

Conservation ethics is a huge, complicated and intellectually challenging subject. In Chapter 8, I skim briefly over the surface, providing the most basic introduction, but it is important. Ethics, or why we conserve wildlife, must guide our decisions and actions if our planning is to make any sense. This is followed by a chapter on what we value or what our obligations might be. I conclude this trilogy of chapters by considering the various approaches to nature conservation management.

Chapters 11, 12, 13, 14, 15, 16, and 17 describe the planning process in considerable detail. The core chapter, 'Objectives for Biological Features', represents the component of planning that has been developed mainly in response to European Natura 2000 and similar legislation. This is an almost uniquely European approach, but it is relevant to wildlife management anywhere in the world. As discussed earlier, these chapters should be read in conjunction with Chapter 2 which will help explain the sequence and relationship between the chapters.

The final Chapters 18 and 19, deal with access, tourism and recreation, and describe a planning process very similar, in most aspects, to that applied to wildlife. Perhaps the most important point is that all aspects of planning on a site must be integrated. Some organisations, erroneously in my opinion, use different teams and different planning approaches for wildlife and people planning. The consequences can be that plans for the same site are contradictory and incompatible, and this approach can foster a culture of division and conflict within organisations.

The book concludes with five case studies. The first is an almost complete management plan, but it omits all the detailed information on the individual projects since a few examples are sufficient. I suggest that you read this at quite an early stage (probably after Chapter 1), and then use it as a reference section while reading Chapters 11, 12, 13, 14, 15, 16, and 17.

Case Study 2 is a complete access section taken from the management plan for Cors Caron NNR. This has been included because it is an exceptionally good example of a site where access provisions prior to preparing the plan were at a relatively low level. The plan takes account of organisational policy which places an emphasis on encouraging the sustainable public use and enjoyment of the site. The case study is mainly relevant to Chapter 18, but it also contains an excellent example of the approach to preparing access objectives described in Chapter 3.

Case Studies 3 and 4 are extracts from management plans. The first demonstrates the relationship between objectives for habitat and species management. This can be a complicated issue that arises in the majority of plans and is best explained by example. Case Study 4 is extremely important. It demonstrates the adaptive management planning process in action. The case study presents the recent history of management, and documents the development of the management objective, for an internationally important population of butterflies.

The final case study introduces computing as a planning tool. Please do not think that because this is the last part of the book it is in any respect less important than any other section. I have put it at the end because, more that any other section, it talks of tomorrow: the future of planning will be dependent on computers. Once we recognise that management is a process, we begin to understand that planning must be dynamic and adaptable. Even on very small, uncomplicated sites, data management can, over time, become a very significant issue. All our decisions should be based on the best available data, and the data collected or generated as part of site management is usually the most useful. Data are only as good as they are accessible. There is little purpose in collecting data if it is hidden in personal notebooks or hopelessly inaccessible filing systems. Computers, and in particular computer databases, provide the obvious solution, a solution which must be entirely relevant at site level but which also generates the information required at corporate levels within an organisation.

Chapter 1
The Need for Planning and Core Management Planning Principles

Abstract Planning is the intellectual or 'thinking' component of the conservation management process. It is in itself a dynamic, iterative process. It is about recognising the things that are important and making decisions about what we want to achieve and what we must do. Planning is about sharing this process with others so that we can reach agreement; it is about communication; it is about learning. It is one of the most important conservation management activities. This chapter considers the need for planning. It begins with an outline of the functions of a management plan. The core of the chapter deals with the reasons for the failure of so many management plans. Planning should be driven by objectives and not issues. Good planning will ensure continuity of management, which is essential, provided, of course, that it is appropriate management. Conservation management will always be influenced by people management and vice versa. It is important that plans are not over-compartmentalised and that the relationship of each section with all the others is recognised. The main reasons for failure, often an inappropriate approach or attitude, are identified and discussed.

1.1 Introduction

With few exceptions, planning is recognised as an essential component of almost all areas of human endeavour. Perhaps, therefore, it is not surprising that most guides to planning offer little in the way of justification or reasons for planning. The following are a few exceptions:

> If something like this is not done there will be constant changes of intention, species, methods, etc., because every officer who takes over a forest will want to try something different and there will be nothing to stop them. (Brasnett 1953)

> All sites managed for nature conservation should have a management plan, the main purpose of which is to ensure that there is continuity and stability of management. Without an effective plan sites are vulnerable to inconsistent management which can result in a waste of resources and, worse, in the loss of the special interest of the site. (NCC 1991)

M. Alexander, *Management Planning for Nature Conservation: A Theoretical Basis & Practical Guide*, DOI 10.1007/978-94-007-5116-3_1,

> Frequently much time and effort is put into management planning for protected areas but the plans are not used – or are unusable. Even in these circumstances there is general agreement about the desirability of such plans; their preparation is supported by most conservation agencies and IUCN wishes to see plans in place for all protected areas (IUCN 1992). Management plans bring many benefits to protected areas and to the organisations or individuals charged with their management – and without them, serious problems can ensue. (Thomas and Middleton 2003)

A rather cynical response to these statements might be: if planning is so important, why are so many sites managed without the support of a planning process? When plans are prepared, why are they so often left unused, lost in computer folders?

Before attempting to answer these questions, there is a need to consider what management planning is and what its functions are.

All management plans should answer six key questions:

- Why are we here?
- What have we got?
- What is important?
- What are the important influences?
- What do we want?
- What must we do?

The functions of a nature conservation site management plan are:

- To identify all the legislation and policies that will govern both the process and outcomes of management.
- To share decision making, whenever appropriate, and to communicate these decisions to all interested individuals and groups.
- To collate all the relevant information about a site and its features.
- To identify or confirm the most important wildlife and natural features.
- To identify all the important cultural features: historic, archaeological, religious, landscape, etc.
- To develop objectives for all the important wildlife features.
- To develop objectives for all important cultural features.
- To identify the range of facilities or opportunities that the site will provide for visitors.
- To identify monitoring and surveillance programmes to ensure that managers are aware of the status of all the important features and the quality of the experience provided for visitors.
- To identify all the management and recording activities required to manage the site.
- To identify and justify all the resource requirements, both human and financial.
- To combine all the above in a cohesive, logical, dynamic and iterative process.

If a plan meets all the above functions it can:

- Help resolve both internal and external conflicts.
- Ensure continuity of effective management.
- Be used to demonstrate that management is appropriate, i.e. effective and efficient.
- Be used to bid for resources.
- Encourage and enable communication between managers and stakeholders, and within and between sites and organisations.

Very occasionally, the preparation of a management plan will be a legal requirement, for example, in the UK, Section 89 (2) of the CROW Act 2000 requires local authorities with an Area of Outstanding Natural Beauty in their area to: 'prepare and publish a plan which formulates their policy for the management of the area … and for carrying out their functions in relation to it.'

Sometimes legislation can be deliberately ambiguous or open to interpretation, for example, the provisions of article 6 of the European Habitats Directive state that: the necessary conservation measures can involve, 'if need be, appropriate management plans specifically designed for the sites or integrated into other development plans'. The words 'if need be' indicate that management plans may not always be necessary (European Communities 2000).

Returning to my original question: why are so many sites managed without plans, and why do plans so often lie unused, forgotten on shelves or lost in computer folders? The answer to the first part of the question lies in the second. So many managers have direct or indirect experience of abysmal management plans, produced at great cost but which deliver nothing, that there is a collective lethargy and aversion for planning. This is surprising at a time when the destructive pressures on the environment that we share with wildlife increase, while the resources available to combat these pressures are decreasing.

There is very little published material that looks critically at management planning, but there are two critiques which provide a very useful perspective. Between them they identify all the most significant failings:

Oliver Rackham (2006), in his book *Woodlands*, expresses concerns about management plans:

> In the 1990s there was a vogue for management plans; every nature reserve in the Kingdom was supposed to have one. Management plans ought to remedy the problem of discontinuity of personnel, but they were approached as a bureaucratic exercise rather than a practical tool.

Fortunately, Rackham concludes his criticism of management plans with a very useful 'list of features of a proper management plan'. The following is an abridged version.

A management plan should:

- Begin with what makes a place special and how it differs from other places.
- Be a public document.
- Be a comprehensive archive of actions and investigations.
- Contain a statement of the core features of management principles to be remembered and maintained through future changes of organisation and policy in the parent body.

Krumpe (2000), in his paper *The Role of Science in Wilderness Planning*, identifies a number of weaknesses associated with the USA planning frameworks, many of which are equally relevant to management planning in general:

- A primary and almost exclusive focus on recreation.
- The quest, for over 20 years, to determine empirically a concrete carrying capacity in terms of the appropriate number of visitors.

- Failure to articulate specific desired future conditions or long-term goals in any but the most general of terms.
- Being issue-driven rather than goal-driven.
- The lack of support and involvement from higher levels of management in the planning process.
- Failure to follow through and systematically complete things that were articulated in the plan.
- Last minute changes by upper level administrators who were not involved in the planning process or knowledgeable about the compromises and tradeoffs that were considered and agreed upon.

The preceding concerns or comments fall into two distinct categories. These are:

- The weaknesses or failures of some planning systems.
- An inappropriate approach or attitude to planning.

1.2 The Weaknesses or Failures of Some Planning Systems

1.2.1 Plans Must Contain Site-Specific Management Objectives

Rackham is clear that a plan should identify, 'what makes a place special and how it differs from other places', and Krumpe writes of the 'failure to articulate specific desired future conditions or long term goals in any but the most general of terms'. Most early management plans (and, unfortunately, many current ones) contain objectives that talk about maintaining or enhancing something, with no indication of what will be maintained or what it will become when it is enhanced. My firm belief is that management plans must contain objectives, and that an objective must be a clear description of something that we want to achieve. Wildlife outcomes are the conditions that we require for habitats, communities and populations of species. It is also very important that objectives are site-specific. Our commitment to maintaining biodiversity must include an obligation to ensure that local distinctiveness is maintained. Generic objectives that can be applied everywhere have very limited value anywhere, and this soon becomes apparent to managers. Once a plan is seen, even in part, as irrelevant it is likely to be abandoned.

1.2.2 A Plan Should Be a Public Document

Most sections in a plan should be in the public domain. Many plans will contain sensitive or confidential information, for example, the location of rare and endangered species. Clearly, this should not be included in a public version of the document (see Chapter 3). If plans have relevance to the widest possible audience there is an increased probability that they will be used.

1.2.3 Archive of Actions and Investigations

It is so important that management plans contain a comprehensive archive of actions and investigations. It is essential that all the significant activities on a site are recorded and that the records are accessible and can be easily interrogated. Recording can become a very expensive activity, and recognising the difference between significant and insignificant information is not always easy. The management planning process is clearly the most appropriate way of identifying or specifying everything that needs to be recorded. In other words, recording should be recognised as an integral component of the management process. The adaptable process is entirely dependent on information.

1.2.4 Continuity of Management

Rackham's final point, 'plans should contain a statement of the core features of management principles to be remembered and maintained through future changes of organisation and policy in the parent body', is a plea for continuity. Continuity of management is essential, provided, of course, that it is appropriate management. Conservation organisations can be even more dynamic than the habitats that they seek to protect: policies change, staff move on, and purpose or direction is lost and reinvented. This is not to suggest that management should never change, but decisions to make changes should take account of the original reason for implementing an action or developing a policy. Also, changes should only be considered when we have better information or when the factors that influence the features change (for example, an alien invasive species may appear on a site).

1.2.5 Focus on Recreation

Krumpe claims that in many plans there is a primary and almost exclusive focus on recreation. This comment is very interesting when viewed from a British perspective. Many European plans can be justifiably criticised for giving insufficient attention to recreation. These issues are discussed in detail in Chapters 18 and 19, on access, tourism and recreation. The main point here is that any management process which fails to deal with all aspects of site management will fail. Conservation management will always be influenced by people management and vice versa. It is also important that plans are not over-compartmentalised and that the relationship of each section with all others is recognised.

1.2.6 Failure to Determine Carrying Capacity

Krumpe's paper, which is about wilderness, stresses the failure of, 'the quest for over 20 years to empirically determine a concrete carrying capacity, in terms of the appropriate number of visitors'. The USA legal definition of wilderness includes the statement: 'land which generally appears to have been affected primarily by the forces of nature, with the imprint of man's work substantially unnoticeable' USA Wilderness Act 1964. So, wilderness is something that nature has created: it is a dynamic and evolving something. Carrying capacity might be defined in two ways:

(a) It could be the point at which the experience enjoyed by visitors to the wilderness is diminished as a consequence of the activities of others. It is extremely difficult, if not impossible, to establish a limit which could be anything other than subjective, a matter of opinion. Tolerance to others will vary enormously from individual to individual and will be influenced by a wide range of experiences and expectations.
(b) The tolerance of the wilderness ecosystem to human activity. This could be the point at which irrevocable change takes place. There are two issues: In a naturally dynamic ecosystem, can we differentiate between change which is the consequence of anthropogenic activity and change which is the consequence of natural processes? And, can we set limits which express the degree of change that is tolerable?

I wonder if the failure to determine carrying capacity in wilderness areas is, in fact, a failure to achieve the impossible.

When management is concerned with obtaining defined outcomes for the features, determining the carrying capacity of features is less complicated. In simple terms, because the condition that is required of a feature is known, whenever a feature is not in the required condition remedial action is necessary. It might not be possible to predict the threshold of carrying capacity for any human activity, but once it is exceeded we will know. This approach will only work if it is possible to monitor both public use and the condition of the feature. In Chapters 15, the use of performance indicators (attributes and factors) will be introduced.

1.2.7 Issue-Driven Versus Goal-Driven Planning

The preceding discussion highlights the value of management by defining outcomes and leads neatly to Krumpe's next point, which is that some plans are issue-driven rather than goal-driven. This is the most significant difference between planning in the Old World and planning in the New World (see Chapter 18). As a consequence of the need to manage a cultural landscape with a very high proportion of valued plagioclimatic communities, and because of the associated legislation, planning in Europe has migrated towards goal-driven or objective-driven management.

1.3 An Inappropriate Approach or Attitude to Planning

1.3.1 Bureaucracy

Once the need for management plans has been recognised or imposed by legislation, the typical bureaucratic approach is to set a target. The target will sometimes include the standard for a plan (and this may be a good and appropriate standard), but the most significant part of the target will be the date by which the plan must be completed. The next step is to identify the most cost-effective way of producing plans, and this is often leads to the employment of external contractors who have no experience or interest in the site. A document is produced, the money has been spent, the target has been met and the plan is then forgotten: forgotten because the site managers have no sense of ownership. The plan makes no contribution except to support the illusion that something must have improved: wildlife has been saved. At least the plan exists, and that, in some minds, is all that really matters. This may come across as extreme cynicism, but, unfortunately, my experience coincides with Rackham's.

A clue that points to a solution to the problem lies in the difference between 'planning' as a process and simply producing a 'management plan' as a finished document: most people do the latter. Some set about creating a document without always considering or understanding its purpose, or even, in extreme cases, that it has a purpose. The perceived barrier between planning and management must be dismantled. People should be taking part in a continuing, iterative management and planning process, where planning is the intellectual, or decision-making, component of management. Management must be adaptive and dynamic to take account of change (see Chapter 6).

Important: site managers should be the site planners. They should, whenever possible, prepare the documentation, or at least supervise its production, and they must be responsible for maintaining the planning process.

1.3.2 Failure to Comply with a Plan

Krumpe also identifies as a problem, 'failure to follow through and systematically complete things that were articulated in the plan'. This will happen for many reasons, including, once again, a failure to recognise planning as a process and a sense of lack of ownership. The latter occurs most often when managers have no confidence in, or fundamentally disagree with, a plan that has been thrust upon them. Another reason for failure to comply may be that some individuals, when newly appointed to a position, take little account of anything that went before them, and this can lead to discontinuity of management. Good planning is particularly important in this context: it is significantly more difficult for anyone to change or abandon management when there is clear justification for, and formal commitment to, an outcome. There will, of course, be occasions when actions articulated in a plan are abandoned for good reason. I am sure that Krumpe is not advocating blind adherence to a plan even when an action ceases to be appropriate.

1.3.3 Lack of Corporate Support

Krumpe raises two issues concerning 'higher levels of management'. He is concerned by the lack of support for planning by upper-level administrators and by the possibility of last-minute changes being made by people who were not involved in the planning process and have no understanding of the compromises and tradeoffs that were considered and agreed. These are, unfortunately, common and widespread issues. I suspect that one of the reasons for this is that, while junior staff are provided with management planning training, senior staff are unlikely to have received this level of instruction and may be unaware of the gaps in their knowledge or skills. I believe that a more significant issue is that organisations often fail to recognise the relevance of conservation management planning to corporate management and planning.

1.4 Management Planning for Nature Conservation – Core Principles

In 2008, the Conservation Management System Consortium (CMSC) recognised a need to review the variety of current management planning protocols and identify core management planning principles that should be applied to any conservation management plan. The CMSC organised a workshop: '*Establishing and Confirming Management Planning Principles on Natura 2000 and Other Conservation Sites*'. This was held from 30 September to 2 October 2008 at Plas Tan y Bwlch, the Snowdonia National Park Study Centre in North Wales, UK.

The following statement outlines the core principles (identified during the workshop) that should be applied when preparing a management plan for nature conservation, and which apply particularly for European Natura 2000 and Ramsar sites. The statement was prepared and endorsed by 21 participants, representing 11 European countries.

1.4.1 General Principles

- Ideally, there should be one comprehensive plan for multi designation sites.
- In addition to planning the management of nature conservation features, plans should also consider stakeholder interests, cultural aspects (including historical, archaeological, religious and spiritual interests), visitor management/tourism, education and interpretation, and social and economic aspects.

- The precautionary principle is important in the context of conservation management and planning. It should guide the planning process and influence the way in which we manage sites, habitats and species.
- Planners should recognise the need to integrate conservation site planning with wider sectorial and land use plans.
- The planning approach should be as uncomplicated as possible (the simpler the better).
- A management plan should be as large as the site requires and no larger.
- Organisational support for the planning process is essential, and this should include a formal approval process.
- Management plans should be easily understood by everyone who has an interest in the site. This will include people who do not have a scientific or technical background. The language used in the plan should, whenever possible, be plain and accessible to all.
- Plans, and in particular plans for large, complex sites, should include a summary. These can be presented as text, but the addition of annotated maps and illustrations will help to explain issues.
- Individuals involved in managing a site should, whenever possible, have an involvement in the planning process and, in all cases, ownership of the plan.
- A record should be kept of all the individuals engaged, at any level, in preparing the plan.
- All consultees and advisors, individual and corporate, should be acknowledged in the plan.
- The plan should contain a glossary of terms.
- Plans must be implemented.

1.4.2 Stakeholder Involvement

Conservation managers must recognise the need to adopt an inclusive approach which takes account of the interests of all stakeholders and, as far as possible, encourages their involvement in all appropriate aspects of management planning and site management. One of the key issues when building and maintaining successful relationships is to have a shared appreciation of what can and cannot be negotiated.

1.4.3 The Planning Process

- Management planning should be a continuous, cyclical, iterative and developmental process (adaptive planning)
- Monitoring must be recognised as an integral and essential component of any planning process. (*Monitoring: Surveillance undertaken to ensure that formulated standards are being maintained.*)
- It is good practice to record all actions undertaken in accordance with a plan.

- Factors must be identified and integrated in the planning process. *(A factor is anything that has the potential to influence or change a feature, or to affect the way in which a feature is managed. These influences may exist, or have existed, at any time in the past, present or future. Factors can be natural or related to human activity in origin, and they can be internal (on-site) or external (off-site).)*
- All planning and management actions should incorporate current best practice and be open to new and innovative ideas.
- Management should be reviewed continually within a time scale that is appropriate to the features. (Fragile and vulnerable habitats or populations will require more frequent attention than robust and secure features.)
- Internal management reviews should be supplemented with formal reviews at predetermined agreed dates. It may, in some cases, be appropriate to hold external reviews.

1.4.4 Information

- Plans require a descriptive section which contains, or provides reference to, the information that will be *needed* to help decide what is important and to undertake the planning process. This is a collation exercise and is generally not dependent on the generation of new information. Further information requirements should be identified during the planning process.

1.4.5 Features as a Focus

- Features should provide the focus for management planning. Management by defining conservation outcomes for features is a reflection of the legal requirement to protect specified features on statutory, and other, sites. This is of particular relevance to Ramsar sites. The desired status for each feature is defined, and these are the *management objectives.*

1.4.6 Objectives

- Objectives should lie at the very heart of a management plan. They are the outcomes of management and the single most important component of any plan. An objective is the description of something that we want to achieve. Wildlife outcomes are habitats, communities or populations (features) at Favourable Conservation Status.
- SMART objectives, as generally applied to business, can, with modifications, be applied to wildlife objectives.

- Objectives for conservation features must not be diminished to accommodate a shortfall of resources. An objective should be an expression of the legal and moral obligations towards features on sites.
- Objectives should, when appropriate, take account of natural and other processes.

1.4.7 The Action Plan

- All management plans should contain an action plan which identifies the total resource requirement.
- The action plan should identify and cost all the activities required to obtain and maintain features at Favourable Conservation Status.
- The action plan should identify priorities for all management activities.
- The action plan should identify all individuals or organisations that will be responsible for implementing the activities.
- The action plan should identify realistic, achievable and effective management actions.

1.5 Conclusion

Planning is the intellectual or 'thinking' component of the conservation management process. It is in itself a dynamic, iterative process. It is about recognising the things that are important and making decisions about what we want to achieve and what we must do. Planning is about sharing this process with others so that we can reach agreement; it is about communication; it is about learning. It is the most important of all conservation management activities.

The emphasis should be rather more on thinking and less on the production of elaborate, verbose documents.

Planning should always come before management. Conservation management is about taking control in order to obtain and maintain desirable conditions. 'Control' does not necessarily mean doing something: it could mean choosing to do nothing. Taking control can have implications for the actions and freedoms of others.

All this could be summarised in three simple words: *thought before action.*

References

Brasnett, N. V. (1953). *Planned Management of Forests.* George Allen and Unwin, London.

European Communities (2000). *Managing Natura 2000 Sites, The provisions of Article 6 of the Habitats Directive 92/43/EEC.* European Communities, Luxembourg.

IUCN (1992). *Caracas Action Plan.* IUCN, Gland, Switzerland.
Krumpe, E. E. (2000). *The Role of Science in Wilderness Planning – A State-of-Knowledge Review.* USDA Forest Services Proceedings RMRS-P-15-VOL-4. U.S. Department of Agriculture, Forest Services, Ogden, UT.
NCC (1991). *Site Management Plans for Nature Conservation, A Working Guide (Revised).* Nature Conservancy Council, Peterborough, UK.
Rackham, O. (2006). *Woodlands.* (The New Naturalists Library) Collins, London, UK.
Thomas, L. and Middleton, J. (2003). *Guidelines for Management Planning of Protected Areas.* IUCN Gland, Switzerland and Cambridge, UK.

Chapter 2
Structure, Preparation & Precautionary Principle

Abstract This is an essential chapter as it sets the scene for the remainder of the book. It begins with an overview of the entire planning process and establishes the key components of any management plan. These can be expressed in plain language as: Why are we here? What have we got? What is important? What do we want? What must we do? This simple outline is followed by a detailed structure and recommended contents for a management plan. There is no 'one size fits all' in management planning: individuals and organisations will invariably wish to tailor a format to meet their specific requirements, but there are advantages to be gained when an organisation adopts a corporate standard. There are a number of issues that have to be considered. These include the size of a plan: it should be as large as the site requires and no larger. Planning should be an inclusive process or a team effort. Planners should understand the difference between outputs, i.e. the incidental by-products of conservation management (for example, a management plan), and outcomes, i.e. the purpose of conservation management (for example, habitats and species in the required condition). Finally, there is an important relationship between conservation planning and the precautionary principle. We cannot afford to take unnecessary chances when managing our natural environment.

2.1 The Structure and Contents of a Plan

This section provides an overview of the entire planning process. Each stage in the process is described in outline. (Most of the text is taken directly from the main chapters.) In all cases, reference is given to the location of the full text.

Any management plan format can be reduced to six key components, or sections, which can be dealt with as a series of questions, which follows, as far as possible, a logical sequence or structure.

M. Alexander, *Management Planning for Nature Conservation: A Theoretical Basis & Practical Guide*, DOI 10.1007/978-94-007-5116-3_2,

Table 2.1 The main sections in a management plan

The main sections in a management plan		
Legislation and policy	1 Why are we here?	All management plans must contain a section on legislation and policy. Together, these provide the foundations that support the plan and act as a guide to the direction that the planning process should follow.
Description	2 What have we got?	Once we know why we are here, the next question is what have we got? Plans require a descriptive section which contains, or provides reference to, all the information that will be **needed** to help decide what is important and to complete all the following sections in the plan.
Evaluation	3 What is important?	Once we know what we have got we can move on to evaluation. This is the process used to identify the important features on a site. (When dealing with the access section, evaluation is concerned with identifying the level of access provisions that are appropriate for a site.)
Factors	4 What are the important influences?	A factor is anything that has the potential to influence or change a feature, or to affect the way in which a feature is managed. These influences may exist, or have existed, at any time in the past, present or future.
Objectives	5 What do we want?	An objective is, or should be, the description of something that we want to achieve.
Action plan	6 What must we do?	The action plan is derived directly from the objectives. When we are clear about what we want to achieve we can decide what we need to do. The action plan will contain individual projects which describe and cost all the work required on a site. This information is used to create various work plans and programmes.
Monitoring & review	Monitoring must be regarded as an integral and essential component of the entire management process. We need to know that we are responding to our policies, achieving our objectives and that management is appropriate.	

2.1.1 *The Structure of a Plan*

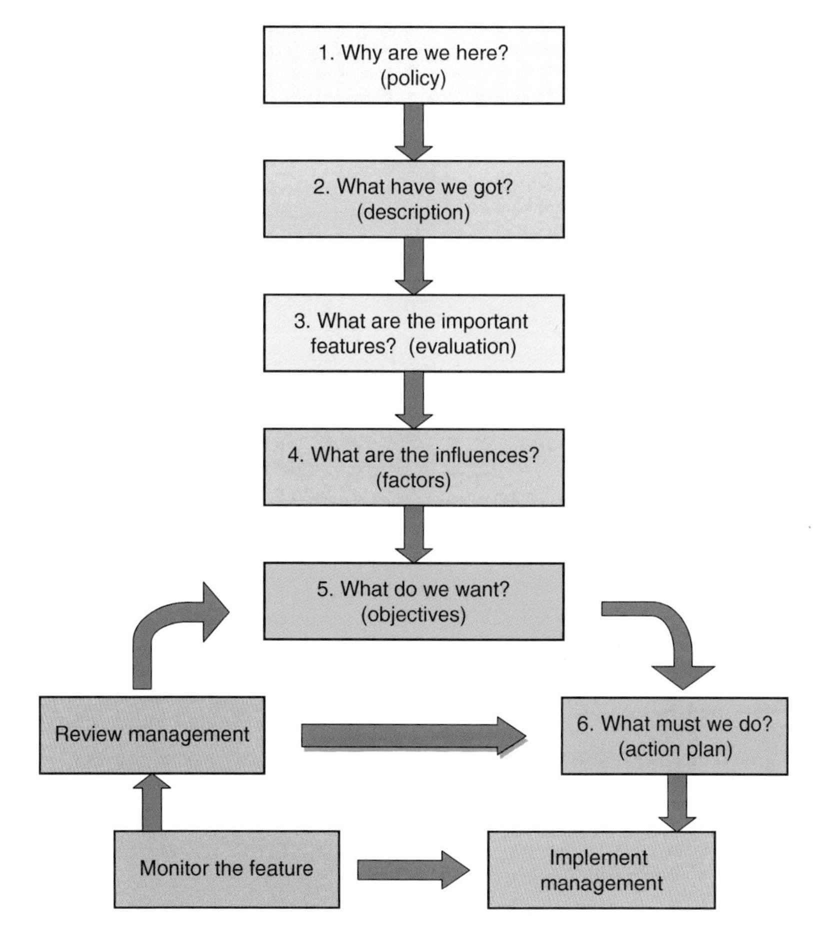

Fig. 2.1 The plan structure for a single feature

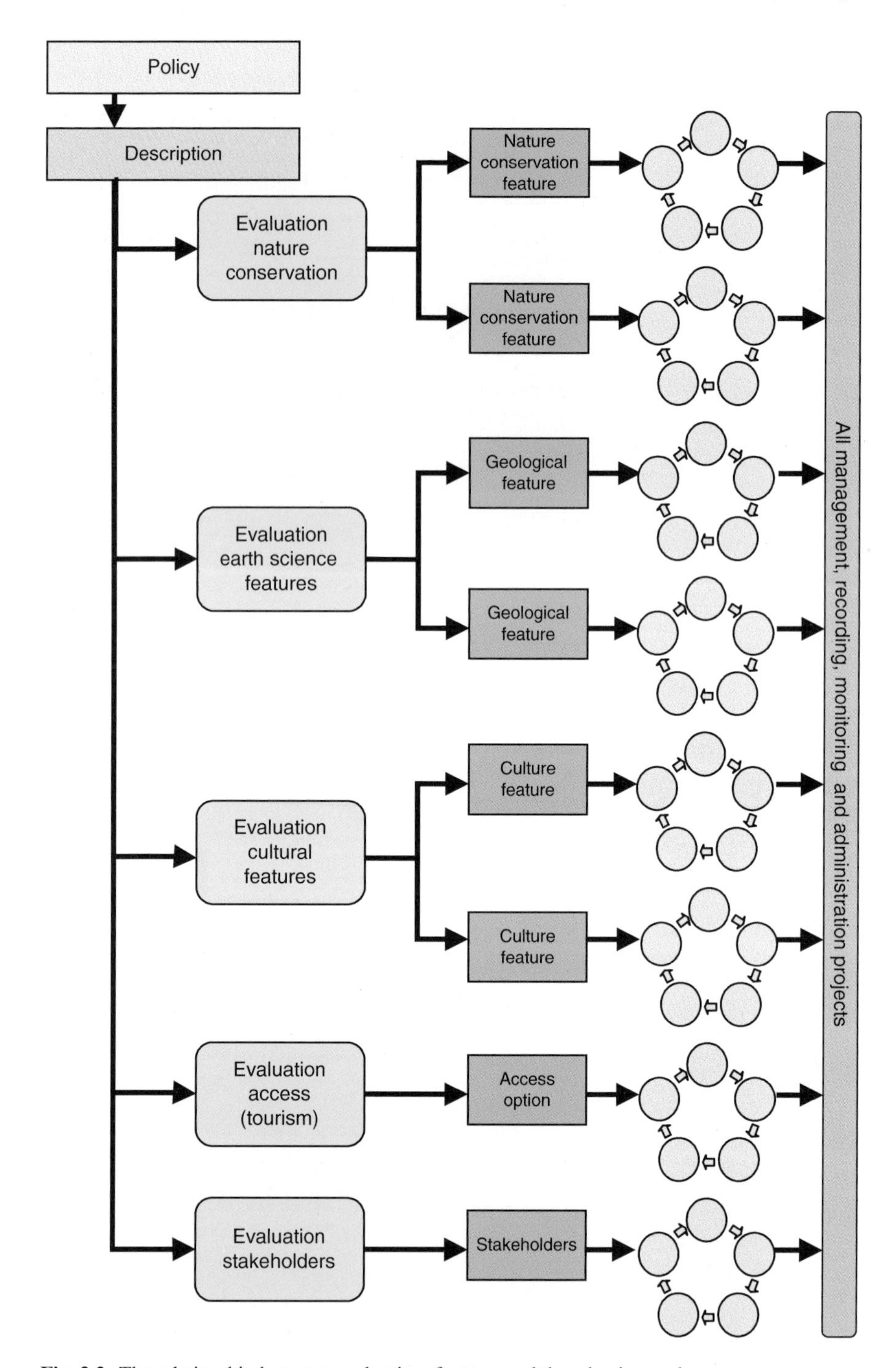

Fig. 2.2 The relationship between evaluation, features and the adaptive cycle

The above diagrams (Figs. 2.1 and 2.2) are a simplified representation of a plan. The cyclical adaptive process is repeated for each individual feature on a site. The actual structure of all but the simplest plans will be more complex than that given below. There is no one size that fits all in management planning. Different organisations, even different site managers, will often wish to develop their own structure and contents to meet organisational requirements or specific site conditions. However, there are clear advantages of attempting to specify a common structure and standard contents within an organisation. This will improve communication: plans are easier to read and assimilate if the structure of the document is familiar to the reader. In addition, a uniform approach will help to establish common standards of planning and to facilitate approval and audit processes. I recommend the plan contents given above. They have a long and proven track record.

2.1.2 The Contents of a Plan

The following example is a simplified version of the list of contents of a plan. For clarity, it is restricted to the nature conservation features and the access section. A complete plan would contain the full range of the interest areas, including, for example, geological, cultural and stakeholder interests. The objectives section for nature conservation features will be repeated for each feature.

1 Plan Summary
2 Legislation & Policy
3 Description
4 Nature Conservation Features
- 4.1 Evaluation
- 4.2 Factors – (preparation of a master list)
- 4.3 Objective
  - 4.3.1 Objective 1
    - 4.3.1.1 Vision
    - 4.3.1.2 Performance indicators
    - 4.3.1.3 Status & Rationale
  - 4.3.2 Objective 2
    - 4.3.2.1 Vision
    - 4.3.2.2 Performance indicators
    - 4.3.2.3 Status & Rationale

5 Access
- 5.1 Evaluation
- 5.2 Options
- 5.3 Objective
  - 5.3.1 Vision
  - 5.3.2 Performance indicators
- 5.4 Status & Rationale

6 Action plan

Important: This section follows the numbering system used in the preceding contents list and not the general numbering sequence used elsewhere in this book.

1 Plan Summary

The purpose of the summary is to give the reader a rapid and clear overview of the entire site. It should be based on the sections in the full description. Some plans contain a site vision in place of a summary. The vision provides a portrait, in words, pictures or maps, of the site when all the objectives have been achieved. This is an assemblage of all the individual visions prepared for each objective. For obvious reasons, it should be one of the last sections to be written in plan.

2 Legislation & Policy (Chapter 11)

All management plans must contain a section on legislation and policy, and this should be completed before most other stages in the planning process. Together, they provide the foundations that support the plan and act as a guide to the direction that the process should follow.

The management of statutory conservation sites can be governed almost entirely by legislation. Even non-statutory sites do not escape the implications of legislation: there will be health and safety legislation, access legislation and a, sometimes bewildering, range of other national and local laws, all requiring compliance.

Policies, or more specifically organisational policies, are a high-level statement of the purposes of an organisation (why it exists). The policy section should begin with the inclusion of all relevant organisational policies. This should be followed by an assessment of the extent to which organisational policies can be met on individual sites.

3 Description (Chapter 12)

The description is fundamentally a collation exercise. All relevant data are located and arranged under various headings. The order in which the headings are organised is of no particular significance. The only reason for a list is to ensure that the contents are reasonably comprehensive.

The description should only include statements of fact. This is not the place for making judgements. The facts are collated and recorded, and, at a later stage, they will provide the basis for evaluation and decision making.

Many planners put too much emphasis on the description. This is often some form of displacement activity, and it may create problems if resources are scarce. Management plans are about communication; they should be as succinct as possible

The full list of contents (given in Chapter 11) will not be appropriate for many sites. The various subsections should be completed only if the information has relevance to site management or the planning process.

4 Nature Conservation Features (Chapter 13)

4.1 Evaluation

Feature assessment or evaluation is simply the means of identifying, or confirming, which of the features on a site should become the focus for the remainder of the planning process. It is about asking a question of each provisional feature in turn: is this feature, in its own right or in association with other features, sufficiently important to be regarded as one of the prime reasons for maintaining the protected area?

For most sites, the presence of conservation features will have been the basis of site acquisition, selection or designation. This means that at some time in the past the site will have been evaluated and the most important features identified.

There are two different approaches to identifying or confirming the important features on a site:

- Selection based on the use of the Nature Conservation Review criteria for identifying important features. This is a derivative of an approach developed in Britain to identify the most important nature conservation sites (Ratcliffe 1977).
- Selection based on the use of the previously recognised status (local, national and international) of a feature. In some ways, this may be regarded as a consensus approach because it takes account of as wide a range of opinion as possible.

4.2 Factors: (preparation of a master list) (Chapter 14)

A factor is anything that has the potential to influence or change a feature, or to affect the way in which a feature is managed. These influences may exist, or have existed, at any time in the past, present or future. Factors can be natural or anthropogenic in origin, and they can internal (on-site) or external (off-site).

The management of habitats and species is nearly always about controlling factors, or taking remedial action following the impact of a factor. Control means the removal, maintenance, adjustment or application of factors, either directly or indirectly. For example, grazing is the most important factor when managing grassland. Grazing can be removed, reduced, maintained, increased or introduced.

Factors are considered at several key stages in the planning process for each feature: the selection of attributes for features, the selection of performance indicators for features and the management rationale. However, an individual factor can have implications for many different features on a site; for some it will be a positive influence, for others negative. To avoid unnecessary repetition, a master list of all the factors is prepared at an early stage in the plan. The list should contain all the factors that have affected, are affecting, or may in the future affect, any of the features on a site. Once a master list has been prepared it can be used to ensure that all the relevant factors are considered for each feature.

4.3 *Objective (Chapter 15)*

An objective is, or should be, the description of something that we want to achieve. These are the outcomes of management. Wildlife outcomes are habitats, communities or populations at a favourable status.
Objectives comprise two components:

- A vision which describes, in plain language, the outcome or condition that we require for a feature.
- Performance indicators which are monitored to provide the evidence that will be used to determine whether the condition that we require is being met or otherwise.

4.3.1 Vision

Writing an objective for a feature is much easier when the vision is based on the definition of Favourable Conservation Status. FCS is an uncomplicated and common sense expression of what we should attempt to achieve for all important features. An objective can be built around the FCS definition by dealing with each section of the definition in turn. Consider the current condition of the feature on the site. If any part, or parts, of the feature appear to be in the required condition, this provides an excellent starting point for deciding what favourable might mean. In situations where features are not in a favourable condition, the question should be: why is the feature unfavourable, and what is the difference between what we see and what we want to see? Experience from other similar places where the feature is considered to be favourable may help, but do not forget the importance of local distinctiveness.

4.3.2 Performance Indicators

A number of performance indicators can be used to quantify the objective and provide the evidence that a feature is at Favourable Conservation Status or otherwise. An objective based on FCS must be provided with two different kinds of performance indicators. These are:

- Quantified attributes with limits. (An attribute is a characteristic of a feature that can be monitored to provide evidence about the condition of the feature.)

– Factors with limits which, when monitored, provide the evidence that the factors are under control or otherwise.

4.3.3 Status & Rationale (Chapter 16)

This section is the process of identifying, in outline, the most appropriate management for the various site features. The procedure which is applied to each feature in turn comprises two distinct phases:

It begins with the identification of the status of the feature and an assessment of current conservation management. We will have some confidence in current management when the feature is considered to be at Favourable Conservation Status and little confidence when it is not. The relationship between factors and the condition of the feature is then considered, along with the implications of the factors to management.

5 Access

The provision of access for local visitors and tourism, and opportunities for recreational use, is an important, if not essential, function of most nature reserves and protected areas. For some sites it will be the most important function. Access planning is concerned with all the provisions that are made for people who visit or use a site for any reason other than for official, business or management purposes.

5.1 Evaluation

The outcome of this section is a clear statement of the level of access, including recreational activities, that is appropriate for a site, or parts of a site. In other words, to what extent can an organisation's access policy be applied to the site?

5.2 Options

Access options are a simple means of indicating the level of access that is considered appropriate for the site, or for zones within the site, following the evaluation.

5.3 Objective

An access objective is a simple and succinct expression of the level and provisions for access that are appropriate for a site.

5.3.1 Vision

The provision of opportunities for people to gain access to sites is not simply about enabling them to enter and wander around the site. There is an obvious need to provide visitors with a very positive experience, and it is possible to describe the experience that they should gain when visiting a site.

5.3.2 Performance Indicators

Performance indicators for access need to be selected with care. They must be measurable and quantified (i.e. so that they can be monitored), and the data should be easy to collect. The number of indicators should be kept to a minimum, but there should be sufficient to provide the evidence necessary to ensure that the quality of the access provisions can be measured.

5.4 Rationale

This section is the process of identifying, in outline, the most appropriate management for the various site features. The procedure comprises two distinct phases:

It begins with the identification of the status of access and an assessment of current access management. We will have some confidence in current management when access provisions are considered to be favourable and little confidence when they are not. The relationship between the factors and the status of access provisions is then considered, and an outline of the management requirements is prepared.

6 Action Plan (Chapter 17)

The action plan contains descriptions of all the work that needs to be carried out on a site in order to meet the objectives. Each individual task or project is identified and described in sufficient detail to enable the individuals responsible for the project to carry out the work. All the basic information for each project (i.e. when and where the work should be completed, who should do the work, the priority, what it will cost, etc.) is aggregated and used to produce a wide range of work programmes, for example, annual programmes, programmes for a specified period, programmes for an individual, financial programmes, long-term programmes, etc. An action plan is prepared for a specified period, usually 5 years. Action plans also provide a structure, and establish priorities, for recording.

2.2 Preparation

2.2.1 *The Size of a Plan*

Small is beautiful. The size of a management plan and, perhaps more importantly, the resource made available for its production must be in proportion to the complexity of the site and also to the total resource available for the safeguard and/or management of the site. Thus, for small, uncomplicated sites, short, concise plans will suffice. A plan should be as small as possible – as large as the site requires and no larger. Unfortunately, the world of wildlife management is littered with sites equipped with large, elaborate, expensively-produced and bound plans, usually paid for by well-meaning donors. This type of plan is often found on sites that have insufficient funds for management.

Even where there may be a long-term intention to prepare a full plan for a site, the process can, and perhaps should, begin as a brief outline or minimal statement. As further information or resources become available the plan may grow.

Plans should, whenever possible, be prepared for an entire site. However, for very large and complicated sites it may be necessary to divide the site into recognisable management units or zones. These units may be based, for example, on tenure, site status, habitat distribution, tourism or public use. Specific plans can be written for each unit but must conform to an overview plan. If possible, the overview should be written in advance of the unit plans.

2.2.2 *Minimum Format*

Some organisations have sought a minimum format for planning which can be applied to all the sites that they manage. Sometimes their reasons are bureaucratic and target driven. They recognise a need to demonstrate that they have written plans but are not prepared to commit appropriate resources and so seek a minimal approach simply to satisfy other bureaucrats. Often, this type of organisation becomes obsessed with the size of a plan and will specify that it must be no greater that a certain length.

More often, conservation organisations, although morally committed and sometimes legally obliged to do their best for their sites, are hopelessly under-resourced. Their search is for a minimum format which represents the least that a site *needs*. There is no universally applicable minimum format for planning. Plans should be tailored to meet the specific requirements of any given site, and that requirement will vary from site to site. Managers usually have far less freedom of choice than they imagine: their ability to make decisions is constrained by legislation, policy and a wide range of other factors. For example, when managing statutory sites, such as Natura 200 sites, the management plan must take account of the minimum requirement to ensure that the features which were the basis of site selection are maintained at Favourable Conservation Status. In some cases, there will be legal or other reasons for providing access to the site, and this will have to be planned. Consequently, it will only be possible to identify shortcuts that will not incur the risk of producing an inadequate plan during the production of the plan, and particularly when preparing the sections of the plan that are concerned with evaluating information or making decisions.

The key to producing a minimum plan is to differentiate between what we need to know and what we want to know.

2.2.3 Who Should Be Involved in the Preparation?

Stakeholder involvement in plan production is covered in Chapter 4.

Management planning must be an inclusive process. Everyone who is involved in the management of the site, or will be in any way affected by management decisions, should at least be consulted and whenever possible and appropriate included in the decision making process. The most important people of all are those responsible for managing the site: the managers must own the plan. That is, they should agree with, or at least appreciate and accept, the reasons behind all decisions. There are many examples where plans have been produced by external consultants, at great expense, but never implemented. The reason for this failure is nearly always the same: the site managers were not fully involved in plan production. Managers will rarely accept the imposition of a plan prepared by others with no experience of managing the site unless they have been fully involved in the planning process.

The preparation of all but the simplest plans should be undertaken as a team effort. No individual will possess sufficient expertise in all the areas that require consideration. It is, however, essential that one person has complete responsibility for the production of the plan. This role should be seen as editorial, and the most appropriate person for this position is the site manager. The author of the plan should have a good knowledge of the site and should understand the practical aspects of management and the interactions between different interests and features.

Unfortunately, it is not always possible for the site managers to set aside sufficient time for planning, and, consequently, organisations often use consultants to write plans. Although this may be far from ideal, it is better to have a plan written by a consultant than no plan at all. Good consultants, who are expert in planning and understand their role, are an invaluable asset. Their employment can be very cost-effective. Most site managers will write only a few plans in their entire career, and they may not have the opportunity to develop planning skills. Experienced consultants will have a thorough understanding of planning, but they should rely on the site staff as a source of expertise about the site. The site managers must be fully involved and consulted regularly throughout the process. The CMS Partnership has developed a technique where the consultants prepare the plan by interrogating the site managers and then obtain their approval for each section. The plan can then legitimately be taken forward with the site manager as co-author.

While the format of a plan follows a logical structure, the production of a management plan need not necessarily follow the precise sequential process. Work on one section will often give rise to something that is relevant to other sections. There are also many good reasons for skipping some sections and then returning to them once others are complete. In the absence of a database, word processors are very useful. A folder is created for the plan with a separate file for each section. This means that moving between the different sections is easy.

One of the best and most effective approaches to preparing a plan is to place raw information into each section of the plan. Often, this will mean pasting in unedited information or preparing rough notes on what information is available. Sufficient information should be included to enable decision making, but there is no need for concern about language or structure at this stage. In this way, it is possible to complete a full plan, in outline, very rapidly. The alternative is to become bogged down in individual sections, and by the time these have been through several drafts or refinements their significance to the rest of the plan will have been forgotten.

2.2.4 *Presentation*

The need for a dynamic or adaptable approach to planning is discussed in Chapter 4. It follows that, if a process is dynamic and subject to review and change, there can be little purpose in producing permanent, expensive documentation. Many organisations create extravagantly bound and lavishly illustrated management plans. These documents are so precious as a result of the effort and cost of production that site managers are very reluctant to modify them, and they quickly become obsolete.

The best possible means of holding and presenting a plan is as a computer document. Ideally, there should be no need to print out the document, but, in reality, many people prefer to read text on paper. (The use of computers in management planning is covered in Case Study 5.)

2.2.5 *Plan Approval*

All organisations that require management plans should adopt a formal approval process. A plan begins life as a draft statement written or approved by the site manager. Stakeholders, and particularly the local community, will have been consulted, and, if appropriate, they will have contributed to some of the decisions. At this stage, the plan can be regarded as a detailed recommendation, with costs, put forward by the reserve manager to the organisation responsible for managing the site. The plan should then be approved, with or without amendments, and returned to the reserve manager. It has now become an instruction. By applying a formal approval procedure, the organisation adopts and accepts responsibility for the plan, including all the resource implications. This is simply good staff management. The reserve managers are fully involved and given appropriate levels of control. Through the approval process, organisations confirm their confidence in their employees and accept full responsibility for their actions. The employees should respond by working to the requirements of the plan.

2.3 Inputs, Outputs & Outcomes

Table 2.2 Inputs, outputs & outcomes

Inputs:	The resources required to manage a site.
Outputs (process):	For example: the production of management plans; survey/surveillance/monitoring; site infrastructure, including provisions for wildlife and visitor management (fences, roads, trails, hides, etc.)
Outcomes:	The condition that we require for habitats and species.

The use of the words 'inputs', 'outputs' and 'outcomes' has become increasingly commonplace in conservation management. This is probably a reflection of their wider association with most management activities. Senior managers and politicians appear to have considerable affection for these words, which they use, often erroneously, when they want to talk about their actual or intended achievements. Some USA papers (McCool and Cole 1998) also use inputs and outputs. Unfortunately, this is very confusing because their 'inputs' are equivalent to our 'outputs', and their 'outputs' (they give 'environmental conditions' as an example) equate to our 'outcomes'. As an alternative to 'inputs, outputs and outcomes', the commercial or business world often uses 'inputs, process and outcomes'. These carry the same meaning but are perhaps less confusing. The messages that these words convey, provided their meanings are understood, are extremely important. Setting aside all other uses, when applied to nature conservation they are defined as follows:

Inputs are the resources that are made available for management. They can usually be divided into finance and staff time. (In addition to staff time, many wildlife organisations rely on voluntary labour.)
Outputs are the consequential by-products of management or the management *process*. For example, management plans are prepared, interpretation is provided, a management infrastructure is developed and maintained, and internal and external boundaries are constructed. Often, outputs are used as a means of assessing whether management is appropriate. Managers will sometimes claim that they have successfully managed their sites because they have achieved a number of outputs. This can be very misleading since it is possible to carry out a wide range of management activities and still fail to obtain favourable status for the conservation

features. One of the worst mistakes that anyone engaged in nature conservation management can make is to claim that a feature is being successfully protected when, in reality, it is not. Outputs are an indication of activity but not necessarily of success.

Outcomes are the end point or purpose of conservation management. They are the status that we require for the features (habitats and species) that we are trying to conserve (Figs. 2.3 and 2.4). Whereas inputs and outputs are usually measured, the measurement of conservation outcomes has, until recently, rarely been attempted. This is partly because managers did not recognise the need, but also because there has been so little guidance available. As is the case with adaptable management, we must be able to determine and quantify the conditions that we require of the conservation features. If we cannot do this, it will be impossible to measure those conditions and, consequently, to judge whether we have achieved required conservation outcomes.

The only means of judging whether or not inputs and outputs are adequate is by considering the outcomes of management. When we are able to do this, and only then, will we be in a position to determine when management is appropriate.

We have come to realise that we *must* measure conservation outcomes as well as outputs. Both measurements are essential if we are to come to any meaningful conclusions about management effectiveness and our ability to safeguard wildlife. This planning process provides a methodology for measuring both outputs and outcomes.

Fig. 2.3 Conservation outcome – a wet pasture

Fig. 2.4 Conservation outcome – a population of a species (gannets)

2.4 The Precautionary Principle

In essence, the precautionary principle is about *not* taking chances with our environment. It moves the 'duty of care' or 'onus of proof' from those who attempt to protect the environment to those who propose changes or development. The principle is almost always associated with the Rio Convention on Biological Diversity. However, its origins were in the German '*Vorsorgeprinzip*', or foresight principle, which appeared in the 1970s and later became a principle of German environmental law. It was adopted by the First International Conference on Protection of the North Sea in 1984, when it was defined as follows:

> Damage to the environment can be irreversible, or remediable only at a considerable cost and over long periods of time, and that, therefore, coastal states and the EEC must not wait for proof of harmful effect before taking action.

At the 1992 United Nations Conference on Environment and Development in Rio de Janeiro, world leaders adopted, and advocated the widespread international application of, the precautionary principle:

> In order to protect the environment, the precautionary approach shall be widely applied by States according to their capabilities. Where there are threats of serious or irreversible damage, lack of full scientific certainty shall not be used as a reason for postponing *cost-effective* measures to prevent environmental degradation.

The Rio Convention is aimed at protecting the world's natural assets. The introduction of 'cost-effectiveness' to the definition is unfortunate as it diminishes the definition by providing an escape route for politicians and developers. However, despite its weakness, the precautionary principle supports most conservation efforts.

Turning to Europe, the precautionary principle was formally adopted by the European Union in the Treaty of Maastricht in 1992. Later, it was applied with legal authority to the Natura 2000 sites. The definition is clearly set out in an official European Commission document, *Managing Natura 2000 sites: The provisions of Article 6 of the 'Habitats' Directive 92/43/EEC* (EC 2000). The article states:

> Member States shall take appropriate steps to avoid, in the special areas of conservation, the deterioration of natural habitats and the habitats of species as well as disturbances of the species for which the areas have been designated, in so far as such disturbance could be significant in relation to the objectives of this directive.

In addition to the article, the document contains two particularly relevant passages:

> This article should be interpreted as requiring Member States to take all the appropriate actions which it may reasonably be expected to take, to ensure that no significant deterioration or disturbance occurs.
>
> In addition, it is not necessary to prove that there will be a real significant effect, but the likelihood alone ('could be') is enough to justify corrective measures. This can be considered consistent with the prevention and precautionary principles.

Despite its formal adoption by European and many other governments, the precautionary principle is extremely controversial. There are concerns expressed by both environmentalists and developers. Many argue that it is an obstacle to innovation and progress. However, most environmental commentators appear to support the principle, and many make the point that it is simple common sense. All life on earth is dependent on a healthy environment: we must not take risks.

The precautionary principle is important in the context of conservation management and planning. It should be adopted regardless of any controversy, and it should influence the way in which we manage sites, habitats and species. If the precautionary principle is applied, the following are some of the more obvious implications for the management of protected areas:

- There is no need for scientific proof in order to restrict human use, or any specific activities, when there is reason to believe that they are a potential threat. Logically, we should, in fact, obtain conclusive evidence to demonstrate that an activity is *not* a threat to the site or to the wildlife before giving consent.
- Unless we have conclusive evidence to demonstrate that conservation features are at favourable conservation status we should assume that they are unfavourable. (If the status of a feature is unknown, we should assume that it is unfavourable.)
- Factors that affect, or may affect, conservation features should not be dismissed until we are confident that they are not a threat.
- We should take steps to control threats (factors) even when there is insufficient scientific evidence to support our concern.
- We must not assume that management will inevitably achieve the desired results. Management can only be considered appropriate when we have conclusive evidence to demonstrate that it is delivering the required outcomes.

References

McCool, S. F. and Cole, D. N. (1998). *Proceedings – Limits of Acceptable Change and related planning processes: progress and future directions;* 1997 Missoula, MT. Gen. Tech. Rep. INT-GTR-371. Ogden, UT, USA.

Ratcliffe, D. A. (ed.) (1977). *A Nature Conservation Review.* Cambridge University Press, Cambridge.

Chapter 3
Language and Audience

Abstract Management plans should be made available to everyone who has an interest in the site. This will include people who do not have a scientific or technical background and may not necessarily have any interest in, or understanding of, wildlife or conservation management. Management plans are about communicating with a sometimes very wide and diverse audience. This suggests that the language used in the plan should, whenever possible, be plain and accessible to all. However, the use of plain language must not be taken as an excuse to be patronising or to diminish the value of the contents. It is also possible, though perhaps not essential, to share with others the values, feelings and enthusiasm that we have for the sites that we manage.

3.1 Audience

Whenever possible, management plans should be made available to the widest possible audience. Occasionally, there will be a need to include sensitive or confidential information, for example, the location of rare and endangered species. Clearly, this should be omitted from a public version of a plan. Everyone who has an interest of any kind in the site, particularly neighbours, local residents and all other stakeholders, should be able to access information which is of interest or relevant to them. Regrettably, this rarely happens and, even when plans are made available to the public, the style of presentation and the language used in the documents can be impenetrable.

M. Alexander, *Management Planning for Nature Conservation: A Theoretical Basis & Practical Guide*, DOI 10.1007/978-94-007-5116-3_3,

3.2 Language

Communication – various definitions:

- A connection allowing access between persons.
- The process of exchanging information and ideas
- The transmission of information so that the recipient understands what the sender intends.

In 2005, I prepared a paper for the annual Welsh Conservation Management Conference. In preparing my paper I interviewed senior representatives of all the major conservation organisations, government and non-governmental, in Wales. One of the most startling observations was that although the interviewees used the same words they often defined even common words or expressions differently. Recently, I asked a group of conservation managers attending a training course if they carried out monitoring on their sites. Most raised their hands. We then discussed and defined monitoring, and I repeated the question. There was no show of hands. Lack of common definitions can lead to confusion, misunderstanding and an inability to share a common purpose. The problems of communication between conservation professionals fade into insignificance when compared with the problems of communication between professionals and others. Usually, professional language is quite inappropriate for general consumption. Most professionals use language that is heavily dependent on abbreviations, acronyms and technical expressions. The following example, taken from the medical profession, illustrates the problem.

Medical language:

> Lansoprazole is effective in the treatment of NSAID associated GUs and DUs and Zollinger – Ellison syndrome and in the eradication of helicobacter pylori. Side effects including LFT alteration, gynaecomastia, petechiae and RF have been reported. Stevens-Johnson syndrome, toxic epidermal necrolysis and erythematous or bullous rashes including erythema multiforme have been reported occasionally. Take 30 mg cap o.d.

In lay terms:

> Lansoprazole is a drug which is very useful in the treatment of ulcers in the stomach and first part of the intestine. Side effects include changes to liver function, breast formation (in men), skin rashes and kidney failure. There are a few very rare but serious other skin conditions. Take one capsule once a day. (Rivett 2005)

The meaning is the same in both versions. This is a clear demonstration that it is possible to use plain language and yet maintain meaning and scientific integrity.

The Susie Fisher Group (2004) was contracted to explore the public understanding of nature reserves in Wales. A small number of local groups were established across the country, and the members of each group were carefully selected in order to ensure a representative cross-section of the local population. The groups discussed a range of terms, ideas and concepts, without any guidance from the consultants. This is a sample of the responses given when the groups were asked to define 'biodiversity':

What does biodiversity mean?

- The hares and rabbits. You see, if there are lots of hares, the rabbits start dying off. If the hares start dying off there is no food and then it is a total circle.
- The environment
- The insides of animals
- Washing powder
- Biology
- It's the interaction of nature with agriculture
- You get a lot of biodiversity in a habitat where there's a lot of species

Biodiversity has become such an important word in our vocabulary that we might assume it is widely understood. It is hardly possible to open a newspaper or watch television without some reference to biodiversity. If we do not even share a definition of a word like biodiversity with the general public, then consider how unfathomable most of our language must be.

If management plans are recognised as a means of communicating our intentions, sometimes to a very wide audience, the use of plain language is essential. Occasionally, there may be circumstances where a plan is prepared entirely by experts for use by experts, but this is rare. Conservation management and planning should be an inclusive activity, and providing stakeholders with access to management plans is possibly one of the best ways of encouraging their involvement. Plans must never be written in a patronising style, but they should not contain difficult or obscure scientific language. For example, scientific species names should be accompanied by a common name whenever possible. Where a common name is widely understood the scientific name may not be necessary. It is, however, important that the quality of the information conveyed in the plan is not diminished as a consequence of using plain language.

3.3 Sharing Enthusiasm

Taking the way in which we communicate a little further, we can improve things by communicating with genuine feeling. If we believe so strongly in the importance of wildlife then perhaps we should also be prepared to share our enthusiasm with others.

> Emotion is the source of all becoming-consciousness. There can be no transforming of darkness into light and of apathy into movement without emotion. (Jung 1968)

> Feelings and emotions are the source of our ideas, inspiration and creativity. (Naess 2002)

Most people involved in nature conservation, and consequently most people who write management plans, will share a love of the natural environment. We take it so much for granted that we often forget to speak about it, and this silence can become inhibiting. It is not always easy to break through these hidden barriers and talk about feelings when the scientific realities are so much safer and easier to quantify. Perhaps sometimes we hide behind the anonymity of scientific jargon because we have no words for our own emotions. At work, we rarely talk about feelings or emotions, and yet, for most people, the reason for their choice of vocation in nature conservation was a deep emotional response to an experience sometime in their lives. Some are motivated by a positive experience and others as a consequence of witnessing disaster or destruction. We disguise our emotions in an attempt to present the illusion of dispassionate objectivity. Clearly, there are times and places when this is important, but, equally, there are times when we need to share our feelings. If no one breaks the silence we will become trapped by conformity. A wide range of influences, particularly peer pressure, encourage us to conform, but simply because ideas of behaviour have become widely accepted it does not mean that there are no better alternatives.

There are several areas in a management plan, none more important than the objectives, which would benefit enormously if the text could also convey some of the values and feelings we have for the very special places that we manage. Through sharing our values with others we might inspire them and help them to gain a deeper appreciation of what we are trying to achieve.

The following examples of visions (the descriptive component of the objective) are intended to inform and, hopefully, to enthuse the reader:

3.3.1 Visions for an Upland Oak Woodland (Fig. 3.1)

The following examples of visions both contain the same factual information, but the messages that they convey are very different.

Vision for an Upland Oak Wood – Simple Version

The entire site is covered by a high forest, broadleaf woodland. The woodland is naturally regenerating, with plenty of seedlings and saplings particularly in the canopy gaps. There is a changing or dynamic pattern of canopy gaps created naturally by wind throw or as trees die.

The woodland has a canopy and shrub layer that includes locally native trees of all ages, with an abundance of standing and fallen dead wood to provide habitat for invertebrates, fungi and other woodland species. The field and ground layers are a patchwork of the characteristic vegetation communities developed in response to local soil conditions. These include areas dominated by heather, or bilberry, or a mixture of the two, areas dominated by tussocks of wavy hair grass or purple moor grass, and others dominated by brown bent grass and sweet vernal grass with abundant bluebells. There are quite heavily grazed areas of more grassy vegetation. Steep rock faces and boulder sides are covered with mosses, liverworts and filmy ferns.

The lichen flora varies naturally depending on the chemical properties of the rocks and tree trunks within the woodland. Trees with lungwort and associated species are fairly common, especially on the well-lit woodland margins.

The woodland does not contain any rhododendron or any other invasive alien species with the exception of occasional beech and sycamore. There is periodic light grazing by sheep and very occasionally by cattle. This helps to maintain the ground and field layer vegetation but does not prevent tree regeneration.

Vision for an Upland Oak Wood – Inspirational Version

In spring, sunlight sifts through the pale, translucent green of the newly emerged leaves sketching bright patterns between the trees. The upland broadleaved woodland covers the entire site. It has a mixture of trees differing in age, size and density, a variety that is maintained by natural processes. Scattered through it is a patchwork of temporary glades that are slowly filled by naturally regenerating tree seedlings and saplings. At the same time, new openings are created, forming a gradually changing mosaic of light and shade, where as much as a quarter of the woodland may be open glades, rides and other canopy gaps. At certain times of the year, sheep may occasionally wander among the trees and glades as they graze. With so much diversity, a whole web of life, from plants and mammals to birds and insects, is woven through the woodland.

Most of the trees and shrubs are locally native broadleaved species such as sessile or hybrid oak, downy or pendulous birch, ash, rowan, holly, elm and hazel. These, together with occasional non-native species such as beech and sycamore, create patterns of dappled green as the mix of trees changes throughout the woodland. The monotonous, deep green of rhododendron, which has invaded some of the surrounding countryside, does not encroach into the woodland. For life in the woodland to flourish, there must be a balance between decay and new growth; dead and dying trees, as well as live trees with holes, hollows and rotten branches, provide the necessary habitat for a rich variety of mosses, liverworts and fungi, and also for specialised insect species.

The field and ground layers make a brilliant tapestry of colours and textures. Some areas are bright with the vibrant greens and muted purples of heather and bilberry. In others there are soft tussocks of wavy hair grass or purple moor grass. There are also swathes covered mainly by brown bent grass and sweet vernal grass, with occasional drifts of pale indigo bluebells in spring. The dense undergrowth helps to maintain the humidity beneath the canopy, which is essential to the survival of many mosses and liverworts. In rocky areas, or where the soil is thin and acidic, these form deep, green carpets. The mix of lichens in all their bright and subtle shades varies throughout the woodland, depending on the rocks and trees that support them. Particularly around the fringes of the wood, where sunlight seeps through, the tree bark is draped with rippled, silver-green clumps of lungwort.

Birdsong resonates through the wood during the breeding season, and there is a faint rustle of leaves where birds such as pied flycatchers, redstart and wood warblers flit between the branches. As the light fades, bats dart silently through the canopy, barely more than shadows in the twilight, and badgers emerge to forage in the growing darkness.

Fig. 3.1 An upland oak wood

3.3.2 Vision for the Condition of a Blanket Bog (Fig. 3.2)

This is a slightly different approach. It describes the feature from the perspective of the experience that a visitor will enjoy on the site:

From a high vantage point, the blanket bog extends as far as your eye can see. At a first glance the bog looks a uniform greenish-brown colour, but a second glance shows a rich mix of reds, browns, greens, yellows and in early summer, the nodding white heads of cotton-grass.

Walk over the blanket bog and you will be aware of the wide range of plants that thrive here. The bog plants grow on a deep layer of waterlogged peat, often several metres thick and made up of the partly decomposed remains of previous bog plants. The surface of the bog is made up of a mixture of small, moss-filled hollows and slightly drier hummocks where heathers grow. You will discover an occasional small bog-pool.

The tallest plants, standing at about knee-height, are cross-leaved heather, which grows in the wetter areas, common heather and cotton-grass. You will also find bilberry, crowberry, cranberry, deer grass, and purple moor-grass.

Sphagnum bog mosses grow below these taller plants. These spongy, water-holding mosses form a low, almost constant and colourful carpet in a variety of greens and reds. Look carefully and you may also see insect-eating sundews and on some of the drier hummocks the fragrant yellow bog asphodel.

The larvae of the large heath butterfly feed on the flower heads of cotton grass and so you may be lucky enough to see some of these rare butterflies on sunny days in early summer.

Occasionally, birds such as hen harrier, merlin and peregrine falcon give spectacular displays as they fly above the blanket bog and surrounding wet heath and acid grassland, which form part of their feeding and nesting areas.

Fig. 3.2 A small blanket bog

3.3.3 Visions for a Species – Guillemots

Two examples of visions for a population of guillemots; both contain the same factual information, but the messages that they convey are very different.

Vision for Guillemots – Simple Version

Skomer Island is a very important breeding site for a large, robust and resilient population of guillemots (Fig. 3.3). The size of the population is stable or increasing (in 2011 the population was 21,688).

The distribution of the colonies, shown on the attached map, is maintained or increasing.

At least 80% of the breeding adults survive from 1 year to the next, and at least 70% of the breeding pairs raise a chick each year. This will help to ensure the long-term survival of the population.

The safe nesting sites and secure breeding environment are protected. There are no ground predators and the impact of predatory birds is insignificant. The size and range of the population are not restricted or threatened, directly or indirectly, by any human activity on the island. The nesting colonies are not disturbed from the sea by boats or other human activities during the breeding season.

Vision for Guillemots – Inspirational Version

The sight and sound of a guillemot colony provides a truly spectacular encounter with seabirds: shimmering clouds of birds skim beneath the cliffs and the constant clamour of their exuberant cries saturates the air. Skomer Island is a secure haven for large numbers of breeding guillemots, allowing the strong and resilient population to thrive. These upright, penguin-like birds, with stark brown and white colouring, lay their eggs on the bare rock ledges of the cliffs. On land they and their chicks are vulnerable, but there are no ground predators on the island to threaten them, and airborne predators have little impact. Though visitors can enjoy wonderful views of the guillemots, the birds are not disturbed by people using the island for any reason, and there is no adverse effect on their numbers, distribution or breeding success. The same is true of the surrounding sea, where no boating or other maritime activities disrupt the lives of the birds at or around the colonies.

In 2011, 21,688 guillemots bred on Skomer, and this number remains constant or may increase. The accompanying map shows how the birds are distributed around the island. In some areas there are only small clusters of guillemots, but in the largest colonies close-packed birds smother the ledges, and the cliffs echo with their deep, resonant calls. None of these colonies is diminishing, and they may continue to grow in size or number. At least 80% of the breeding adults survive from 1 year to the next, and over 70% of breeding pairs raise a chick each year, helping to safeguard the future of these magnificent birds. With so many successfully breeding birds the colonies are alive with activity. When the eggs are hatched, the air hums with the sound of rapid wing beats as birds return to the cliffs with the silver glint of fish just visible in their beaks. At fledging time the urgent cheeping of chicks adds to the cacophony of noise as the near-flightless youngsters tumble from the cliffs to find the parent birds in the water below.

There are many circumstances where this type of language will not be appropriate. One of the more important principles of planning is, ‘small is beautiful’. Our obligation should be to produce fully functional plans at the least possible cost in time and other resources. Plans should also be as concise as possible. However, having made these points, if a plan is intended for a wide public audience inspirational language may be necessary. There is no reason why we should not produce different versions of a plan providing, of course, that the meanings are entirely consistent. There may be the need for a functional version, which adopts a minimalist approach where the language is precise and uncluttered.

Fig. 3.3 Guillemot

References

Jung, C. G. (1968). *Collected Works of C. G. Jung, Vol. 9, Part 1*, 2nd ed. Princeton University Press, Princeton, USA.

Naess, A. (2002). *Life's Philosophy: Reason and Feelings in a Deeper World*. The University of Georgia Press, Athens and London.

Rivett, A. (2005). Personal communication.

The Susie Fisher Group (2004). *Nature reserves and the natural history concepts which underpin them*. Unpublished report, Countryside Council for Wales, Bangor.

Chapter 4
Local Communities and Stakeholders

Abstract A stakeholder is any individual, group or community living within the influence of the site or likely to be affected by a management decision or action, and any individual, group or community likely to influence the management of the site. Conservation managers must recognise the need to adopt an inclusive approach which takes account of the interests of stakeholders and, as far as possible, encourages their involvement in all aspects of management and planning. One of the key issues when building and maintaining successful relationships is to have a shared appreciation of what can and cannot be negotiated. Relationships with stakeholders are so important that they should be dealt with explicitly within the planning process. It is possible to prepare an objective which establishes the desired relationship with stakeholders and to identify within an action plan all the essential activities necessary to meet the objective.

4.1 Background

In earlier years, protected area management had a reputation for excluding people or severely restricting human activities (Borrini-Feyerabend et al. 2004). This may have been true for some sites and it may represent the views held by some pioneering conservation managers, but, as a generalisation, it is an unfair criticism. It is true that many managers saw a separation between wildlife and people, and this may occasionally have led to exclusive attitudes and practices. However, this can be overemphasised: the difficulties that can exist between managers of protected areas and stakeholders are the consequence of a combination of many different factors. Among the most significant is poor communication, leading to misunderstanding and intolerance (unfortunately, an all too common human failing). Over the past decades, certainly since the 1970s, there has been a move or paradigm shift towards a much more inclusive approach to managing protected areas (Borrini-Feyerabend et al. 2004).

M. Alexander, *Management Planning for Nature Conservation: A Theoretical Basis & Practical Guide*, DOI 10.1007/978-94-007-5116-3_4,

Managers of sites established by government agencies face a particular problem. The sites are most often established as an assertion and expression of 'common values' through the democratic process. That is, at a high level society decides to protect wildlife for the common good, and one of the methods employed is the establishment of protected areas. Unfortunately, at site level this can be, or may be perceived as, a threat to the interests of local people. Conservation managers have come to realise that they must involve, and consider the interests of, stakeholders (particularly local communities) in the management of protected areas. Society will tolerate the right of individual owners to do more or less what they like on their own property, often without question, providing their actions are legal and meet any planning regulations. However, when that land is managed for wildlife, and especially when the management is carried out by a government agency, society presumes a right to be involved, to be consulted and to influence management. If we want to maintain conservation areas, we must be prepared to adopt inclusive approaches; we must work with other people. We should also recognise that the capacity to appreciate and enjoy wild places and wildlife, and not simply to regard them as an essential resource, is often restricted to individuals who do not have to depend on these areas for their livelihood.

Sites are never isolated from their surroundings; it is usually only possible to safeguard them with the co-operation of others. Protected area managers should recognise that local stakeholders can make a very significant contribution towards managing a site. Once stakeholders gain a sense of ownership there are many different ways in which they can help: for example, local knowledge and traditional skills are often essential, especially when these complement good science.

Stakeholders, local communities and indigenous people can gain substantial benefits from the presence and management of protected areas. Some of these benefits are obvious and include opportunities for activities such as fishing, hunting, grazing, reed harvesting, recreational use and ecotourism. Other less obvious benefits include the protection and maintenance of spiritual and cultural values associated with a site, and the maintenance of ecosystem functions, for example, flood control and improved water quality.

Any mention of stakeholders, or anything else that might have implied stakeholder or local community involvement, was, until very recently, absent from management planning guides (NCC 1981, 1983, 1988; Alexander 1994, 1996; Eurosite 1999). By the late 1990s, management planning guides began to put greater emphasis on stakeholder involvement. *Measures of Success* (Margoluis and Salafsky 1998), a USA guide to management planning, stresses that local stakeholders must participate in the plan. One of the earliest European planning guides which included guidance on planning the management of 'relationships with the local community' was *The CMS Management Planning Guide for Nature Reserves and Protected Areas* (Alexander 2000b). In 2002, Eurosite organised a workshop on stakeholder involvement in management planning. One of the conclusions of that workshop was:

> Stakeholder involvement is a method that can help in protecting and managing effectively nature conservation sites. Involving stakeholders is not a goal in itself. It should be part of a complete set of activities. Stakeholder involvement is not in all cases needed or equally

important. Involving stakeholders can be time and money consuming, so consider whether or not it will really help you.

The Eurosite report, while recognising the importance of stakeholder involvement, adopts a very pragmatic approach: it is something that we do in order to improve site management. This is an important point. Many people seem to think that stakeholder involvement is an end in itself: they forget the wildlife.

The IUCN *Guidelines for Management Planning of Protected Areas* (Thomas and Middleton 2003) contains a chapter on 'involving people'. It provides a list of the benefits of involving people in management planning:

- Increased sense of ownership
- Greater support for the protected area
- Greater public involvement
- Links planning for conservation with planning for development
- Provides a mechanism for communication

This very succinct list sums up the recognised benefits of including stakeholders in management planning.

In sharp contrast to earlier Eurosite publications, their guide, *Management Planning for Protected Areas* (Idle and Bines 2005), offers an approach to planning 'arising from the need to involve stakeholders'. This guide appears to hand decision making over to stakeholders. It also puts considerable emphasis on the use of 'professional facilitators' to guide the development of the plan. While these facilitators are completely impartial, with no hidden agendas, their employment is likely to add greatly to the cost of producing a plan. Although the guide is intended for European Natura 2000 sites, surprisingly, it does not appear to recognise the legal nature of the features on these sites.

In summary, the move away from an exclusive, towards a more inclusive, approach to conservation management has been followed by a similar change in management planning guidance. Most current guides recognise that stakeholders' involvement is crucial to the success of a plan. One guide goes further and, in addition to advocating stakeholders' involvement, also recommends that an objective for ensuring continued stakeholder participation is included in the plan. More recently, some people have suggested that the entire process can be handed over to stakeholders. I believe that stakeholder involvement should mean working in partnership with others to seek solutions that protect our sites and the wider environment, enhance the quality of life and provide benefits for stakeholders and the wider society, and have the least possible impact on the freedom of individuals. The level of stakeholder involvement will vary according to local circumstances. There may, occasionally, be sound justification for a community to manage its own sites.

Unfortunately, poor relationships with stakeholders are not uncommon and may occasionally lead to serious conflict. Whatever the reasons for the divisions (and the sometimes disastrous and expensive consequences), conservation managers can only ever control or change their own attitudes and actions. Where others are concerned, managers can only hope to influence them through reasonable discussion that treats their opinions with respect.

4.2 Stakeholder Definitions

There are many different definitions of 'stakeholder', but the following is reasonably typical:

> The term 'stakeholder' is generally used to define a person, group or organisation who has an interest in an issue, service or resource, or who is, or would be, affected by it. The current definition used in conservation circles is 'those likely to be affected by a decision or likely to affect the implementation of a particular decision'. (Caldwell and Evison 2005)

Some definitions divide stakeholders into different categories:

Example 1 (Rientjes 2000)

Primary stakeholders
- Stakeholders whose permission, approval or (financial) support is required
- Stakeholders directly affected by the plan or activity
- Stakeholders who will benefit
- Stakeholders who will suffer loss or damage

Secondary stakeholders
- Stakeholders who are indirectly affected

Tertiary stakeholders
- Stakeholders who are not directly involved but can influence opinion

Example 2 (Baker Associates 1997)

- Direct partners – priority relationships where there is ongoing contact
- Participative groups and significant others – those with an interest in the situation, important to the process, some ongoing contact
- Statutory agencies
- General consultees – community groups, general public

These divisions can be very contrived or intended for specific applications. Any approach to creating divisions of this kind will have both positive and negative consequences. People often find that dividing a topic under specific headings helps to ensure that all areas are considered: in other words, the headings act as a series of prompts. Conversely, whenever divisions are created they very rarely provide unambiguous 'boxes' for all possible categories. The consequence is that some people will spend an inordinate amount of time trying to decide on the most appropriate divisions. The divisions should be regarded as an aid and not an encumbrance, and should be tailored to meet any specific situation.

For the purpose of site management, and this chapter, the definition of stakeholder is extended to mean:

A stakeholder is any individual, group, or community living within the influence of the site or likely to be affected by a management decision or action, and any individual, group or community likely to influence the management of the site.

This definition includes 'local community'. There will rarely be one single, clearly identifiable community. Individuals can be part of several different communities. A simplistic view of communities will regard spatial boundaries as the only definition. However, even within a clearly defined area there can be several quite distinct communities, often overlapping. For example, religious divisions often exist within a community. Other divisions will include age, occupation and political inclination. These sections are sometimes in conflict and may not agree on all issues. This means, of course, that from a site manager's perspective it will rarely, if ever, be possible to obtain the approval of everyone.

4.3 Stakeholder Involvement in Plan Preparation

The Ramsar Convention Bureau (2000) claims that experience has shown that it is advisable to involve local and indigenous people in a management partnership when:

- The active commitment and collaboration of stakeholders are essential for the management of a site, for example, when the site is inhabited or privately owned.
- Access to the site is essential for local livelihood, security and cultural heritage.
- Local people express a strong interest in being involved in management.
- Local stakeholders have historically enjoyed customary/legal rights over the site.
- Local interests are strongly affected by the way in which the site is managed.

4.4 Facilitators

Before making any attempt to engage with stakeholders the need to involve a facilitator should be given some consideration. Skilled and competent facilitators are often an essential prerequisite to successful negotiation. Some publications (The Ramsar Convention Bureau 2000; Idle and Bines 2005) recommend the use of facilitators or coordinators when involving stakeholders in any participatory process. An obvious consideration when using facilitators is that people may assume they have been employed to address confrontational issues, and discussions could be hampered by preconceived ideas that at least one of the parties is expecting disagreement. This may not be the case, but when people expect confrontation they often find it. Whether going it alone or involving facilitators organisers should:

- Ensure that all the stakeholders understand the role of the facilitator or organiser.
- Regularly verify that all stakeholders agree on the basic objectives of the initiative.
- Ensure the involvement of representative groups and individuals from all significant sectors of the community and ensure that no key participants are excluded. (Assist in the establishment of representative groups if they do not already exist.)
- Ensure stakeholders share ownership of the process.

- Ensure that key parties have a clear understanding of each other's needs, responsibilities and limitations.

There will always be problems associated with stakeholders. Some people, because of their position in society or their background, will be articulate and confident communicators. They can dominate discussion, even when their particular interests are of little consequence. Other people, who may have a crucial role or interest, fail to be heard because they lack confidence or perhaps because they feel that they cannot influence decisions or make a difference. It is the responsibility of site managers to ensure that everyone is given an equal opportunity to participate and that proceedings are not dominated by a minority.

4.5 What Is Negotiable?

This is perhaps the most significant question of all. The answer, of course, is that the limits on negotiation, i.e. what can and cannot be negotiated, will vary from circumstance to circumstance. It is essential, from the onset of any collaborative venture or discussion, that both parties understand what is and what is not negotiable. The worst possible mistake is to give stakeholders the impression that something is negotiable only for them to find out later that it is not. Unfortunately, this happens all to often, usually when, albeit with the best of intentions, managers or planners rush into consultation with stakeholders before they have taken the time to understand for themselves what constraints may be imposed on their ability to negotiate or compromise. Probably the best examples come from statutory sites where managers have a legal obligation to protect the qualifying features. For example, there is a legally imposed obligation to obtain and maintain Favourable Conservation Status for all Natura 2000 sites, and this obligation has been extended to include Ramsar sites. Clearly this is not negotiable: there are no choices or decisions that can be shared with others. However, the way in which a site is managed should be negotiable. Sometimes, but not always, there will be several different management options. The most likely penalty of compromise is that management will be less efficient or it will take longer to achieve the desired results.

Whenever there is a case for involving stakeholders in the planning process, the first stages in the plan should be drafted before making any formal contact. The legal status of the site, along with any other legal obligations and constraints, must be clearly understood. As far as possible, the information that is relevant to planning should be collated, and any gaps that might be filled with information from stakeholders should be identified. The objectives should be drafted and the extent, if any, to which they may be modified is noted. The full range of potential management options should be identified. Taken together, this information will provide the basis for shared decision making.

4.6 The Stakeholder Section in a Plan

So far in this chapter I have very briefly considered the involvement of stakeholders in site management and then, in greater detail, I have discussed their involvement in preparing a plan. Stakeholder and community interests can have considerable implications, both positive and negative, for site management, and they can impose significant obligations on the site manager. Public interest, at all levels, must be taken into account. Conservation managers must recognise that other people may have many different, and sometimes opposing, interests in the site. It is essential that these interests are safeguarded wherever possible. There may be a justifiable need for compromise, providing, of course, that the prime objectives of management are not jeopardised. Maintaining communication and, whenever necessary, consultation with stakeholders is essential, at the very least to keep them informed of any developments that may affect them. In order to safeguard wildlife successfully, conservation managers need to adopt a flexible approach that will allow them to respond to the legitimate interests of others, to adapt to the ever-changing political climate, to accommodate uncertain and variable resources, and to survive the vagaries of the natural world.

For these reasons, the Conservation Management System Consortium planning guides (Alexander 2000, 2005) recommend, and provide guidance on, preparing a section in a management plan which deals explicitly with stakeholder relationships. This is such an important consideration that it deserves the same attention as any other area within the plan. Site managers must organise or plan their involvement with stakeholders: simply responding and taking remedial action when things go wrong is not good enough. Relationships must be established and maintained. This takes time and effort, and it should be planned. This is a very simple section to complete, but the planning principles used in all other sections of the plan are just as important here. It is only possible to identify activities when we know what we are trying to achieve. This means that a plan should contain an objective for stakeholder relationships, followed by the activities that are necessary to meet the objective.

When preparing a plan for isolated, remote sites where there are few, if any, people and little external interest, this section can be dealt with in a few paragraphs, but for some sites with sizeable resident populations, or sites surrounded by densely populated areas, this section can be larger than the rest of the plan. As is the case for all sections in the plan, this section should be as large as it needs to be and no larger.

4.6.1 Description/Stakeholder Analysis

A stakeholder analysis is simply a systematic approach to identifying all relevant stakeholders. The information will later be used to prepare an action plan which will define the circumstances when the stakeholders will be consulted and how they should be involved. All stakeholders or groups of stakeholders will require attention at some time, but usually not all at the same time or for the same purpose.

All the stakeholders must be identified. One way of doing this is to consider the reasons for wanting or needing to engage with stakeholders, and also why they would want or need to be contacted. Earlier in this chapter, I gave examples of different ways in which organisations define or categorise stakeholders, and recommended that categorisation should be tailored to meet the requirements of an organisation or of any specific situation. The Rientjes (2000) approach is a good general guide to this process. Stakeholders are divided into three groups, each with different interests. Regardless of how the stakeholders are divided, each category can comprise:

- Professional staff from various organisations, including businesses.
- Representatives of organised groups: these can be local or national, for example, a fishing society, county wildlife trust or community council.
- Individuals.

The following is a *modified* version of the Rienjes approach:

Primary stakeholders

- Stakeholders whose permission, approval or (financial) support is required. (This will include statutory consultees.)
- Stakeholders directly affected by site management.
- Stakeholders who will benefit.
- Stakeholders who will suffer loss or damage.

Secondary stakeholders

- Stakeholders who are indirectly affected.

Tertiary stakeholders

- Stakeholders who are not directly involved, but can influence opinion.

Once the stakeholders have been identified, the following information should be recorded:

- The most appropriate means of communicating with each individual or group.
- The sort of engagement they require, if any.
- Their contact details (or, when dealing with a group, details of a representative).
- How they want to be contacted, for example, mail or email.
- Their interests or the issues they want to be involved in.
- Their relationship with the protected area. (This will include an extremely diverse range of interests, for example, dog walking, bird watching, fishing, grazing or other agricultural rights).

Fig. 4.1 Stakeholder involvement

4.6.2 Objectives for Stakeholders

Please note: a fuller account of preparing objectives is included in Chapter 15.

The relationships that organisations choose to have with stakeholders are entirely a consequence of their policies. A simple objective for relationships with stakeholders and the local community could be to achieve a state where:

> Mutual understanding, co-operation and respect optimise benefits for stakeholders and make a positive contribution towards protecting the site.

This is possibly too universal, i.e. it could be applied more or less anywhere. An objective, in this context, is a simple statement of the ideal state for relationships with stakeholders. It may not be an obtainable state in the short term, but it will provide a consistent direction for all developments in this area.

4.6.3 Performance Indicators and Monitoring

In common with all objectives, there must be some means of measuring achievement. It is not an easy task to select performance indicators that will measure the quality of relationships with stakeholders. One obvious approach is to identify a series of monitoring projects to ensure compliance with any management projects identified in the rationale. For example, the rationale could identify the need to contribute towards the provision of environmental education in the local schools. This activity must be planned, and a compliance-monitoring project identified, to ensure that the work is carried out as required. Monitoring compliance will tell us that a planned action has taken place, but it will not enable any evaluation of how effective the action has been towards meeting our objective of improving relationships with the community. It should, however, be possible to make direct, though possibly subjective, measurements. For example, it may be possible to gauge stakeholder opinion by recording the number of complaints or compliments received and noting any trends. Informal liaison will provide a proactive approach. It may even be appropriate, in some circumstances, to use formal interview or questionnaire techniques.

4.6.4 Status and Rationale

This section is concerned with justifying the allocation of resources and time to obtaining and maintaining good relationships with stakeholders. The first step is to consider the implications of status. Status is quite simply the difference between what we want and what we have. If relationships are excellent, where excellent is defined by the objective, then any current management activities are probably appropriate. Conversely, if relationships are poor a change of management is required.

The rationale is concerned with identifying and describing, in outline, the management activities considered necessary to obtain and maintain an appropriate relationship with stakeholders. Management activities may include, for example, liaison, provision of environmental education, consultation, compensation and direct aid. Given that managers should always seek ways of improving relationships and involving stakeholders, there are two key questions which should guide this section:

- What opportunities are there to obtain benefits for the site and its wildlife by improving community relationships?
- How can local people benefit from the presence of the site?

4.6.5 Management Projects

This section is a continuation of the rationale in which the need for, and the nature of, possible management has been discussed. The function of this section is to describe in detail all the individual projects that must be carried out in order to meet the stakeholder objective.

Planning Individual Projects
(See Chapter 17)

Recommended Further Reading

Borrini-Feyerabend, G., Kothari, A. and Oviedo, G. (2004). *Indigenous and Local communities and Protected Areas: Towards Equity and Enhanced Conservation.* IUCN, Gland, Switzerland & Cambridge, UK.

Eagles, P. F. J., McCool, S. F. and Haynes, C. D. A. (2002). *Sustainable Tourism in Protected Areas: Guidelines for Planning and Management.* IUCN Gland, Switzerland and Cambridge, UK.

Eurosite (2002). *Report on the 67th EUROSITE Workshop – Nature Management Planning and Stakeholder Involvement.* Eurosite, Tilburg, The Netherlands.

Ramsar Convention Bureau. (2000). *Ramsar Handbooks for Wise Use of Wetlands.* Ramsar Convention Bureau, Gland, Switzerland.

Thomas, L. and Middleton, J. (2003). *Guidelines for Management Planning of Protected Areas.* IUCN Gland, Switzerland and Cambridge, UK.

References

Alexander, M. (1994). *Management Planning Handbook.* Countryside Council for Wales, Bangor, Wales.

Alexander, M. (1996). *A Guide to the Production of Management Plans for Nature Conservation and Protected Areas.* Countryside Council for Wales, Bangor, Wales.

Alexander, M. (2000b). *Guide to the Production of Management Plans for Protected Areas.* CMS Partnership, Aberystwyth, Wales.

Alexander, M. (2005). *The CMS Guide to Management Planning.* The CMS Consortium, Talgarth, Wales.

Borrini-Feyerabend, G., Kothari, A. and Oviedo, G. (2004). *Indigenous and Local Communities and Protected Areas: Towards Equity and Enhances Conservation.* IUCN, Gland, Switzerland and Cambridge, UK.

Caldwell, N. and Evison, S. (2005). *The role of National Nature Reserves in the Countryside Council for Wales' Approach to Stakeholder Involvement.* Unpublished report, Countryside Council for Wales, Bangor.

Eurosite (1999). *Toolkit for Management Planning.* Eurosite, Tilburg, The Netherlands.

Idle, E. T. and Bines, T. J. H. (2005). *Management Planning for Protected Areas, A Guide for Practitioners and Their Bosses.* Eurosite, English Nature, Peterborough, UK.

NCC (1981). *A Handbook for the Preparation of Management Plans for National Nature Reserves in Wales.* Nature Conservancy Council, Wales, Bangor, UK.

Margoluis, R. and Salafsky, N. (1998). *Measures of Success: Designing, Managing, and Monitoring Conservation and Development Projects.* Island Press, Washington, DC.

NCC (1983). *A Handbook for the Preparation of Management Plans.* Nature Conservancy Council, Peterborough, UK.

NCC (1988). *Site Management Plans for Nature Conservation, A Working Guide.* Nature Conservancy Council, Peterborough, UK.

Ramsar Convention Bureau (2000). *Ramsar Handbooks for Wise Use of Wetlands.* Ramsar Convention Bureau, Gland, Switzerland.

Rientjes, S. (2000). *A Practical Guide for Communicating Nature Conservation.* European Centre for Nature Conservation, Tilburg, The Netherlands.

Thomas, L. and Middleton, J. (2003). *Guidelines for Management Planning of Protected Areas.* IUCN Gland, Switzerland and Cambridge, UK.

Chapter 5
Survey, Surveillance, Monitoring & Recording

Abstract Monitoring, surveillance and recording are all activities concerned with the collection and management of information. They are an indispensable and integral component of management planning: without information there can be no planning. '*Survey*' is simply making a single observation to measure and record something. '*Surveillance*' is repeating standardised surveys in order that change can be detected. This is quite different to, but often confused with, monitoring. Surveillance is used to detect change but does not differentiate between acceptable and unacceptable change. '*Monitoring*' is surveillance undertaken to ensure that formulated standards are being maintained. Monitoring should be an essential and integral component of management planning: there can be no planning without monitoring and no monitoring without planning. There should be a direct relationship between the accuracy of the conditions that management can deliver and the level of accuracy that a monitoring project is designed to measure. The development of any monitoring strategy should be based on the availability of resources and on a risk assessment. '*Recording*' is concerned with making a permanent and accessible record of significant activities (including management), events and anything else that has relevance to the site. Recording management activities must be given the highest priority: if something is worth doing it must be worth recording. Recording is an expensive activity, and it must be planned with exactly the same rigour as all other aspects of reserve management.

5.1 Definitions

If we read almost any publication on conservation management, the words 'monitoring', 'survey' and 'surveillance' will be found frequently. Very few authors define what they mean by these words: there appears to be an assumption that there are universally accepted definitions. Unfortunately, nothing could be further from the truth: the standard dictionary definitions of these words are not adequate for the purposes of conservation planning, and there are no other widely accepted, definitive definitions. The following

M. Alexander, *Management Planning for Nature Conservation: A Theoretical Basis & Practical Guide*, DOI 10.1007/978-94-007-5116-3_5,

definitions will be applied throughout this book. This is not an attempt to lay claim to the meaning of these words, but it is important to establish meanings that can be clearly understood within the framework of this book. No doubt other authors will use different definitions. The definitions to be used in this book are shown below in bold type.

Survey: Making a Single Observation to Measure and Record Something

The standard dictionary definitions, for example, 'look carefully and thoroughly at', or 'to view comprehensively and extensively', are not really adequate. In common use, 'survey' is generally taken to mean a once-only observation, and it usually also implies that a record is made. Indeed, unless a record of some kind is made there can be little purpose in 'looking carefully and thoroughly at'.

Surveillance: Making Repeated Standardised Surveys in Order That Change Can Be Detected

Once again, dictionaries do not help: they usually make a link with criminal activity, for example, 'observations of a suspected spy or criminal'. Other definitions written specifically for dealing with the natural environment have been more helpful. For example:

- 'Surveillance: A continued programme of (biological) surveys systematically undertaken to provide a series of observations in time.' (JNCC 1998)
- 'Surveillance: Repeated surveys (or counts) designed to detect changes in the abundance of species (or features) with known precision.' (Rose and Mclean 2003)
- 'Surveillance, which is a repeated survey, often used to detect trends in habitats, populations and environmental change.' (Hurford and Schneider 2006)

The definition in this book avoids using terms such as 'biological' and 'species' because surveillance (and survey) are also used in the context of public use, tourism and cultural features. The need for an appropriate and constant level of precision is implied by the inclusion of 'standardised' in the definition of surveillance.

Monitoring: Surveillance Undertaken to Ensure That Formulated Standards Are Being Maintained (JNCC 1998)

Monitoring, according to common use, can mean almost any kind of measurement, including survey, census and even research. In fact, it has such a broad range of meanings that without a clear definition it is almost useless for planning purposes. 'Monitor' is derived from the original Latin word 'monere' which means 'warn'. Wildlife managers need a warning when things are going wrong and confirmation when things are satisfactory. This means that they have to decide what conditions they require for a feature: this is the objective or formulated standard. They must then make repeated measurements to ensure that their objective is being met. There are many definitions of monitoring which are similar to the JNCC (1988) version, but none are quite as suitable for management planning.

Recording: Making a Permanent and Accessible Record of Significant Activities (Including Management), Events and Anything Else That Has Relevance to the Site

There is surprisingly little in the management planning literature about recording and there are no definitions. This is probably because the word has such an obvious meaning.

5.2 Survey

Some planning guides suggest that surveys are an essential precursor to management planning (NCC 1988). Others express a different opinion:

> There is a tendency for organisations to enter a 'decision-making paralysis' when faced with having to make decisions on tricky subjects without adequate information. Managers (and planners) rarely consider they have enough information and generally have to accept this situation: possible lack of information should not become an excuse for delaying the production of the plan. (Thomas and Middleton 2003)

> A competent plan can be developed from relatively simple descriptions of the physical, biological and socio-economic characteristics of an area. More sophisticated data add to the confidence of the manager or planner, but they rarely justify a dramatic change of plan. The absence of site-specific information is not normally a good reason for postponing management in favour of more research. (Keller 1999)

There can be no doubt that management plans should be based on the best *available* knowledge. Decisions (planning is making a series of decisions) made in the absence of sufficient and reliable data are potentially dangerous. However, a failure to make a decision or to take an action can be even more dangerous.

Surveys can be very expensive, particularly in respect of the time required. There are many examples where the cost of pre-plan data collection has exceeded the resource available to manage the site for the duration of the plan. All the resources available for managing sites should be allocated through a structured, logical planning process which identifies and prioritises the work required to manage a site, and data collection should be no exception. There will always be things that are not known. An intelligent approach differentiates between those things that we need to know and those that we would like to know. It then prioritises the different needs. In short, it is the planning process that identifies the need for data and provides a justification for surveys. It is also the planning process that identifies and prioritises the need to maintain inventories or to ensure that the site description is always up to date.

5.3 Monitoring

This section is concerned with describing the relationship between monitoring and planning. It is not about the science of monitoring or the design of monitoring projects.

5.3.1 Background

The 1983 NCC planning guide, which was the precursor of most European guides, does not mention 'monitoring' in the main text. An example of a plan given in the guide contains a project to 'monitor any change in factors', but this is included without explanation. It would appear that 'monitoring' was not recognised as part of the UK planning process at that time. A few years later, when the NCC published a new planning guide in 1988, 'monitoring' had become recognised as a component of the plan. There was no direct link with the features at that time, but a monitoring project was included in each project group. (A project group comprised a number of projects which were linked to each operational objective.) This approach remained unchanged until 1996 when the Countryside Council for Wales published *A Guide for the Production of Management Plans for Nature Reserves and Protected Areas* (Alexander 1996). This guide made the connection between features, attributes (performance indicators), limits and monitoring. Monitoring was recognised as an integral and essential component of management planning.

The recognition of the relationship between monitoring and planning has not been universal. The *Eurosite Toolkit for Management Planning* (Eurosite 1999) does not give any significant attention to the need for monitoring. There is only one mention of monitoring in the entire section on preparing a plan: 'When selecting the operational objectives, managers may also wish to take into consideration how the site will be monitored and may choose to relate the objectives to the parameters which will be used for monitoring'. At a later stage, in the section on audit, it is recommended that auditors ask the question: 'Are the effects (of management) being monitored – are changes in the biological/physical systems of the site, including both the impacts of management and natural processes being systematically recorded?' At best, Eurosite are recommending surveillance and not monitoring (the issue of the lack of a universally accepted definition for monitoring was discussed earlier). Eurosite, in common with many other organisations, suggest that the 'effects of management' can be monitored, but they do not explain what this means or how it should be done. The IUCN *Guidelines for Management Planning of Protected Areas* (Thomas and Middleton 2003) treats monitoring as a completely separate exercise, so much so that the guide refers to another IUCN guide, *Evaluating Effectiveness: A Framework for assessing the Management of Protected Areas* (Hockings et al. 2000), which deals with the whole issue of monitoring and evaluation of protected area management.

In complete contrast, the IUCN guide, *Sustainable Tourism in Protected Areas* (Eagles et al. 2002), contains a chapter devoted entirely to monitoring. It begins with:

> Monitoring is an essential component of any planning or management process, for without monitoring, mangers know nothing about progress towards the objectives they have set or have set themselves. Monitoring is the systematic and periodic measurements of key indicators of biophysical and social conditions.

Turning to the New World, in the paper which introduced the dominant USA management process, Limits of Acceptable Change (LAC) (Stankey et al. 1984), the ninth and final step in the process is: 'implement actions and monitor conditions'. The monitoring programme focuses on indicators and compares their condition with a previously described standard. McCool (1996) describes 11 principles of LACs. The ninth principle, 'monitoring is essential to professional management', is set out below.

> Monitoring, in an informal sense, has historically been a component of the protected area manager's job. However, monitoring has generally been conducted informally, with little systematic planning and implementation. Monitoring is defined as the period and systematic measurement of key indicators of biophysical and social conditions. It performs two major functions in the LAC process. First, it allows managers to maintain a formal record of resource and social conditions over time. In serving this function, data points can inform managers of changes in these conditions rather than relying solely on informal perceptions of changes that might have occurred. This is particularly important in situations where managers change frequently or where effects are slow to develop. Second, it helps assess the effectiveness of management actions. Thus, monitoring helps managers understand, in a relatively objective way, if the action addressed the problem.

McCool and Cole (1998), in a subsequent review of LACs, propose a significant change to the LAC process. They recommend the addition of a new first step that involves 'defining goals and desired conditions'. Later, they identify 'inadequate attention to monitoring protocols' as an issue that requires attention, and recommend that more attention should be given to monitoring protocols early in the LAC process.

Other USA systems, 'Visitor Impact Management' (Graefe et al. 1990) and the 'Visitor Experience and Resource Protection' planning system (Manning et al. 1995), include monitoring as a central component of planning. The 'Protected Area Visitor Impact Management System' (Farrell and Marion 2002) replaces the use of 'indicators' and 'monitoring' with the 'views of an expert panel'. In reality, they have replaced one approach to monitoring objectives with another.

Some planning systems which acknowledge the need for integrated monitoring create a separate, almost independent, section for monitoring within the management plan. The planning process described by Margoluis and Salafsky (1998) is an example. Considerable emphasis is placed on the need for, and the methods of, monitoring. They recommend the development of a 'monitoring plan': this is undertaken once their goals, objectives and activities have been identified. There is a link between monitoring and their goals and objectives, but it is tenuous and does not appear to recognise a need to monitor the feature directly. They include the following goal as an example:

To Conserve the Grassland Savannah Ecosystems of Karimara National Park

Their goal is accompanied by six objectives, although these are not objectives as defined in this book.[1] They are in fact objectives for activities: things that they wish to control.

- 20% of the park revenue goes to local communities
- Waste from local hotels does not pollute the park
- Tour operators adhere to park guidelines
- There are no uncontrolled fires to provide pasture for cattle in the park
- There is no grazing of cattle in the park
- Hunting is reduced by 90%

This approach is similar to many that originate in the USA: it is a factor driven process. Although they use goals, they do not define the required condition for the feature in any detail. The most important factors are identified, management activities intended to control the factors are described and then monitoring strategies are developed. The following are examples of their monitoring strategies:

Objective: *Reduce illegal hunting by 90% by the end of the project.*

Monitoring strategies:

i. *Compare the number of elephants and rhinos killed before and after project interventions.*
ii. *Measure the change in the number of encounters with hunters inside the park over time.*
iii. *Compare the levels of illegal hunting by trophy hunting operators in Karimara National Park to a neighbouring park.*

This approach appears to be based on several assumptions:

- There is no need to describe the condition required of the park ecosystem.
- All the factors that are affecting, or might affect, the ecosystem have been identified and activities have been identified for their control.
- If all the anthropogenic factors are controlled the ecosystem will be in good order.
- Sufficient information can be obtained through monitoring the factors to provide the evidence that the feature is in the required condition.

In this particular example, they do not monitor 'all aspects of illegal hunting'. They measure three distinct elements, and this leads to another assumption: if there are fewer elephants and rhinos killed, fewer encounters with hunters, and the levels of hunting are no different to a neighbouring park, hunting will have been reduced by 90%.

[1] An objective is, or should be, the description of something that we want to achieve. These are the outcomes of management. Wildlife outcomes are habitats, communities or populations at a favourable status.

(What if there are fewer animals killed because there are fewer left to be killed? What if there are fewer encounters with hunters because they have learned to avoid detection? What if the neighbouring park is badly managed?)

These are all high risk assumptions: we can never be certain that all the factors have been identified, monitoring factors provides only part of the evidence required to assess the condition of a feature, and evidence based on measuring a sample of the impacts of a factor is unlikely to be conclusive.

The examples used by Margoluis and Salafsky are taken from hypothetical scenarios. They are clearly intentionally simplified to illustrate the concepts, but it is difficult to judge a concept when the examples are possibly oversimplified.

To summarise: With the exception of PAVIM, I have not found any published material which challenges the need for monitoring in management planning. Earlier European planning systems (i.e. pre 1990) and, surprisingly, a few recently revised systems pay little or no attention to monitoring. Generally, most current management planning systems acknowledge the value of monitoring, and some recognise the need to integrate monitoring in the planning process.

5.3.2 Monitoring and Adaptive Planning

Monitoring should be an essential and integral component of management planning: there can be *no planning without monitoring.* The adaptive planning process and all other functional management planning processes are entirely dependent on an assessment of the status of the features, and this is obtained through monitoring. Monitoring is 'surveillance undertaken to ensure that formulated standards are being maintained'. The 'formulated standards' are the 'objectives with performance indicators', and these are a product of the planning process. Therefore, there can be *no monitoring without planning.*

5.3.3 Performance Indicators

The integration of monitoring in the adaptive planning process occurs when the objectives for the features are formulated. An objective must be measurable, and this is achieved by including performance indicators that are directly linked to, and part of, the objective. This process is fully described in Chapter 15. Two different kinds of performance indicators are used to monitor an objective. These are:

- Quantified attributes[2] with limits which, when monitored, provide evidence about the condition of a feature.

[2] An attribute is a characteristic of a feature that can be monitored to provide evidence about the condition of the feature.

– Factors[3] with limits which, when monitored, provide the evidence that the factors are under control or otherwise.

Some of the attributes are selected because they provide the evidence that is needed to assess the condition of the feature and others because they are indicators of change. The latter are directly related to a factor, i.e. where the factor is the agent of change it is the attribute that changes in response to the factor. This means that some of the attributes are selected and used because they provide an indirect means of monitoring the factors. Ideally, this indirect approach (i.e. monitoring the attributes that change rather than directly monitoring the factors) should be used for all factors, because management decisions need to be based on changes to the feature and not simply on an assumption that if a factor changes we can predict with certainty the way in which a feature will change.

We should only rely on setting limits or thresholds for a factor if we have a complete understanding of the relationship between a factor and its impact on a feature. Ideally, factors would only be monitored directly when this relationship is understood. Monitoring factors can, very occasionally, be much easier than monitoring attributes, for example, the height of a water table, or the presence of an invasive species. Although the actual method of measuring these can be very easy, the difficulty lies in establishing the limits for the factors. How low can the water table drop before the bog is damaged? Unfortunately, our knowledge of conservation biology is often very limited, and there will be many occasions when the only option is to establish surveillance projects for the factors. This will help us to develop an understanding of the relationship between the factor and a feature, and will eventually lead to the establishment of a surveillance project.

Monitoring or establishing surveillance for many factors can be extremely difficult or prohibitively expensive. For example, the most significant factor as far as all species are concerned is the quality of the habitat or habitats that support them. Management decisions are often unavoidably based on practical, but quite dangerous, assumptions, i.e. we assume that if a habitat is in more or less good order then the species that it supports will be secure. Although the habitat requirements of a few flagship species are known, for many species, particularly invertebrates, these relationships are poorly understood. When the habitat that supports a species is contained within a site and can be managed as a feature there is a chance that our assumption is defensible. Unfortunately, for many mobile species this is not the case. For example, a population of bats that breeds on a site may occupy a feeding territory containing habitats that extend for many kilometres beyond the site boundaries. The consequences of these difficulties is that although the need to monitor (either directly or indirectly) all the significant factors is obvious and generally accepted this is rarely achieved in practice.

[3] A factor is anything that has the potential to influence or change a feature, or to affect the way in which a feature is managed. These influences may exist, or have existed, at any time in the past, present or future.

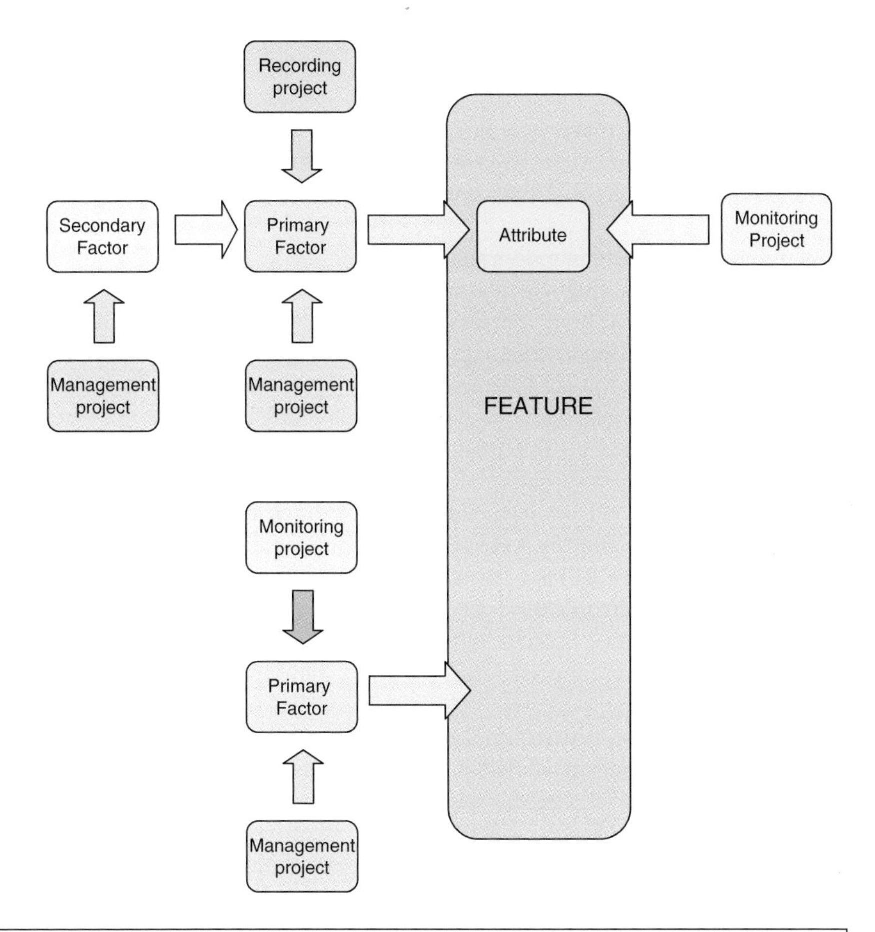

- *Primary factors* will have a direct influence on the feature: as the factor changes so will the condition of the feature.
- *Secondary factors* will not have a direct influence on the feature but will influence primary factors or our ability to manage the primary factors.
- *An attribute* is a characteristic of a feature that can be monitored to provide evidence about the condition of the feature. Attributes change in response to the influence of a factor or factors.
- *Management projects* are the activities or interventions that are employed to control the factors. All management projects are recorded.
- *Monitoring* a feature can be *direct,* when an attribute is monitored, or *indirect*, when a factor is or monitored.
- *Recording* projects are generally applied to factors where the range of acceptable levels is unknown, i.e. they cannot be monitored.

Fig. 5.1 The relationship between features, factors, attributes, management, monitoring, & recording

5.3.4 Establishing Monitoring Projects

So far in this chapter, monitoring has been defined, recognised as an integral component of planning and its location in the plan has been identified. The following notes are intended to guide the selection of appropriate monitoring methodologies. To repeat the point that was made at the beginning of this chapter, this is not monitoring for reporting purposes and it is certainly not about research (hypothesis testing). This is monitoring which is specific to management planning.

Monitoring projects should not be unnecessarily complicated: 'Monitoring need not be highly complex and expensive – if too expensive, it often would be better to put the money into additional land acquisitions' (Noss et al. 1997). Most conservation managers, even those who fail to carry out any monitoring, will readily accept that monitoring is extremely important. When sites are not monitored the most common claim is there are insufficient resources. However, I suspect that another, and more significant, reason is a misguided perception of what monitoring means. Many people believe that monitoring is always a demanding, scientific activity that requires high levels of expertise and is consequently expensive and time-consuming. There is no point in pretending that this is not sometimes the case. When managing important, fragile or threatened habitats and species it may, occasionally, be necessary to obtain very accurate and precise information, but this should be the exception and not the rule.

A decision must be made about how accurate a monitoring project needs to be. There should be a direct relationship between the accuracy of the conditions that management can deliver and the level of accuracy that a monitoring project is designed to measure. 'Accuracy is the ultimate measure of the quality of an estimate' (Ratti and Garton 1994). Nature conservation management is a crude and often clumsy process and, given the tools and levels of control that are available, attempting to fine-tune the quality of a habitat can be a futile activity. We should also question the need to obtain precisely-defined outcomes. Do they make any sense when managing semi-natural or plagioclimatic habitats which were originally created as the by-products of farming or other human activities? The quality of semi-natural communities would have varied enormously in the past. They responded to a wide range of factors, including market demand, poverty, mechanisation and war. There was no constant state. So why do some people believe that we need precisely-defined, constant states today? The management of habitats – grassland is a good example – can be as serendipitous today as it always was. Nature conservation organisations, particularly in the voluntary sector, have variable and unreliable resources. In addition, their ability to obtain grazing is often dependent on other people, graziers and farmers who themselves are influenced by changing agricultural policies and legislation. For example, there is a limited market demand for some important grazing animals and particularly for animals past a specified age.

In the UK, domestic and European legislation can be a serious constraint, for example, fallen stock regulations, the 6 day rule on animal movements, pony passports, and the loss of local abattoirs (PONT 2007).

So, even if we believe that there is justification for precisely-defined outcomes for semi-natural habitats they are generally not obtainable. When managing the habitats that have suffered least from anthropogenic influence the outcome is, or should be, determined as far as possible by natural processes. In these situations, can there be any sense in seeking precisely-defined outcomes? Allowing for legislation, the preceding arguments are applicable to the vast majority of protected areas. Whatever the conditions that we want to obtain they will be variable and to some extent unpredictable. If we also acknowledge global climate change and the consequential potential for habitat change it should be even more obvious that we can only provide an approximation of what we wish to achieve and that we will have to continually revise our objectives.

It is essential that monitoring projects are affordable. There is no purpose whatever in developing expensive monitoring regimes, or planning individual monitoring projects, if the resources required to undertake the work are not available. This is a common problem: even government conservation agencies have sometimes fallen into the trap of developing rather ideal monitoring strategies based on hopelessly expensive methodologies. The usual consequence is that features on a few sites are monitored to a very high standard while the remainder are completely neglected. The development of any monitoring strategy should be based on the availability of resources and on a risk assessment. What can we afford to do, which features are the most vulnerable (i.e. most likely to change) and which need remedial management (i.e. those which should change)? Ideally, all features should be monitored to a minimum standard, even if the minimum is based entirely on expert opinion. Once the minimum is achieved for all features, the information can be used to identify the need for, and to prioritise, any additional, or more detailed, monitoring for the most vulnerable features.

Most experts, including experienced reserve managers, should be competent to assess the status of many features without relying on detailed data collection and analysis. Their assessment should always be based on a written and agreed objective with performance indicators. This will ensure consistency between visits and between assessments made by different individuals. The experts should, in addition to making the assessment, give an indication of the level of confidence in their decision. If their confidence level is above a predetermined threshold, for example 80%, there may be no justification for any further monitoring. Where there is limited confidence in an expert view this could be the justification for monitoring based on detailed data collection.

Examples of simple or cost-effective monitoring projects, which are not dependent on highly-skilled labour, are given in Case Study 1.

Fig. 5.2 Surveillance

5.4 Surveillance

Surveillance: making repeated standardised surveys in order that change can be detected. This is quite different to, but often confused with, monitoring. Surveillance lacks the 'formulated standards' that are so important in monitoring. Surveillance is used to detect change but does not differentiate between acceptable and unacceptable change. Surveillance is often used when monitoring is not possible because the 'formulated standard' or specified limits for the attributes and factors are unknown. In these circumstances, surveillance projects can be a precursor to monitoring projects. By measuring and recording changes we can begin to understand the limits within which a factor or attribute can vary without giving any cause for concern. Where monitoring is a very specific and targeted activity, surveillance can have a broader function and can be used to detect a much wider range of changes. Surveillance can be a site-specific activity or part of a wider national, or sometimes international, programme, such as The Wetland Bird Survey (WeBS). This is a scheme used to census non-breeding water-birds in the UK.

There is nothing that can replace the sort of informal surveillance that is only possible when reserve managers maintain a continual presence on a site. Their experience and familiarity with the features and factors often means that they will

Fig. 5.3 Thrift *Armeria maritima 1980*

Fig. 5.4 Sea campion *Silene uniflora 1999*

Fig. 5.5 Yorkshire fog *Holcus lanatus* 2008

recognise very small changes that would be difficult to detect even with more sophisticated surveillance, changes that would certainly be missed by tightly focused monitoring projects.

Photo-surveillance, often confused with photo-monitoring, is a relatively cheap and very effective way of maintaining a record of changes on a site. The main advantage is that it is not targeted or specific, and, as a consequence, it can be used to detect unpredictable changes. Of course, the fact that it is not targeted is also a disadvantage, especially if too much reliance is placed on the results. Photo-surveillance, as with all surveillance, is an excellent means of maintaining a general awareness of change. It can supplement, but not replace, monitoring.

Figures 5.3, 5.4, and 5.5 demonstrate the value of photo-surveillance. In this example, they clearly show how coastal vegetation has changed over a period of almost 30 years. A succession of different species has dominated the community.

5.5 Recording

The concept of an integrated recording, reporting and planning system for conservation management is not new (NCC 1988). Unfortunately, managers rarely place a sufficiently high priority on this aspect of their work. This is quite surprising, because the collection of information about wildlife is often the first activity that engages people who eventually go on to become conservation management professionals. They bring into the profession a 'recording ethic', but they do not always record relevant information.

Recording management activities must be given the highest priority: if something is worth doing it must be worth recording. One of the most irritating problems that reserve managers have to face is knowing that, at some time or other, some form of management actions were taken, but they do not know when or what. They may be aware of the results, but where these are favourable the management cannot be repeated, and if they are unfavourable there is a danger that the same mistakes will be made again.

When management activities are carried out by a third party, as the consequence of a management agreement, for example, the work must be recorded. This is sometimes called '*compliance monitoring*'. It is a means of checking that planned work is actually completed.

The maintenance of records on a site is occasionally a legal requirement, for example, compliance with health and safety legislation. The advent of a litigious society has placed a considerable burden on the managers of all public access sites. Safety checks have become routine, and these activities must be recorded.

Once managers appreciate the need to maintain records they are often faced with the dilemma of having to decide what should and should not be recorded.

They must guard against the experience described by Graham Burton, Senior Reserves Manager RSPB, in his introduction to *Ecological Monitoring of Protected Areas* (Eurosite 2003):

> I decided to carry out an audit of all the monitoring and recording being carried out at the protected areas that I was responsible for. It will come as no surprise that almost every site had some projects that collected data with no apparent purpose. Ecological data is a precious resource, gathered at high cost to any organisation. It is, therefore, essential that the data collected is either central to the management of protected areas or a vital part of a wider recording scheme.

Recording is an expensive activity and it must be planned with exactly the same rigour as all other aspects of reserve management. Whenever a management activity is planned a system for recording the work must also be established. This will ensure that nothing of significance goes unrecorded.

It is essential that managers avoid irrelevant or unnecessary recording. There is a need to recognise the crucial difference between the information that is needed for site management or protection and information that managers want to collect. 'Want' is often driven by personal interest, and many reserve mangers are driven by a passionate interest in wildlife. There are many examples where every single bird that is seen on, or flying over, a reserve is meticulously recorded, despite the fact that this information is not in any way relevant to managing the site features. Clearly, if managers had unlimited time these activities should be encouraged. Unfortunately, many sites have extremely good records of things that we do not need to know and poor, or even no, records of the things that we need to know. The prime function of any protected area must be the protection of the wildlife or conservation features that were the basis of site acquisition, selection or designation and any other features of equal status discovered post acquisition. Casual recording, valuable though it can be, must be relegated to the 'if only we had spare time' category.

Information and records are only as good as they are accessible. Good data management is essential, but this can be quite a challenge, especially on large sites or when there is a need to share information over several sites. The obvious solution is to use a computer database. This will be discussed in Case Study 5.

Recommended Further Reading

Elzinga, C. L., Salzer, D. W., Willoughby, J. W. & Gibbs, J. P. (2001). *Monitoring Plant and Animal Populations: A Handbook for Field Biologists*. Blackwell Science, Inc. Malden, Massachusetts, USA.

Hill, D., Fasham M., Tucker G., Shewry, M., Shaw, P. (2005). *Handbook of Biodiversity Methods – Survey, Evaluation and Monitoring*. Cambridge University Press, Cambridge, UK.

Krebs, C. J. (1999). *Ecological Methodology*, 2nd ed. Addison-Welsey Longman, Menlo Park, California, USA.

References

Alexander, M. (1996). *A Guide to the Production of Management Plans for Nature Conservation and Protected Areas*. Countryside Council for Wales, Bangor, Wales.

Eagles, P. F. J., McCool, S. F. and Haynes, C. D. A. (2002). *Sustainable Tourism in Protected Areas: Guidelines for Planning and Management.* IUCN, Gland, Switzerland and Cambridge, UK.

Eurosite (1999). *Toolkit for Management Planning.* Eurosite, Tilburg, The Netherlands.

Eurosite (2003). *Ecological Monitoring of Protected Natural Areas – Providing Guidance and Best Practice*. Eurosite, Tilburg, The Netherlands.

Farrell, T. A. and Marion, J. L. (2002). *The Protected Area Visitor Impact Management (PAVIM) Framework: A Simplified Process for Making Management Decisions*. Journal of Sustainable Tourism, 10(1). Portland Press Ltd. London, UK.

Graefe, A., Kuss, F. R., Vaske, J. J. (1990). *Visitor Impact Management: The Planning Framework*. National Parks and Conservation Association. Washington, DC, USA.

Hockings, M., Stolton, S. and Dudley, N. (2000). *A Framework for Assessing the Management of Protected Areas*. IUCN, Gland, Switzerland and Cambridge, UK.

Hurford, C. and Schneider, M. (eds.). (2006). *Monitoring Nature Conservation in Cultural Habitats*. Springer, The Netherlands.

JNCC (1998). *A Statement on Common Standards Monitoring.* Joint Nature Conservation Committee, Peterborough, UK.

Keller, G. (1999). *Guidelines for Marine Protected Areas*. IUCN Gland, Switzerland.

Manning, R. E., Lime, D. W., Hof, M. and Freimund, W. A. (1995). *The Visitor Experience and Resource Protection (VERP) Process*. The George Wright Forum 12(3).

Margoluis, R. and Salafsky, N. (1998). *Measures of Success: Designing, Managing, and Monitoring Conservation and Development Projects*. Island Press, Washington, DC.

McCool, S. F. (1996). *A Framework for Managing National Protected Areas: Experiences from the United States*. Maritime Institute of Malaysia, Kuala Lumpur, Malaysia.

McCool, S. F. and Cole, D. N. (1998). *Proceedings – Limits of Acceptable Change and Related Planning Processes: Progress And Future Directions*; 1997 Missoula, MT. Gen. Tech. Rep. INT-GTR-371. Ogden, UT, USA.

NCC (1983). *A Handbook for the Preparation of Management Plans.* Nature Conservancy Council, Peterborough, UK.

NCC (1988). *Site Management Plans for Nature Conservation, A Working Guide.* Nature Conservancy Council, Peterborough, UK.

Noss, R. F., O'Connell, M. A. and Murphy, D. D. (1997). *The Science of Conservation Planning, Habitat Conservation Under the Endangered Species Act*. Island press, Washington, DC, USA.

PONT (2007). *Grazing animals constraints report*. Unpublished internal document, PONT, Brecon Wales, UK.

Ratti, T. R. and Garton, E. O. (1994). *Research and Experimental Design* – contribution 565, University of Idaho Forestry, Wildlife and Range Experiment Station, Idaho, USA.

Rose, P. and McLean, I. (2003). *The Future Role of JNCC in Biological Surveillance and Monitoring*. JNCC, Peterborough, UK.

Stankey, G. H., McCool, S. F. and Stokes, G. L. (1984). *Limits of Acceptable Change: A new framework for managing the Bob Marshall Wilderness.* Western Wildlands, 10(3), 33–37

Thomas, L. and Middleton, J. (2003). *Guidelines for Management Planning of Protected Areas.* IUCN Gland, Switzerland and Cambridge, UK.

Chapter 6
Adaptive Management, Adaptive Planning, Review and Audit

Adaptive management is not really much more than common sense. But common sense is not always in common use.
(Holling 1978)

Abstract Planning should be a continuous, iterative and developmental process. Adaptive management can be applied to any site, regardless of size. It is, in its more complex form, an approach to experimental management that enables changes to be linked to cause and to management operations. This chapter considers the main versions of adaptive management and introduces a minimal version: a basic approach, but with some significant differences. It is not experimentation but a simpler system based on monitoring and then, if necessary, modifying management. The cyclical, adaptable management process allows site management to: respond to natural dynamic processes; accommodate the legitimate interests of others; adapt to the ever-changing political and socio-economic climate; and, in the long term, succeed, despite uncertain and variable resources. There is a continuum, from trial and error to full scale active adaptive management, and somewhere within this range there lies a version of adaptive which is appropriate for any given place and time. Management reviews are an integral and essential component of the adaptive management process. Audit is not strictly a component of the management planning process but a complementary activity that sits alongside management planning. Audit is the procedure for assessing whether a site is being managed to the standard required by an organisation and ensuring that the status of the site features is accurately reported.

M. Alexander, *Management Planning for Nature Conservation: A Theoretical Basis & Practical Guide*, DOI 10.1007/978-94-007-5116-3_6,

6.1 Introduction

Fig. 6.1 Hafod Garegog

There are very few certainties in life, but we can be sure that our natural environment and the wildlife that it supports will change. It has always changed. The values that we apply to our environment and its components will change, as will the condition of habitats and populations. The management actions that we take will also change. As a generality, we must learn to accept, and even welcome, change.

People usually talk about writing a management plan when, in fact, they should talk about a planning process. We need to abandon the production of static management plans and adopt a dynamic, iterative and adaptive planning process. While it is obvious that management activities will change with time, so planning – the intellectual, or decision-making, component of management – must also be dynamic and able to take account of change.

Far too much emphasis is placed on the idea that it is somehow possible to prepare a definitive site plan that will last forever, and an enormous amount of time has been wasted in this pursuit. The end-product of these attempts is usually an extremely expensive document that spends its life gathering dust, forgotten on a shelf. Even where there may have been an initial intention to review the plan at intervals (usually 5 years), this is forgotten, and then, at some time in the future, a decision is made to rewrite this long-obsolete document and to produce yet another dust trap. This may sound harsh and over-critical but, unfortunately, it happens all too often.

It is not unusual for a plan to be out of date within months of production, some even before they are finished. Sites, habitats and species are dynamic and constantly changing, as is our knowledge and expertise. It is not uncommon for new features, species, or even habitats, to be discovered after a plan has been prepared. For example, when the first management plan was written for Hafod Garegog, a woodland National Nature Reserve in North Wales, the recognised features were the oak-birch-bilberry woodland, *Quercus petraea-Betula pubescens-Dicranum majus* woodland, and a population of silver-studded blue butterflies *Plebejus argus* (Oliver and Hellawell 2001). Areas of wet heath were included within the site boundary but regarded as incidental features, and not of any particular importance. Shortly after the plan was completed, an important population of lesser horseshoe bats *Rhinolophus hipposideros* was discovered. This was followed by the discovery of a population of small red damsel flies *Ceriagrion tenellum* and, most surprising of all, an area of extremely rare flood plain woodland. The manager also realised that although the wet heath had not been originally regarded as an important feature it supported the population of silver-studded blue butterflies. The heath is fragile and requires intervention management to ensure its survival. Consequently, the heath was added to the list of important site features.

This example is unusual in that the discoveries were of important features, but, nevertheless, it is commonplace, even predictable, for new features to appear or be discovered as our experience of managing a site increases. The advent of climate change will, without any doubt, be accompanied by movement and changes in the distribution of species. In the future, plans will probably have to change or be adapted even more rapidly.

Adaptive management emphasises the need to change, or adapt, the management section or action plan. However, most other sections of the plan must also evolve. The description should obviously be developed to account for new knowledge. This will have consequences for most other areas in the plan. For example, new factors will appear which will require attention. It is important that we learn to accept that a plan is never completed and can never be complete. Ideally, a plan should, at any time, meet all the requirements of site management and be based on the best available science, knowledge and experience. Where there are gaps in our knowledge there will be gaps in the plan.

Anyone with any experience of planning will, of course, realise that plans must be updated. The usual response is to prepare a plan for a specific period and then review and rewrite. Unfortunately, despite these good intentions, the review is often delayed. The plan then becomes irrelevant and is ignored. Why does this happen? Each year since 1990 I have delivered management planning courses or workshops. I always begin by asking the participants about their reasons for attending. One of the most common responses is, 'I have been asked to revise a management plan that has been out of date for several years'. When asked why the plan has not been maintained, the usual answer is that there was no time available. With very few exceptions, conservation management is seriously under-resourced: there is insufficient money and staff time. Plans take time to prepare and review. Consequently, when a manager is required to set aside a considerable period of time to do this work, and when they already suffer an impossible work schedule, their response is predictable.

Planning is a continuous, iterative and developmental process. This means that a plan can, and perhaps should, begin life as a simple outline statement. It can then grow over time, with emphasis on the most important sections, until it fully meets the requirements of the site. Thereafter, it is constantly kept up to date. This will avoid the need to set aside long, and often unavailable, periods of time for plan writing and rewriting. It will also mean that the plan is always up to date. The Countryside Council for Wales uses a forward-rolling 5-year-cycle for plans. This means that each year the plan is rolled forward by one financial year, thus ensuring that the plans are always valid for 5 years.

By recognising that planning is an on-going component of site management, and by spreading the workload, it is possible to maintain up-to-date plans. We must have the confidence to put off until tomorrow what we do not *need* to do today.

6.2 Background – Europe and the Old World Perspective

The adaptive approach to managing nature reserves, and other places, for nature conservation has been applied in Britain (in its simplest form and to varying degrees) since at least the early 1960s, but the concept was not formalised, and there was no standard definition, or even recognition, of the process. It was just the way in which many conservation managers managed their sites. As Holling (1978) claimed, adaptive management is not really much more than common sense. It is important that we understand how nature reserves were perceived at that time. The National Parks and Access to the Countryside Act 1949 provides a clear, two-part definition of a reserve. The second section deals with 'preserving flora, fauna or geological features', but the first, and dominant, section is a reflection of the supremacy of science or, more significantly, scientists in the formulation of the act and subsequent nature conservation management:

> Meaning of 'nature reserve'; (a) of providing, under suitable conditions and control, special opportunities for the study of, and research into, matters relating to the fauna and flora of Great Britain and the physical conditions in which they live, …

Many reserve mangers were scientists, or were supervised by scientists. They came to conservation equipped with a knowledge of scientific protocols and procedures. In short, scientific thinking dominated the way in which species and sites were managed. Managers were aware that they were pioneers: they understood their limitations and also the need to learn. It was not surprising that they intuitively adopted the core elements of adaptive management. They recognised the need to identify 'problems'. A problem was often a reflection of their uncertainty or their inability to protect a species or habitat. Various solutions, in terms of approaches to management, were identified, trialled and applied. These pioneers understood the importance of recording or monitoring the impact of management, and they realised that many of the problems were common to most sites and situations. It became obvious that more research was needed and that the growing bank of knowledge had to be shared more widely, since it was becoming increasingly clear that the development of management techniques was not necessarily site-specific. The one component of the USA adaptive approach that was given least attention was the need to work with stakeholders and in particular local communities. Unfortunately, an awareness of these issues has not always been reflected in practice.

Despite its acceptance in common practice, the earliest UK management planning guides failed to recognise planning as a dynamic process (Wood and Warren 1976, 1978; NCC 1983). The emphasis was on the production of a document – a statement of intent – that was valid for a specified period. There was no suggestion of review or development of any kind.

The first UK guide to hint that planning is a process was the NCC *Site Management Plans for NatureCconservation – a working guide* (1988 revised 1991). It contains a diagram of the 'planning system and its control' which depicts a review process comprising three stages:

- Short term control of the annual work plan.
- Medium term control – review of projects and their progress (annual).
- Long term control – review in depth (5 or 10 years).

The problem is that, although these guides contain an elaborate diagram, there is no supporting explanation.

A French planning guide, *Plans de gestion des reserves naturelles* (Ministere charge de l'Environmement 1991), includes two feedback loops in the planning process: an annual loop, which ensures compliance with work programmes, and a longer term (5 year) review, to ensure that management is effective. This guide was produced in collaboration with NCC and was a derivative of the NCC 1991 approach to planning.

The Countryside Council for Wales *Management Planning Handbook* (Alexander 1994) was a development of the NCC (1991) guide. It contains a detailed explanation of an annual review which simply looks at compliance and, if necessary, adjusts the work programme for subsequent years. The handbook also describes a 'long-term review', where the interval between reviews is determined by the nature of the individual site: 'The prime function of the review is to ensure that the long-term objectives and options, as stated in the plan, are still pertinent, and that the prescriptions have been, and will continue to be, effective in achieving the desired objective'. The problem with this handbook is that, although it emphasises the need to associate monitoring projects directly with each objective, the objectives are rather vague, e.g. 'to maintain and enhance '.

The Eurosite *Toolkit: Management Planning* (1999) adopted a similar approach to CCW's. It recognises the need for an annual review, which is a compliance and performance review, and a 5 year review (Eurosite plans are written for a 5 year period). The latter is a performance review which considers the achievement of objectives, and it is suggested that, 'following the 5 year review it will probably be desirable to produce a complete revision of the plan'.

Long-term review: The long-term review, recognised in most European planning guides, is a major review undertaken at predetermined intervals. The length of the interval can, in exceptional cases, be as little as 1 year and would not usually exceed 10 years. The more dynamic or threatened a site the shorter the interval or planning cycle becomes. This does not mean that objectives are restated or that it is necessary to rewrite the entire plan. The prime function of the review is to ensure that the objectives, as stated in the plan, remain relevant and that management has been, and will continue to be, effective in achieving the desired objectives. Whenever an adaptive planning approach is applied, the long-term review as a stand-alone activity ceases to have any relevance. In a modified form, it becomes a component of the 'adaptive management' process. The adaptive cycle, which I describe later in this chapter, specifically identifies the need for periodic reviews of the objectives.

Short-term or annual review: The annual review is not strictly a component of adaptive management, but it remains an essential stage in any planning process. The main purpose of the annual review is to ensure that the site is being managed in accordance with the approved management plan. It is important to ensure that any serious, unexpected events or trends that could affect management are taken into account. The review is also an opportunity for site managers to present the preceding year's work to others in the organisation.

The management team responsible for the site should undertake the review. The structure of the team carrying out the review will vary from organisation to organisation and from site to site. The essential point is that the team should be able to make an objective appraisal of the year's work and reach agreement on the next annual work plan.

The group must ensure that all high-priority projects have been completed and that all lesser projects have been reported on. In the case of the former, they should seek a satisfactory explanation to account for any projects that have not been completed. Shortfalls in achievement and performance should be noted on each appropriate project form, and any necessary amendments should be made to the project register record and/or the next annual plan. If there are serious problems and it becomes clear that, for example, an organisation is not providing sufficient resources for a site, priorities may have to be reassessed and/or the operational objectives redefined. Any additional resources that become available for use on a site should be dealt with in a similar way, by reassessing priorities.

Adaptable management: Readers familiar with US adaptive management (Holling 1978) may be wondering why, at this stage, I introduce a different term: adaptable. The first edition of this book talks of 'adaptable' rather than 'adaptive' management.

A Management Planning Handbook (Alexander 2000a), prepared for the Uganda Wildlife Authority, and *A Guide to the Production of Management Plans for Protected Areas* (Alexander 2000b) first introduced the concept of 'adaptable management'. This was followed in 2002 by *New Guidelines for Management Planning for Ramsar Sites and Other Wetlands* (Ramsar resolution VIII.14), which also included the concept of adaptable management. In that document, the initial intention had been to use the term 'adaptive' in place of 'adaptable'. However, the Ramsar Scientific and Technical Panel was concerned that 'adaptive management' had become the subject of some controversy and that, since there are so many (occasionally contradictory) definitions, the meaning of 'adaptive' had become obscured. A year earlier, Elzinga et al. (2001) made a similar statement; they claimed that because 'adaptive management' had been adopted as a buzzword its definition and meaning had become muddled by widespread use.

In the first edition of this book I followed the Ramsar lead and used the term 'adaptable'. At that time, I believed that by using 'adaptable' I could avoid the confusion and controversy that Elzinga had highlighted. On reflection, I think that I probably contributed to the confusion. Through common use and frequent variation for specific purposes, and to reflect organisational preferences, 'adaptive' is now applied to a wide range of different approaches. What I described as 'adaptable' was, in fact, a minimal, but legitimate, version of adaptive.

6.3 Background – USA, Australia and the New World

Adaptive management is an approach to managing ecological systems[1] that is most frequently encountered in North America and Australia. Over the past decade, it has begun to appear in Europe where, regrettably, it is often poorly understood and misrepresented. As a concept, it can be traced back to Frederick Taylor in the early 1900s. However, it is Holling, an ecological theorist, and his colleagues who developed adaptive management at the University of British Columbia's Institute of Resource Ecology in the late 1960s, who can lay claim to the first significant publication (Holling 1978; Walters 1986). Professor Kai N Lee's influential book, *Compass and Gyroscope – Integrating Science and Politics for the Environment* (Lee 1993), was one of the more important landmarks in the development and application of adaptive management. Lee developed the ecological model, originally conceived by Holling and others, by emphasising a social or stakeholder dimension. Later, in 1999, Lee published an article, *Appraising Adaptive Management,* in *Conservation Ecology* (Lee 1999). This is one of the most useful individual papers published on adaptive management. It is certainly one of the easiest to understand. Lee makes a critical observation which is so important if we are to understand why 'adaptive' from his, and from the USA, perspective is about people:

> Cultivating an ecosystem in order to foster its wild state is paradoxical... This paradox has been resolved by turning around the objective: to think of ecosystem management as *managing the people* who interact with the ecosystem.

[1] It is important that managing ecological systems in this context is not confused with the Convention on Biological Diversity (CBD) 'ecosystem approach'. See Chapter 7 which deals with the CBD approach in some detail.

A literature search will reveal countless versions of 'adaptive' management, for example:

> Adaptive Management is an approach that involves learning from management actions, and using that learning to improve the next stage of management. (Holling 1978)

> Adaptive management is "learning by doing" with the addition of an explicit, deliberate and formal dimension to framing questions and problems, undertaking experimentation and testing, critically processing results, and reassessing the policy context that originally triggered investigation in light of the newly acquired knowledge. The concept of learning is central to AM. It is a process to accelerate and enhance learning based on the results of policy implementation that mimics the scientific method: experimentation is the core of adaptive management, involving hypotheses, controls and replication. It is also irreducibly socio-political in nature. (Stankey et al. 2005)

> Adaptive Management is a rigorous approach for learning through deliberately designing and applying management actions as experiments. (Marmorek et al. 2006)

> A process in which management activities are implemented in spite of uncertainties about their effects, the effects of management are measured and evaluated, and the results are applied to future decisions. (Elzinga et al. 2001)

There are three divisions of adaptive management (Walters and Holling 1990):

6.3.1 Evolutionary Adaptive Management

This is dismissed as irrelevant by some authors. They see it as being effectively little more than a version of trial and error (Wilson et al. 2007). This is a very simplified approach but is probably much better than doing nothing. In essence, it is about learning from experience. This implies that, at very least, managers should record management activities and include an assessment of management effectiveness. This is sometimes described as 'observational monitoring'; it is, or should be, a formalised or organised version of trial and error.

6.3.2 Passive Adaptive Management

This approach focuses on the implementation or application of management techniques or policies which have a successful track record or are believed to represent 'best practice'. Implementation is followed by monitoring, review or evaluation of the results, and, if necessary, the technique is modified and the cycle repeated. This approach will identify effective management techniques. Most of the various definitions of adaptive management insist that experimentation is an essential component. Adaptive management treats human interventions (management) as experimental probes (Lee 1993). Consequently, the need for replication and the use of controls is also implied. Without controls it will not be possible to differentiate between changes that would have happened had the management not been applied or changes that are a result of other unidentified factors. This process clearly provides a structured approach to learning.

An example of *passive* adaptive management is the application of burning to manage maritime dry heath,[2] where a simple objective might be to maintain a diverse and regenerating community dominated by dwarf shrubs. Best practice management is applied. This is believed to be burning, in small patches distributed across the entire community, followed by grazing with ponies. The stocking levels and timing would also be specified. A control area, where there is *no* management treatment, is established. The heath is burned, grazed and monitored. If the specified required conditions are met, and the trial management can be replicated, the management regime is adopted. If otherwise, the management experiment is modified. This could, for example, be a change of timing between burns, a change in stocking levels, or both.

6.3.3 Active Adaptive Management

This shares the basic concept described above, but differs in that it is specifically designed to test various hypotheses (Lee 1993). It involves the application of a range of management techniques and should eventually identify 'best practice'. This approach is more likely to identify new and innovative techniques or policies and could highlight the most efficient practices as well as the most effective.

The application of an active adaptive management approach to the example of a dry heath, given above, implies that a range of both tried and new methods of managing the heath should be applied at the same time. These could include the obvious, for example, using different grazing animals, mowing in place of burning, and, of course, a wide range of variations in the timing and intensity of the various management trials. The experimental management areas will be accompanied by sufficient control areas.

The dry heath example describes the process of adopting passive and active adaptive approaches in an uncomplicated situation, where the outcome, i.e. healthy heath, is easily defined and where there is a direct relationship between management interventions and the desired condition of the heath. However, the adaptive process is most often associated with the management of very large ecosystems. Lee (1993) describes its application to the Columbia Basin, a catchment with 1,200 miles of river extending over an area the size of France. In these circumstances there are a wide range of interests and values. For example, the Columbia Basin has to accommodate important legal tribal fishing rights associated with the salmon fishery, the world's largest hydro-electric power system, farming, flood control, water supply, recreation and much more. This diverse range of interests is accompanied by an equally complex range of community interests which had to be integrated in the adaptive process. It should, therefore,

[2] Maritime dry heath is generally dominated by dwarf shrubs: varying proportions of heather, bell heather and, in some areas, western gorse *Ulex gallii*.

not be a surprise to learn that Lee is credited with introducing a social or stakeholder dimension to adaptive planning.

As a footnote to this very brief background to adaptive planning, there is a new, emerging approach labelled 'adaptive co-management', which is intended for the 'governance of social-ecological systems'. Adaptive co-management is the iterative learning component of adaptive management combined with collaborative management, in which collective rights and responsibilities are jointly shared. I am not sure that there is a need for this new label when the approach appears to be covered by the definition of active adaptive management.

6.4 Characteristics of Adaptive Management:

I have looked at a wide range of definitions, and the following is an attempt to identify the characteristics that are common to most, but by no means all, definitions:

Management is treated as experimentation.

Experimentation has three components: a clear hypothesis, a way of controlling factors that are (thought to be) extraneous to the hypothesis, and opportunities to replicate the experiment to check its reliability (Lee 1999).

1. There should be clear and specific objectives.
2. Monitoring with a feedback link to management or policies is essential.
3. Learning is an explicit objective.
4. There is a social or stakeholder dimension.
   (This implies that communication is also an important component.)
5. Management is not delayed by uncertainty.
   (Uncertainty is an inevitable factor which limits nature conservation management. We do not know enough to manage ecosystems.)
6. Adaptation is essential.
   (This means changing management interventions or assumptions in response to new information gained through focused monitoring.)

Some definitions also suggest that adaptive management should:

- retain a focus on statistical power and controls
- use computer or conceptual models
- use embodied ecological consensus to evaluate strategic alternatives

There are many areas of potential confusion because there are so many different approaches to adaptive management. Some use different words to describe the same thing. Some approaches rely on the development of conceptual models. Most appear to focus on the activities associated with management, while a minority include goals or objectives as central to the system. Although, at first glance, they look very different, in essence there are more similarities than differences.

6.5 Adaptive Management – Discussion

It is important to recognise that nature reserves, sites, species and habitats have been successfully managed for a very long time without formalised versions of adaptive management. So, should we adopt any version of adaptive management? The answer is yes and no, but mainly (though not always) no to the application of the more complex adaptive approaches and particularly to active adaptive management. Lee (1999) made two extremely important statements:

> Adaptive management is difficult to sustain.

> Adaptive management has been more influential, so far, as an idea than as a practical means of gaining insight into the behaviour of ecosystems utilised and inhabited by humans.

There are relatively few US examples of passive or active adaptive management which have stood the test of time. Even the example of the Columbia basin experience, used by Lee (1993) as the focus for his book *Compass and Gyroscope*, was not maintained for more than a few years.

The most significant book published recently on adaptive management, *Adaptive Environmental Management* (Allan and Stankey 2009), contains a selection of adaptive management case studies representing, presumably, some of the best examples of passive or active adaptive management. In the conclusion to the book, the authors make it clear that most of the case studies are characterised as 'works in progress'. This is not intended as a criticism but recognises that adaptive management is, and must always be, an on-going process. But, has it been going on for long enough for this approach to be adopted without very careful consideration? I am not sure that either passive or active adaptive management have achieved the status they need, or possibly deserve: they lack a substantial and sustained track record.

There is an issue that will apply to the use of passive or active adaptive management on many nature reserves, particularly statutory conservation sites. For example, on a NATURA 2000 site the managers are usually required to obtain Favourable Conservation Status or optimal status, for the recognised conservation features. Central to the definitions of favourable and optimal is an obligation to ensure that habits and species are not declining. This means that it will not often be possible to set aside sufficient land for large-scale experimental approaches, and it could be difficult, if not impossible, to establish large areas for high risk controls, i.e. areas that would not be managed and, as a result, could be damaged. The restricted ability to use controls and the risk of damage to unique systems is a recognised and significant limitation to the application of active adaptive management. We learn from the North American experiences that adaptive management should be applied at an appropriate spatial scale. Adaptive management is also usually described as a site-specific approach and, under ideal circumstances, this clearly makes sense, but this does not imply that passive or active adaptive management cannot be used in small European countries which have few, if any, large sites.

Fig. 6.2 Dry coastal heath, Llyn Peninsular, North Wales

There is an additional consideration. Earlier I quoted Lee's (1993) description of the application of active adaptive management to the Columbia Basin, a catchment with 1,200 miles of river extending over an area the size of France. France, by European standards, is a large country: 547,030 km^2. Once again, I will return to the example of dry coastal heath that was used to illustrate the differences between passive and active adaptive management. My experience with dry heath is mainly from Wales, a country of 20,779 km^2: a fraction of the size of the Columbia Basin. We have to be very careful when comparing areas. Lee points out that it is the social structures and interdependent use that defines the 'size' of an ecosystem. Accepting the limitations of making comparisons, within Wales dry heath occupies an area of approximately 8,900 ha, with over 70% of this being maritime (Fig. 6.2). This is tiny compared to the Columbia basin, and with very different, but no less complex, social structures. However, it is surely large enough for the application of an active adaptive approach. Is there any significant conceptual difference between applying adaptive management over a habitat, which is in part naturally fragmented, across an entire European country and a river catchment fragmented by 19 major hydro-electric dams? The management issues and the factors that influence our ability to maintain maritime dry heath are more or less common throughout Wales. Management that is appropriate in one area will, with some minor modification, be relevant throughout the dry heath habitat.

A possibly imperfect, but entirely realistic, answer to this dilemma is to recognise that different management regimes, and in particular the development of new and innovative approaches, may have to take place over a range of sites. The application of a diverse range of different management approaches over a range of similar sites could be applied as surrogates for large-scale active adaptive management. It would be important that each individual site is managed in compliance with the principles of at least minimal adaptive management. If this sounds familiar, it should, because, although we may not have described it as adaptive, this is what we have been doing, with considerable success, for quite a long time. Unfortunately, a fair criticism could be that we have not always given sufficient attention to sharing knowledge, and, consequently, we have not been able to learn from the successes and failures of others.

The application of active adaptive management in some European countries will probably be more relevant at a strategic or country-wide perspective, where it can be used to develop policies and formulate activities that will guide local management.

Conservation managers must always ground their plans and aspirations in the real word of resource constraints. While society demands more and more of the natural world – more access, tourism, recreation and a host of other environmental services – politicians, and others who represent society, are becoming less able, or prepared, to allocate appropriate resources to conserving biodiversity. In times of financial recession maintaining biodiversity ceases to be recognised as a necessity and becomes perceived as a luxury. The consequences are that, even when we think that we should be applying passive or active adaptive management, the costs (and no one will deny that these approaches can be extremely expensive) can be prohibitive. Conservation managers must do their very best, even in a seriously impoverished financial situation, and should recognise that many of the most basic components of adaptive management can be applied even when the complete approach is beyond the limitations of their resources. Elzinga et al. (2001) recognise that it is possible, sometimes even desirable, to apply a simplified version of 'adaptive' which is based on observational monitoring.

I believe that *there is a continuum, from trial and error to full scale active adaptive management, and somewhere within this range there lies a version of adaptive which is appropriate for any given place and time.* A minimal, or occasionally passive, approach will be most suitable when dealing with single-purpose sites, for example, National Nature Reserves in Britain and similar sites in Europe, where land use is primarily, and often exclusively, for nature conservation. On complex sites, with a multitude of stakeholders, cross-boundary issues and poorly understood ecosystems, active adaptive management will possibly be most appropriate (Allan and Stankey 2009).

The range of adaptive approaches can also vary from place to place, time to time, and from feature to feature within a site. There may be features, where we have considerable successful management experience, where we believe that any form of experimental management is not necessary, but for others multiple experimental management trials will be necessary.

Both passive and active adaptive management are 'scientific approaches'. That is, they rely on scientific experimentation. No one with any experience of conservation management can fail to recognise the importance of science to conservation:

without good science there would be no conservation. But, does this mean that conservation management should always be a scientific process; is it only scientists who can manage species and places for nature conservation? Good conservation mangers should be multi-talented. They need to have at least a basic understanding of biology, ecology, earth sciences and often a more than superficial understanding of history and other cultural or heritage issues. They need advanced technical and practical skills. They must understand the legal implications and obligations associated with their work. Perhaps above all, they must have highly developed interpersonal skills. The management of places for nature conservation is more often concerned with people management – establishing and maintaining relationships – than anything to do with direct species, or habitat, management. Inclusivity, and the need to work with stakeholders, is central to adaptive management.

It is important that conservation management comes of age and is recognised for what it is: an activity guided by science, but not itself a scientific process. Modern farming is never regarded as science, yet it is vastly more sophisticated, and reliant on science, than most conservation management. We need to develop and apply adaptive management approaches that are not entirely dependent on the active application of complex and expensive scientific techniques. For example, in laboratory science, when testing hypotheses, inferences should be reliable 95 % of the time. In conservation management we must accept 'more likely than not' (Lee 1999). Most conservation organisations recognise that monitoring is essential, but, in reality, most fall far short of their intentions. The main reason, or excuse, is the lack of resources or skilled staff to carry out the work. However, another reason is that most monitoring is far too complex and designed to achieve much more than 'more likely than not'. The adoption of simplified monitoring, that is affordable and does not require sophisticated or specialised skills, would be much better than no monitoring. Perhaps the role of science in conservation should be rather more about enabling others.

6.6 Adaptive Management – A Minimal Approach

This is the process that the Ramsar Convention on Wetlands and I have described as adaptable management (Ramsar 2002; Alexander 2000a). I must make it clear that this minimal approach is a modified version of the adaptive management process as described by Holling, Walters, Lee and many other North American and Australian authors. It is not my intention to challenge their ideas in any way, and I acknowledge that it is their pioneering work that provides the foundations and guidance for my interpretation. Adaptive management, in all its forms, has not generally been a significant feature in European conservation, although we have used our own language to describe similar approaches. If there is to be a wider application of adaptive management, there is a need to describe the approach in a way that is more relevant to the condition of our natural heritage and to our legislation, for example, NATURA 2000. Much, and in some countries most, of what we manage for nature conservation is plagioclimatic or semi-natural, and sites can be very small – the fragmented remnants of original ecosystems – but we place considerable value on these wonderful places. They represent generations of human interaction (albeit incidental) with nature. The products are landscapes that are best regarded as cultural entities.

Lee (1999) is clear that ecosystem management is about *managing the people* who interact with the ecosystem. As a concept, this is not difficult to understand when it is applied to the management of ecosystems where the intention is to maintain, or restore, near pristine natural conditions. However, conservation management to maintain a cultural landscape can be a very different process. An oversimplified rule is that in the most natural ecosystems human influences are negative factors that must be controlled, but, conversely, in a cultural landscape, comprising plagioclimatic habitats, natural processes can be the negative factors.

Fig. 6.3 A cultural landscape, Tal y Llyn, North Wales

We may have aspirations for a Europe where large areas of land are freed from cultural constraints and where unfettered nature once again rules. This will not happen in the lifetime of my generation, and it will only become a reality for succeeding generations if we are able, at least, to maintain the populations of species and the habitats that will provide the foundations of a different future.

I provide my own definition of 'minimal adaptive management' for use in this book, as I do not want to 'hijack' the approach described as 'evolutionary' by Wilson. I believe that a modified version of trial and error is a useful, and entirely relevant, management approach. It is useful because it is a process where management is implemented in spite of uncertainty. In reality, we can never be certain that management is entirely appropriate, but, provided that management is directly linked to a clear, quantified and measurable objective for a feature, the condition of the feature can be monitored before and after the implementation of management. This version of adaptive is not experimentation. It is not dependent on replicating management actions or establishing control plots. It is a system for managing sites and features based on monitoring performance indicators. Johnson (1999) describes this simplified version as 'monitor and modify', and not as adaptive management. Elzinga et al. (2001) define their approach as, 'a process in which management activities are implemented in spite of uncertainties about their effects, the effects of management are measured and evaluated, and the results are applied to future decisions'. They describe their system as a simplified version of 'adaptive' based on observational monitoring.

The application of this minimal approach to adaptive management can be illustrated by returning to the example of maritime dry heath. A simple objective might be to maintain a diverse and regenerating dry heath community, dominated by dwarf shrubs. Best practice management is believed to be burning in small patches distributed across the entire community, followed by grazing with ponies. The stocking levels and timing could also be specified. The heath is burned, grazed and monitored. If the specified required conditions are met then management is continued; if not, management is modified. This could, for example, be a change of timing between burns, a change in stocking levels or both.

6.6.1 Size or Spatial Considerations

There is a rather obvious presumption that adaptive management should encompass a specified and defined area. Many authors describe adaptive management as an approach suitable for ecosystems. Lee (1993) suggests that adaptive management is most urgently needed in large ecosystems. The Convention on Biological Diversity definition of an ecosystem can refer to any functioning unit, at any scale, which is determined by local circumstances. A unit could be a veteran tree, a small woodland, a forest or the entire biosphere. The concept of an ecosystem is a human construct used to describe the natural world, and we define ecosystems according to the scale of human interests and our decision-making powers. I believe that some form of adaptive management can, and should, be used at any scale, ecosystem or otherwise. In Europe and many other parts of the world, areas managed for nature conservation can range from large ecosystems to small fragmented areas of woodland or individual meadows. None need be excluded from a basic adaptive approach.

6.6.2 Stakeholder/Community Involvement

'Adaptive', in all its guises has become, for many, a surrogate for an inclusive approach to management, where communities or stakeholders are involved in the process. Stakeholder involvement is central to all versions of adaptive management, but stakeholder management does *not* always need to be adaptive. While we recognise that there will be areas of conservation management which will not be open to negotiation, for example a legal obligation to protect the features on a European Natura site, whenever possible, management and planning should be an inclusive process. It is the planning process on a site that provides the best opportunities for negotiation, communication and the involvement of stakeholders and local communities. The relationship between planning, management and stakeholders is discussed in Chapter 4.

6.7 Adaptive Planning

Up to this point, I have written about adaptive management, but this book is concerned with management planning. Much that has been written about adaptive management focuses on its application to the management projects. In short, adaptive management is the process of identifying appropriate or best practice management regimes. It is possible to extend the process to encompass the entire planning and management process, and I believe that we should. It is not only the management techniques or policies that will need to be developed and adapted but also our objectives: the things that we are trying to achieve will also change over time, for any number of unpredictable reasons. Although the original concept was conceived as an approach for managing nature or ecosystems, it can also be applied to all aspects of countryside management, including recreation, access and relationships with stakeholders. Whenever we want to achieve or maintain something, and we can articulate this 'something' as an objective, we can use an adaptive process to ensure that we meet our objectives. Planning is an essential component of management: it is the intellectual component, the time and place when we make the most important decisions, and a way of communicating, justifying resources and seeking approval. Adaptive management should not be something that we do as an aside, or adjunct, to conservation planning. It should be embedded in, and provide a conduit for, the entire range of interdependent processes that constitute good management planning.

The minimal approach to adaptive management planning has the potential to deliver greater security for the wildlife and places that we mange. When applied at this most basic level, the adaptive process should not incur any significant additional cost, but the benefits are considerable.

A minimal adaptive approach can be used to:

- Demonstrate that management is appropriate and effective, and that resources are well spent
- Ensure that we learn from experience
- Ensure continuity of effective management
- Encourage and enable communication between managers and stakeholders, both within and between sites and organisations

6.7.1 The Minimal Adaptive Process – Step by Step

Table 6.1 The adaptive cycle

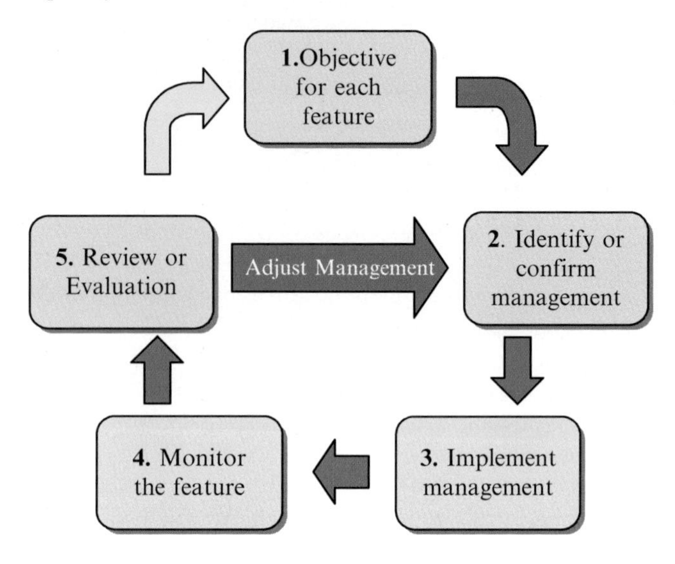

1. Prepare an objective for each feature

This is quite different to some adaptive cycles, which begin with a management 'problem' that requires attention. In common with the Ramsar approach and an Australian version this cycle begins with an objective.[3] That is, it begins with a decision about what we want to achieve for a feature. An objective is prepared for each feature, and performance indicators (which will be monitored) are identified for each objective. Objectives lie at the very heart of the planning process, and are perhaps the most important component. A full account of how objectives are formulated is given in Chapter 15.

2. Identify or confirm management – the rationale

This is the stage where the decision must be finalised about which version, or which components, of adaptive management will be used. The outcome of this stage will be the techniques or approaches that we believe are best practice or most likely to deliver the objective.

[3] The most fundamental problem in conservation management is how to obtain and maintain the condition defined by the objectives.

The approach in the rationale will change according to whether it is being applied in the first management cycle or in subsequent cycles:

(a) When adaptive management planning is introduced for the first time:
An assessment is made of the *condition* of the feature. (The condition of the feature is the difference between the state described by the objective and the actual state of the feature at the time that an assessment is made.) If a feature is in the required condition there is reason to assume that past or current management is probably appropriate. Conversely, if a feature is unfavourable there is reason to believe that management is inappropriate.
This is followed by considering, in turn, the factors that have changed, are changing, or may have the potential to change, the feature. The management required to keep the factors under control must then be identified. (Conservation management is always about directly or indirectly controlling the influence of factors, or about the remedial management required following the impact of a factor.)

(b) When adaptive management planning is at the end of the first, or any subsequent, cycle:
An assessment of the *status*[4] of the feature will have been completed in Stage 5 (see below). This means that we should know whether current management is effective or otherwise. If the feature is favourable, and the factors are under control, management is considered to be effective. If the feature is unfavourable, and / or the factors are not under control, management is either ineffective or has not been in place for long enough. If possible, management efficiency should also be considered at this stage.

At stage 2 there is nothing to be gained by reinventing procedures. Managers may draw on a wide range of sources, often external, and make use of the best available information, evidence, expertise and experience to inform the decision-making process. Information generated as part of an internal process on one site will often be the source of external information for other sites. Do not assume that because a management activity worked at some time in some place it will necessarily work elsewhere. It may be reasonably easy to identify an appropriate type of management, but it will usually be much more difficult to quantify the intensity or frequency of that management.

Management is implemented for a period of time. The length of the management period is determined by two main factors:

- The predicted rate of change of the feature. Some features can change very rapidly, while for others change is extremely slow.
- The more important factor relevant to the minimal approach is our level of confidence in the planned management activities. Confidence levels will be high when adequate expertise and experience are available and low when this is not the case. For example, a fragile habitat, managed for the first time by inexperienced staff, will require close surveillance and monitoring each year. As experience and confidence grow the period can lengthen.

[4] The difference between status and condition, and the full implications of status, are described in Chapter 14.

Ideally, monitoring should be carried out prior to the commencement of management activities (the initial assessment of conservation status) and thereafter at intervals which match the management period, or more frequently. Monitoring, in this context, is a very specific activity: the performance indicators (quantified when the objectives were prepared) are measured. This is one of the most critical aspects of the adaptive process. Monitoring is linked directly to both the objectives for a feature and the associated management activities. A detailed description of the rationale is given in Chapter 16.

3. Implement management

This is when the action plans are implemented. Management which is believed to be the most appropriate is applied. All management activities must be carefully recorded.

4. Monitor the feature

The performance indicators, which were identified when the objective for the feature was formulated, are monitored. The indicators will provide evidence about the status of the feature that is sufficient to allow us to evaluate management. Chapters 5 and 15 provide guidance on the use of monitoring in the context of management planning.

5. Review or evaluation

The results of monitoring, along with reports of management activities and any other relevant observations (including external information), are considered. The first question should always be: is there any reason to change the objective? Even when management is concerned with obtaining specified outcomes defined by legislation, there will occasionally be a need for revision. Objectives will need to change for many different reasons. For example, we may have got it wrong in the first instance, or the status of a species can change with time. At best, an objective is an expression of something that we believe we want at any specific time: it can only be a reflection of our values, knowledge, experience, science and the evidence available at that time. All of these things will, and must, change. If there is a need to change the objective in any significant way this can, of course, have implications for many of the planning stages. Each will have to be considered in sequence and, if necessary, revised.

If the objective does not require revision, move on to consider the status of the feature. (This will have been disclosed by monitoring the performance indicators, and any additional relevant information will also have been taken into account.) There are two main questions: what is the condition of the feature, and are the factors under control?

If the feature is in a favourable condition and the factors are under control, then we can assume that management is appropriate, i.e. it is effective. If the feature is unfavourable and/or one or more of the factors are not under control, management should be reconsidered and, if necessary, changed. In some circumstances, we may conclude that a particular management regime has not been in place for long enough for the required changes to have taken place, in which case, providing there are no signs of deterioration, we could continue with the existing management for a longer period. A further essential consideration is that both the type and intensity of management can vary with time, depending on the condition of the feature. The condition of a feature when management commences can be extremely unfavourable. The management required to move a feature from an unfavourable condition to a favourable condition can be regarded as 'recovery management', while the management required to maintain a feature in favourable condition would be 'maintenance management'. Clearly, there can be very significant differences between the two types of management.

The adaptive management process is both cyclical and repetitive (Table 6.2). Cyclical management systems appeared in the USA literature in the 1950s. Lindblom (1959) talks of 'cyclic incrementalism'. Adaptive management recognises that wildlife managers may be unsure of their objectives and management requirements. However, each time a management cycle is completed the management activities are tested, new knowledge is obtained and skills are improved.

Table 6.2 A cyclical, iterative & developmental process

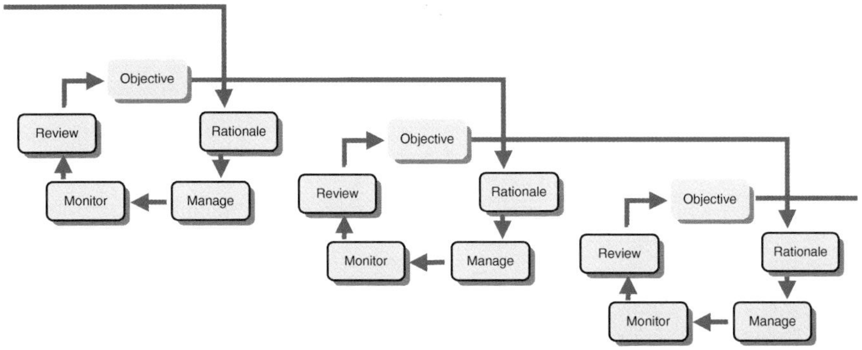

There is no point in claiming that adaptive management, and particularly the approach described in this book, is a proven system. Professor Kai N Lee said much the same thing in 1999. We have not been doing it for long enough. The response of habitats to management is slow, and there are no long-term examples of adaptive management. The longest application of an imperfect adaptive approach in the UK, that I am aware of, has only been in place for 15 years (see Case Study 4). Only time will tell, but if the approach is less than satisfactory it can be changed or adapted.

6.8 Audit

Audit is not strictly a component of the management planning process but a complementary activity that sits alongside management planning.

Eurosite published an approach to 'Site Conservation Assessments (audit)' in their *Toolkit: Management Planning* (1999): 'A tool for reviewing management, preferably under a management plan, by assessing the performance of the managing organisation of the site'. This process is based entirely on an examination of the degree of compliance with the Eurosite planning guide. Their planning guide does not recommend quantified objectives or other measurable expressions of what the plan is intended to achieve for the conservation features. Consequently, the Eurosite audit is restricted to measuring compliance with procedures.

The CMS guide to management planning (Alexander 2005) introduced 'audit' as an addition to, or a replacement for, the long-term review. The audit combined two essential processes: an assessment of compliance and of management effectiveness. This was probably a mistake because, although both processes can, or even should, be carried out at the same time, they fulfil quite different functions. Compliance audit should be concerned with ensuring that organisational standards are met, while the assessment of management effectiveness should focus on management outcomes, for example, the status of the site features and the quality of the experience enjoyed by visitors. An assessment of outcomes is, of course, central to the adaptable management process. Thus, where sites are managed through an adaptable process, audit need only be concerned with compliance. This is more or less the Eurosite approach.

6.8.1 *The Audit Procedure*

Organisations will develop their own procedures to meet specific organisational requirements. The following example outlines a procedure that could be adapted for use in many circumstances.

Function of audit

- To assess whether or not a site is being managed to the standard required by the organisation or department responsible
- To ensure that the status of the site features is accurately reported

Timing of audit

The outcome of an audit should influence the timing of the following audit. When a site scores highly at audit, i.e. there is compliance with the required standards or protocols and the audit team are confident that these standards will be maintained, the interval between audits can be extended. The converse should also be true. In any situation, the maximum permitted period between audits should be specified, and should probably be no longer than 5 years.

Personnel

The audit team can comprise:

- Auditors – these should be external consultants or independent staff from an auditing group/department
- The site manager(s) must always attend
- Personnel responsible for managing site staff
- Other relevant staff may also be invited to attend as required

Procedure

An audit will comprise two stages:

- An examination of the management plan, the adopted project planning and recording system, the results of monitoring and surveillance programmes, safety documentation and any other relevant information. It is generally good practice for the audit team to read the documentation prior to the audit.
- A site visit/inspection.

Reporting

A draft audit report should be sent to the site managers to allow them to comment on its accuracy. This will then be returned to the audit team and an amended audit report, including observations and recommendations, will be sent to the site managers and to appropriate senior staff. Management responses must then be returned to the audit team. A final report will be issued to senior staff and other responsible officers. This report will identify agreed management responses and actions, together with officers responsible for ensuring these are undertaken and deadlines for action.

Recommended Further Reading

There is a wealth of literature on adaptive planning. Many different organisations have redefined, or modified, the process to meet their own specific requirements, and much of this has been published. Often, there are contradictions, and it is easy to become confused by the different use of language. In this chapter, I have focused on a minimal approach to adaptive management. This is because I know that it can be used and that it has worked in the situations where I have direct experience. I do not have first-hand experience in the application of large-scale active adaptive management schemes, but the following publications are useful in giving a broader insight:

Allan, C. and Stankey, G. H. (eds.), (2009). *Adaptive Environmental Management: A Practitioner's Guide*. Springer, The Netherlands.

Holling, C. S. (1978). *Adaptive Environmental Assessment and Management*. John Wiley and Sons, New York.

Lee, K. N. (1993). *Compass and Gyroscope: Integrating Science and Politics for the Environment.* Washington, DC: Island Press.

Lee, K. N. (1999). *Appraising adaptive management*. Conservation Ecology 3(2).

References

Alexander, M. (1994). *Management Planning Handbook.* Countryside Council for Wales, Bangor, Wales.

Alexander, M. (2000a). *A Management Planning Handbook*. Uganda Wildlife Authority, Kampala Uganda.

Alexander, M. (2000b). *Guide to the Production of Management Plans for Protected Areas*. CMS Partnership, Aberystwyth, Wales.

Alexander, M. (2005). *The CMS Guide to Management Planning*. The CMS Consortium, Talgarth, Wales.

Allan, C. and Stankey, G.H. (eds.), (2009). *Adaptive Environment Management* . Springer, The Netherlands.

Elzinga, C. L., Salzer, D. W., Willoughby, J. W. and Gibbs, J. P. (2001). *Monitoring Plant and Animal Populations: A Handbook for Field Biologists*. Blackwell Science, Inc., Malden, MA, USA.

Eurosite (1999). *Toolkit for Management Planning.* Eurosite, Tilburg, The Netherlands.

Holling, C. S. (1978). *Adaptive Environmental Assessment and Management.* John Wiley and Sons, New York.

Johnson, B. L. (1999). *The role of adaptive management as an operational approach for resource management agencies.* Conservation Ecology, 3(2), 8.

Lee, K. N. (1993). *Compass and Gyroscope: Integrating Science and Politics for the Environment.* Island Press, Washington DC, USA.

Lee, K. N. (1999). *Appraising adaptive management*. Conservation Ecology, 3(2).

Lindblom, C. E. (1959). *The Science of 'Muddling Through'*. Public Administration Review, 19, 79–88.

Marmorek, D. R., Robinson, D. C. E., Murray, C. and Grieg, L. (2006). *Enabling Adaptive Forest Management*. National Commission on Sustainable Forestry, Vancouver, BC.

Ministere charge de l'Environmement (1991). *Plans de gestion des reserves Naturelles*. Direction de la Protection de la Nature, Paris.

NCC (1983). *A Handbook for the Preparation of Management Plans.* Nature Conservancy Council, Peterborough, UK.

NCC (1988). *Site Management Plans for Nature Conservation, A Working Guide.* Nature Conservancy Council, Peterborough, UK.

NCC (1991). *Site Management Plans for Nature Conservation, A Working Guide (Revised).* Nature Conservancy Council, Peterborough. UK.

Oliver, D. M. and Hellawell, T. C. (2001). *Management plan for Hafod Garegog National Nature Reserve.* Unpublished document, Countryside Council for Wales, Bangor, Wales.

Ramsar Convention Bureau (2002). *New Guidelines for Management Planning for Ramsar Sites and Other Wetlands, Ramsar resolution VIII.14.* Ramsar Convention Bureau, Gland, Switzerland.

Stankey, G. H., Roger, N., Clark and Bormann, B. T. (2005). *Adaptive Management of Natural Resources: Theory, Concepts, and Management Institutions*. Department of Agriculture, Forest Service, Portland, OR, USA.

Walters, C. (1986). *Adaptive Management of Renewable Resources*. Macmillan, New York.

Walters, C. J. and Holling, C. S. (1990). *Large-scale Management Experiments and Learning by Doing.* Ecology, 71, 2060–2068.

Wilson, A. L., Dehaan, R. L., Watts, R. J., Page, K. J., Bowmer, K. H. and Curtis, A. (2007). *Proceedings of the 5th Australian Stream Management Conference. Australian Rivers: Making a Difference*. Charles Sturt University, Thurgoona, New South Wales Australia.

Wood, J. B. and Warren, A. (ed.), (1976). *A Handbook for the Preparation of Management Plans – Conservation Course Format, Revision 1*. University College London, Discussion Papers in Conservation, No. 18, London.

Wood, J. B. and Warren, A. (ed.), (1978). *A Handbook for the Preparation of Management Plans – Conservation Course Format, Revision* 2. University College London, Discussion Papers in Conservation, No. 18, London.

Chapter 7
The Ecosystem Approach

Abstract This chapter introduces the ecosystem approach to management. This is a series of 12 principles that can be applied to planning and management. It is not necessarily about managing ecosystems; the principles can be used to guide biodiversity management in any situation. In order to avoid unnecessary confusion, the concept of an 'ecosystem' is discussed. Most of the early thinking about ecosystem management originated in the USA. This generally assumed that a natural system is sustainable if anthropogenic factors are removed or controlled. The ecosystem approach has not, until recently, been a feature of European conservation, but Europe has recognised and applied most of the principles implied by this approach for a very long time. The most commonly applied, and widely recognised, definition of an ecosystem approach comes from the Convention on Biological Diversity (CBD). This chapter will introduce the 12 principles promoted by CBD, which are all given varying levels of attention in other chapters of the book. The principles provide guidance that all management planners should, at least, consider, but it must be recognised that an ecosystem approach does not mean abandoning the established tried, tested and effective methods of conservation. The ecosystem approach to management must be adaptive: this book is about an adaptive approach to management and planning. Finally, there is no single way to implement the ecosystem approach, as it depends on local, provincial, national, regional or global conditions.

7.1 Introduction

This chapter explains the ecosystem approach to management. The most important point to bear in mind when reading this chapter is that the ecosystem approach is not a management system but a series of 12 principles that can be applied to planning and management. The term 'ecosystem approach' is, unfortunately, misleading, because many people believe, erroneously, that it is about managing ecosystems.

M. Alexander, *Management Planning for Nature Conservation: A Theoretical Basis & Practical Guide*, DOI 10.1007/978-94-007-5116-3_7,

I must stress that the application of the ecosystem approach is not restricted to ecosystems: it can be used to guide biodiversity management in any situation.

7.2 What Is an Ecosystem?

Before I discuss the ecosystem approach, I think that, in order to avoid any confusion, I should make an attempt to introduce the concept of an ecosystem. The term and concept were first introduced in 1935 by A G Tansley in his paper '*The Use and Abuse of Vegetation Concepts and Terms*' (Tansley 1935). Since that time, so much has been written and taught about ecosystems that it is easy to become lost in the morass of information (Golley 1994). One of the best definitions that I have seen comes from a children's website, Geography4Kids.com:

> The word ecosystem is short for ecological systems. An ecosystem includes all of the living organisms in a specific area. These systems are the plants and animals interacting with their non-living environments (weather, Earth, Sun, soil, atmosphere). An ecosystem's development depends on the energy that moves in and out of that system. As far as the boundaries of an ecosystem, it depends upon how you use the term. You could have an entire ecosystem underneath a big rock. On the other hand, you could be talking about the overall ecosystem of the entire planet (biosphere). An ecosystem can be as small as a puddle or as large as the Pacific Ocean. That ecosystem includes every living and non-living thing in the area. It is several small communities interacting with each other.

One of the most commonly used definitions is from the Convention on Biological Diversity website, http://www.cbd.int/ecosystem:

> A dynamic complex of plant, animal and micro-organism communities and their non-living environment interacting as a functional unit.
>
> The concept of an ecosystem is a human construct used to describe the natural world, and we define ecosystems according to the scale of human interests and our decision-making powers.

The CBD definition can refer to any functioning unit, at any scale, which is determined by local circumstances. A unit could be a veteran tree, a small woodland, a forest or the entire biosphere. We can easily understand that the entire biosphere of planet Earth is an ecosystem, because all the elements, living and otherwise, interact to maintain the global ecosystem, but at a smaller scale it is rather more contrived. A guiding principle is that, 'an ecosystem has strong interactions amongst its components and weak interactions across its boundaries' (UK National Ecosystem Assessment 2010). An example that may help to demonstrate this is the interactions between the organisms in a lake (strong interactions) compared to their interaction with the surrounding land (weak interactions). We might think that an offshore island could easily be defined as an ecosystem, with strong interactions among its isolated components (habitats and populations). But, consider a small Atlantic island off the coast of Wales, with a population of Manx shearwaters

Puffinus puffinus (Fig. 7.1). These seabirds nest on the island, but feed in the Irish Sea and spend the winter in the South Atlantic, feeding, en route, in the North Atlantic. This example illustrates why we must not take an over simplistic view of ecosystems. To varying degrees, other than the global ecosystem, none are completely isolated: there will be movements in and out. Mobile, and more specifically migratory, species are an obvious case, but there are also movements of the non-living elements, for example, nutrients from seabird guano. In addition, the factors that can influence wildlife on the island can be local or on-site (for example, predatory species); they can be offsite (for example, oil pollution); or they can be global in origin (for example, increasing levels of carbon dioxide in the atmosphere). The shearwater example also illustrates the need to work both at a local level, in this case on the island, and at a much wider, even global, scale. Species, sites, places and what we describe as ecosystems cannot be managed or protected in isolation. Wildlife cannot survive as small, isolated islands in a desolate, ravaged landscape.

Fig. 7.1 Manx shearwater

The UK National Ecosystem Assessment website (uknea.unep-wcmc.org) provides an excellent guide to the various categories of ecosystems:

There is not a universal categorisation of ecosystems, but they can be defined as areas which share similar features amongst the factors of:

- climatic conditions;
- geophysical conditions;
- dominant use by humans;
- surface cover (based on type of vegetative cover in terrestrial ecosystems or on fresh water, brackish water, or salt water in aquatic ecosystems);
- species composition;
- resource management systems and institutions.

7.3 The Ecosystem Approach – Background

Most of the early thinking about ecosystem management originated in the USA (Stankey et al. 1984). Stankey has been publishing since the early 1970s. His approach, endorsed by many new world scientists, is most relevant to wilderness ecosystems. It assumes that a natural system is sustainable if anthropogenic factors are removed or controlled. The latter, i.e. control, dominated their thinking, hence the development of limits of acceptable change (LACs).[1] Conservation management becomes a sequence of risk assessments: how much of a certain human influence, or influences, can an ecosystem tolerate without suffering any permanent damage? The only things that managers would then need to monitor would be the various influences or factors. This may be a good idea, but it assumes that it is always possible identify all the factors and that we will understand how these factors can change the ecosystem. In reality, this can never be done with any certainty.

By contrast, in Europe we have been rather more concerned with expressing the desired outcome of management, i.e. quantified and measurable objectives. This is mainly because of the value that we place on semi-natural communities and habitats, the need for species-specific management and the impact of European and domestic legislation. The difference between Old World and New World conservation had led to a divergence, particularly at the site and species management level. This might explain why the ecosystem approach has not, until recently, become a feature of European conservation. We have, however, recognised and applied most of the principles implied by this approach for a very long time, but we have sometimes used a different language to describe them.

Various ecosystem approaches have been described and defined by many different individuals and organisations in many different ways. The most commonly applied and widely recognised definition comes from the Convention on Biological Diversity (CBD). For most people, the words 'ecosystem approach' are assumed to mean the CBD approach. The CBD definition of the ecosystem approach is given above. In interpreting this, it should be understood that:

- It was designed to balance the three objectives of the Convention on Biological Diversity: conservation, sustainable use and the fair and equitable sharing of the benefits arising out of the human use of genetic resources. It recognises that humans, and their cultural diversity, are an integral component of many ecosystems.
- It is a strategic approach for the integrated management of land, water and living organisms that promotes nature conservation and sustainable use in an equitable way.

[1] Please see Chapter 18 for a description of LACs.

- No individual component of a system is treated in isolation from any other component. Integration should facilitate a much more strategic approach to tackling the factors that influence biodiversity and help prioritise the effective and efficient use of resources.
- It is an approach that should encourage synergies rather than conflicts between social, economic and environmental objectives.

It is important to understand that an ecosystem approach does not mean abandoning the established tried, tested and effective methods of conservation. It is about widening our current perspective, not changing it completely. We still need to designate and manage protected areas, and we still need to target priority species and habitats. We will need to improve the quality of current protected areas, increase their size, enhance connections and create new sites. Nature reserves and other special conservation sites are fundamental to a successful ecosystems approach. They are the reservoirs of species, providing the essential resources for maintaining biodiversity, along with all the benefits that it provides.

In order to respond to uncertainties and the potential for unpredictable outcomes, the ecosystem approach to management must be adaptive. (Adaptive management is grounded in the admission that we do not know enough to manage ecosystems.) Adaptive management must be planned. It is entirely dependent on the formulation of clear objectives that define our outcomes, the implementation of management and, most importantly, monitoring. This cyclic and repetitive approach is about learning by doing (see Chapter 6). Ecosystem management reflects the precautionary principle: measures may need to be taken even when some cause-and-effect relationships are not yet fully established scientifically (Chapter 2). There is no single way to implement the ecosystem approach, as it depends on local, provincial, national, regional or global conditions.

7.4 The Key Principles of the Ecosystem Approach to Management

Some professional conservationists hold the view that the CBD principles are an overwhelmingly politically correct statement of the obvious, which could weaken our ability to protect wildlife. On the other hand, politicians and government organisations usually respond favourably to the principles. Given the status of the CBD, the ecosystem approach is rarely questioned and is generally accepted as an authoritative expression of fact that cannot be challenged. The principles are expressed in extremely cautious terms. This is an inevitable product of international conventions, where decisions are made through consensus. International conventions can do little more; they seek the agreement of all representatives when making statements and formulating policies.

Weaknesses aside, the principles provide guidance that all management planners should at least consider. As with all guidance, we can take from it the sections that are relevant to our particular circumstances and set aside the remainder.

The following principles are complementary and interlinked.

Principle 1: The objectives of management of land, water and living resources are a matter of societal choices.

> Different sectors of society view ecosystems in terms of their own economic, cultural and society needs. Indigenous peoples and other local communities living on the land are important stakeholders and their rights and interests should be recognized. Both cultural and biological diversity are central components of the ecosystem approach, and management should take this into account. Societal choices should be expressed as clearly as possible. Ecosystems should be managed for their intrinsic values and for the tangible or intangible benefits for humans, in a fair and equitable way.

See Chapter 4 for a detailed account of the relevance of stakeholders to management planning. A stakeholder is any individual, group or community living within the influence of the site or likely to be affected by a management decision or action, and any individual, group or community likely to influence the management of the site. It is far easier to express an intention to take account of societal choices than it will ever be to apply this principle in practice. The different sectors in any society will have varied, and often extremely contradictory, views and aspirations. Even obtaining a balanced view, one that has not been skewed by the extreme ideas of a vociferous minority, can be extraordinarily difficult. The majority, and these are generally the people that really matter, are usually silent. Regardless of all difficulties and impediments, conservation managers must recognise the need to adopt an inclusive approach that takes account of the interests of stakeholders and, as far as possible, encourages their involvement in all aspects of management and planning.

'*Ecosystems should be managed for their intrinsic values and for the tangible or intangible benefits for humans, in a fair and equitable way.*' This is one of the most important statements in the entire suite of principles, but it is buried in the subtext, and, surprisingly given its significance, it is not one of the key principles. (See Chapter 8 for a detailed section on intrinsic value.)

Principle 2: Management should be decentralized to the lowest appropriate level.

> Decentralised systems may lead to greater efficiency, effectiveness and equity. Management should involve all stakeholders and balance local interests with the wider public interest. The closer management is to the ecosystem, the greater the responsibility, ownership, accountability, participation, and use of local knowledge.

Managers should seek to balance local interests with the wider public interest. These are fine sentiments, often used to justify decentralisation. However, there are many examples where localised public interest contradicts the wider interests of society. For an ecosystem approach to work, we will need to regulate human activity: this is never popular with the public. Local decisions will have to be based on strategic plans. Otherwise, there is a threat that too much decentralisation will diminish any hope of a coordinated national or international approach. Nature conservation and the maintenance of biodiversity are, or should be, global concerns.

Principle 3: Ecosystem managers should consider the effects (actual or potential) of their activities on adjacent and other ecosystems.

> Management interventions in ecosystems often have unknown or unpredictable effects on other ecosystems; therefore, possible impacts need careful consideration and analysis. This may require new arrangements or ways of organization for institutions involved in decision-making to make, if necessary, appropriate compromises.

This is a very obvious and simple principle which is much easier said than done. We really do not know enough about natural systems to have complete confidence in our ability to manage them, and we certainly cannot predict the implications of management for other sites. The precautionary principle (Chapter 2) should always be applied. So, if there is a risk that intervention on one site may have serious consequences for another we need, at the very least, to work with the managers of the threatened site. It is quite difficult to find examples in the more natural ecosystems where these threats occur. An example could be coastal engineering solutions applied at one site to prevent natural erosion, thereby starving a second site of its sand supply.

It is when we interrupt the natural process in an ecosystem that we are most likely to impact on adjoining systems. For example, the Jaldapara Wildlife Sanctuary is a protected park situated at the foothills of Eastern Himalayas in West Bengal. It has vast areas of floodplain grassland interspersed with patches of riverine forest, marsh and swamp. The sanctuary was established in 1941 for the protection of the Indian one-horned rhinoceros *Rhinoceros unicornis*. The habitat that supports the rhinoceros is tall grassland. This is an ephemeral community, created when monsoon rains flood the river Torsa, which flows through Jaldapara. The grassland rapidly becomes scrub and, eventually, riverine forest. This persists until flooding changes the landscape and creates new opportunities for grassland to establish.

The floods are, from a human perspective, destructive, threatening events that happen every year, but, periodically, they can be devastating. People react by attempting to control the floods by, among other measures, building dams. Management intervention in the river ecosystem above Jaldapara has already interrupted and diminished the impact of flooding. Where nature once provided, gangs of women now clear scrub by hand. This is expensive and probably not sustainable. A new, large dam is currently being planned for the Torsa, up river of Jaldapara. This may be good for people, but it will be devastating for the grasslands and, of course, for the rhinos.

Fig. 7.2 *Rhododendron ponticum,* an invasive and destructive species

In the semi-natural world of cultural landscapes, there are so many examples where the management of one site has implications for others. Probably the most common are the consequences of inappropriate management or lack of management intervention. This is best illustrated when alien invasive species threaten several adjacent sites or ecosystems. For example, rhododendron *Rhododendron ponticum,* (Fig. 7.2) an extremely invasive and destructive species, can only be effectively controlled when tackled over its entire range within any locality. The range can easily extend over several sites and ecosystems. The conditions which prevail when rhododendron has been cleared are perfect places for rhododendron seeds to germinate.

Where the management of one ecosystem is impacting on another the obvious question must be the size or geographical definition of the separate ecosystems. Should they be managed as a single system? Clearly, when ecosystems are defined as management areas attention must be given to Principle 7: 'The ecosystem approach should be undertaken at the appropriate spatial and temporal scales.'

Principle 4: Recognizing potential gains from management. There is usually a need to understand and manage the ecosystem in an economic context.

Any such ecosystem-management programme should:

- Reduce those market distortions that adversely affect biological diversity;
- Align incentives to promote biodiversity conservation and sustainable use;
- Internalise costs and benefits in the given ecosystem to the extent feasible.

The greatest threat to biological diversity lies in its replacement by alternative systems of land use. This often arises through market distortions, which undervalue natural systems and populations and provide perverse incentives and subsidies to favour the conversion of land to less diverse systems.

> Often those who benefit from conservation do not pay the costs associated with conservation and, similarly, those who generate environmental costs (e.g. pollution) escape responsibility. Alignment of incentives allows those who control the resource to benefit and ensures that those who generate environmental costs will pay.

Principle 5: Conservation of ecosystem structure and functioning, in order to maintain ecosystem services, should be a priority target of the ecosystem approach.

> Ecosystem functioning and resilience depends on a dynamic relationship within species, among species and between species and their abiotic environment, as well as the physical and chemical interactions within the environment. The conservation and, where appropriate, restoration of these interactions and processes is of greater significance for the long-term maintenance of biological diversity than simply protection of species.

The management and planning approach which I describe and promote in this book places considerable emphasis on the concept of 'Favourable Conservation Status' (FCS) as a means of defining what we want to achieve for habitats and species. There is no legal basis for using FCS at site level in Europe, but in England and Wales it is a policy applied to sites protected through domestic legislation (the Sites of Special Scientific Interest). The conservation of ecosystem structure and function is central to the definition of Favourable Conservation Status. (See Chapter 15 for a full account of FCS and how structure and function can be addressed in a management plan.)

Principle 6: Ecosystems must be managed within the limits of their functioning.

> In considering the likelihood or ease of attaining the management objectives, attention should be given to the environmental conditions that limit natural productivity, ecosystem structure, functioning and diversity. The limits to ecosystem functioning may be affected to different degrees by temporary, unpredictable or artificially maintained conditions and, accordingly, management should be appropriately cautious.

This is best explained by example. Consider a raised bog which has a very long history of exploitation: ditches have been cut throughout the site, the hydrological integrity damaged and large quantities of peat have been removed. The result is a rather desolate bog with large areas of mature scrub and a dry surface incapable of sustaining the bog flora (Fig. 7.2). Some of the most derelict areas provide habitats for populations of rare species of invertebrates and birds, including nightjars *Caprimulgus europaeus*. Nightjars are birds of heathland, moorland, open woodland with clearings, and recently felled conifer plantations.

Despite the extremely unfavourable condition of the site, it is still recognised as an internationally important raised bog. Its status and designation impose an obligation to restore the raised bog. If this is successful, the habitat suitable for the invertebrates and the nightjars will be destroyed. Any attempt to protect the 'temporary artificially maintained conditions' would seriously jeopardise the legal obligation to obtain the management objective, which would be to restore, and thereafter maintain, the raised bog at a favourable conservation status.

Fig. 7.3 A damaged bog

Principle 7: The ecosystem approach should be undertaken at the appropriate spatial and temporal scales.

> The approach should be bounded by spatial and temporal scales that are appropriate to the objectives. Boundaries for management will be defined operationally by users, managers, scientists and indigenous and local peoples. Connectivity between areas should be promoted where necessary. The ecosystem approach is based upon the hierarchical nature of biological diversity characterized by the interaction and integration of genes, species and ecosystems.

This is a very important principle. Many people believe that the ecosystem approach can only be applied to very large areas of land; they often talk of the landscape level. This principle makes it clear that the ecosystem approach can, and should, be applied at any appropriate scale.

The 'temporal scale' is slightly more difficult to understand. It does not suggest that we can manage an ecosystem for a given period of time and then move on. There are approaches to planning, mainly advocated by some authors from the USA, that describe conservation management as 'conservation projects' (Margoluis and Salafsky 1998). The implication, intended or otherwise, is that, in common with all projects, there is a beginning and an end. I believe that the management of places or ecosystems for nature conservation should, whenever possible, be recognised as a very long term commitment. In Chapter 14, I introduce recovery and maintenance management. This is in recognition of the fact that both the type and intensity of management can vary with time, depending on the condition of the feature. The condition of a feature when management commences can be extremely unfavourable. The management required to move a feature from an unfavourable condition to a favourable condition can be regarded as 'recovery management', while the management required to maintain a feature in favourable condition would be 'maintenance management'. A common failing of conservation management, particularly when it is treated as a series of projects, is to assume that once recovery management has been successful we can forget about the site. However, there is no escaping the fact that conservation management is occasionally opportunistic and short-term: clearly it is better to provide short-term opportunities for wildlife than none at all. There are also ephemeral features: for example, many sand dune systems have a naturally limited life span; they can be part of a dynamic coastal process which creates and eventually erodes a system.

The 'spatial scale' is easier to deal with. We might be tempted to think, 'the bigger the better', and, clearly, there is some truth in this. The conservation of ecosystems will only be possible if we are able to control all the factors, positive and negative, that impact on and change, or have the potential to change, the ecosystem. (These are the management interventions described in Principle 3.) We can no longer avoid the fact that global climate change is probably the single most significant factor that we will have to confront. This can only be dealt with strategically at a global level, though most of the actions will have to take place at a local level. Some factors, for example, the use of destructive pesticides, are best tackled at a country level by, for example, legislating against their use. Factors can operate at a regional level: alien invasive species are an obvious example. Finally, some factors can be site, or ecosystem, specific. So, what are the implications? Successful ecosystem management will only be possible, at any level, if we can adopt a tiered approach at the global level. This requires that, at each level in the tier, management or controls are appropriate for that tier. Tiers can be anything from global to the smallest manageable unit. Remember the CBD definition of an ecosystem: 'A unit could be a veteran tree, a small woodland, a forest or the entire biosphere'.

So often, when organisations or governments consider adopting the ecosystem approach their response is to delineate selected areas, which they call ecosystems. (The concept of an ecosystem is a human construct, used to describe the natural world, and we define ecosystems according to the scale of human interests and our decision-making powers.) This perpetuates the myth that isolated sites, islands of biodiversity, can, in the long term, survive in a desolate, artificial landscape. 'The rate of extinction once a species manages to colonize any island is affected by island size' (MacArthur and Wilson 1967).

The 'spatial' scale is often defined by specific distinct habitat types. The adoption of management at an ecosystem level offers opportunities for setting aside the traditional divisions between habitats along with any idea that habitats should be static and unchanging. Most habitats in the Old World exist in any given place and time as a consequence of a range of natural and anthropogenic factors. As the factors have changed, and will inevitably continue to change over time, so should the distribution, structure and composition of the habitats in any ecosystem. The transitional areas between habitats provide opportunities for so much wildlife, and these areas can only prosper in dynamic ecosystems.

The ecosystem approach, when applied to large areas of land or to a tiered system, could provide dynamic landscapes, better able to respond to natural and anthropogenic factors. This would enable the establishment of robust and resilient ecosystems, which have the potential to optimise opportunities for biodiversity and secure a future of wildlife. These areas will inevitably be different to anything that has happened in the past, but should we try to triumph over evolution and, for the first time in the history of our countryside, prevent any further change? Can we realistically fossilise our countryside with reference to a baseline which represents some arbitrary point in time? This is not to suggest that this approach should be universally applied. Even if we work strategically at an ecosystem level, work on the ground has to be at an appropriate scale. Whatever the approach adopted, we will always need to manage the components that comprise ecosystems. We will also need to hold some places in their current condition, or at a status that we define as favourable. These places are a record of our cultural history. They provide the reservoirs of wildlife, habitats and populations of species that are essential if we are to obtain improvements elsewhere.

The sub-text promotes the need for connectivity: this is the re-establishment of linkages between isolated fragments of habitat. Although connectivity is an essential consideration when planning and managing sites, it is not easy to decide where it should be included in a plan. I think that it is best regarded as a factor or influence, because if it is an issue on a site it will have a significant impact on the management of the features. (See Chapter 14 for a full account of why factors are so important in the planning process.)

Principle 8: Recognizing the varying temporal scales and lag-effects that characterize ecosystem processes. Objectives for ecosystem management should be set for the long term.

> Ecosystem processes are characterized by varying temporal scales and lag-effects. This inherently conflicts with the tendency of humans to favour short-term gains and immediate benefits over future ones.

The problem is that we do not know what long term means. At a time when our ability to visualise the future is obscured by the reality of climate change, 'long term' becomes 'shorter term'. The main point here, however, is the potential contradiction between this principle and the next, which recognises that change is inevitable and advocates *adaptive management*. Biodiversity or ecosystem objectives, which lie at the very heart of adaptive management, are, quite simply, an expression of something that we want to achieve. They are, and can only be, a reflection of our values, knowledge and expertise at the time of writing and must be reviewed, and if necessary changed, at intervals.

Principle 9: Management must recognize the change is inevitable.

> Ecosystems change, including species composition and population abundance. Hence, management should adapt to the changes. Apart from their inherent dynamics of change, ecosystems are beset by a complex of uncertainties and potential "surprises" in the human, biological and environmental realms. Traditional disturbance regimes may be important for ecosystem structure and functioning, and may need to be maintained or restored. The ecosystem approach must utilize adaptive management in order to anticipate and cater for such changes and events and should be cautious in making any decision that may foreclose

> options, but, at the same time, consider mitigating actions to cope with long-term changes such as climate change.

Please refer to Chapter 6, which deals with adaptive management.

Principle 10: The ecosystem approach should seek the appropriate balance between, and integration of, conservation and use of biological diversity.

> Biological diversity is critical both for its intrinsic value and because of the key role it plays in providing the ecosystem and other services upon which we all ultimately depend. There has been a tendency in the past to manage components of biological diversity either as protected or non-protected. There is a need for a shift to more flexible situations, where conservation and use are seen in context and the full range of measures is applied in a continuum from strictly protected to human-made ecosystems.

Please see Chapter 8.

Principle 11: The ecosystem approach should consider all forms of relevant information, including scientific and indigenous and local knowledge, innovations and practices.

> Information from all sources is critical to arriving at effective ecosystem management strategies. A much better knowledge of ecosystem functions and the impact of human use is desirable.

Please see Chapters 12 and 19.

Principle 12: The ecosystem approach should involve all relevant sectors of society and scientific disciplines.

> Most problems of biological-diversity management are complex, with many interactions, side-effects and implications, and therefore should involve the necessary expertise and stakeholders at the local, national, regional and international level, as appropriate.

All too frequently, stakeholders are reluctant to recognise expert opinion, and organisations can be tempted to set aside expertise in order to appease local stakeholders. Given that the first principle of the ecosystem approach advocates 'societal choice', the intention of this principle must be to emphasise the need for expertise.

References

Golley, F. B. (1994). *A History of the Ecosystem Concept in Ecology. More Than the Sum of Its Parts.* Yale University Press, New Haven/London.

MacArthur, R. H. and Wilson, E. O. (1967). *The Theory of Island Biogeography.* Princeton University Press, Princeton, NJ, USA.

Margoluis, R. and Salafsky, N. (1998). *Measures of Success: Designing, Managing, and Monitoring Conservation and Development Projects.* Island Press, Washington DC.

Stankey, G. H., McCool, S. F. and Stokes, G. L. (1984). *Limits of acceptable change: A new framework for managing the Bob Marshall Wilderness.* Western Wildlands 10(3), 33–37.

Tansley, A. G. (1935). *The use and abuse of vegetation concepts and terms.* Ecology 16(3).

Chapter 8
Ethics and Conservation Management: Why Conserve Wildlife?

> Is it not enough for you to feed on the good pasture, that you must tread down with your feet the rest of your pasture; and to drink of clear water, that you must muddy the rest of the water with your feet? (Ezekiel 34:18)

> The world grows smaller and smaller, more and more interdependent… today more than ever before life must be characterized by a sense of Universal Responsibility not only nation to nation and human to human, but also human to other forms of life. (His Holiness the Dalai Lama 1987)

> Someday children coming upon the picture of a tiger will view it the way we view dinosaurs, wondering if such creatures ever really existed. But the extinction of tigers – and the gorillas and the wolves and the whales – will be different. We will have exterminated these species, unthinkingly, without purpose, without remorse. (Roszak 2001)

Abstract This chapter deals with one of the most important aspects of conservation management: why we do it. It is not an easy chapter: there are no clearly defined, or widely accepted, rights or wrongs. It is entirely a matter of opinion or belief. My intention, therefore, is simply to provoke thought and debate. When sites, particularly legally protected areas, are managed by organisations that have policies to guide management there may be little reason to consider conservation ethics when preparing a plan (though anyone engaged in nature conservation should at least be aware of the ethical considerations). Where there is no formal guidance, legislation or policy, planners must understand why they are managing the site. It is only by understanding 'why' that we are able to decide what we are trying to achieve and what we must do. Human values are considered, with an emphasis on scientific values and conservation ethics. One conclusion is that scientific values, if they exist, must be supplemented with the full range of other human values. The biocentric/anthropocentric (ecosystem services) divide represents perhaps the most significant issue in conservation ethics. Some authors suggest that this has done more harm than good. Norton (1991) offers a 'convergence hypothesis' and argues that the outcome, i.e. environmental protection, will be a consequence of either ethical position. There is, however, at least one significant difference: the burden of proof.

M. Alexander, *Management Planning for Nature Conservation: A Theoretical Basis & Practical Guide*, DOI 10.1007/978-94-007-5116-3_8,

An anthropocentric approach implies that conservationists should prove that a habitat or species has value to people, whereas a biocentric approach requires developers to justify their position. There is no consensus, no single and universally accepted conservation ethos. Conservation managers should be aware of the breadth of the debate and attempt to develop a personal ethical position.

8.1 Introduction

Fig. 8.1 Low tide on a rocky shore

A search for ethical enlightenment is like being a child at the seaside, turning over rocks at low tide on an unfamiliar shore. Each new rock reveals a diverse, exciting host of organisms, some vaguely recognisable, most completely new. At first, we are dazzled by the seemingly endless discoveries, so delighted by the novelty that we are ready to forget everything that has gone before. We ignore the familiar things that we know and understand; our obsession with the new distorts our perspective and can distort our beliefs. This is a dangerous place, and I am aware that my personal search for an ethical position is both naive and poorly informed. I know that new ideas (new, at least, to me) can, for a time, disproportionately influence my beliefs. The consequence is that as my understanding evolves so does the emphasis or precedence that I give to various topics and issues. I do not claim any expertise. I am a novice. I cannot claim to have a balanced or impartial view. Nothing could be further from the truth. So if, like me, you are a newcomer to the world of environmental ethics, regard what I have to say as a very small sample from under the first rock that you turn, and understand that the shore is littered with thousands upon thousands of bigger rocks.

What I do know with some certainty is that it is only through trying to understand our relationship with nature that we can begin to make sensible, and perhaps even wise, decisions about nature conservation. The questions we need to resolve are why we should conserve nature and what we should conserve. The answers to these questions should guide politics and the formulation of policies that lead to the development of legislation and conservation strategies. Logically, sites should be managed within a strategic framework, where protected areas are recognised as some of the many tools used to conserve nature and to safeguard our natural environment. Some statutory conservation sites, for example Natura 2000 in Europe, are selected and designated as a consequence of a political process which we can only assume represents a democratic expression of society's values. In addition to responding to legislation, and particularly to the legal status of the site, all management plans should be guided by policy (see Chapter 11). Policies, or more specifically organisational policies, are a high-level statement of the purposes of an organisation. These policies may have been adopted voluntarily, imposed by legislation, or a combination of both. They should reflect the values of the organisations responsible for managing sites and provide at least an indication of why they manage sites and what, in broad terms, they want to achieve. If this is the case, i.e. there is a clear expression of guiding values, either as a consequence of legislation or organisational policy, few, if any, decisions based on values or ethics need to be made in the site management plan.

By now you may be wondering why this chapter has any relevance to a book on planning. It is important because sites are often managed by organisations that do not have, or have not expressed, any policies. We also work in areas that are not governed by any external policies. In the absence of clear guidance, all our actions as site managers should be underpinned by a personal ethos. The easiest way to justify the case for conservation management to be guided by ethics is to consider the alternative. Should we make even the most simple and seemingly inconsequential decisions without values or ethics?

Fig. 8.2 A wild and remote coast

Consider this example: It was decided on an extremely beautiful, wild and remote coastal site to provide information signs for visitors (Fig. 8.2). This was important because visitors needed information to help them find their way around and discover when and where they could see the wildlife, including birds and seals. Because the sea cliffs are very dangerous signs were also needed to address serious health and safety considerations. A commercial consultancy was contracted to provide the signs, but, unfortunately, they did not appear to have any guiding ethos or, indeed, sensitivity. The signs that they produced contained all the necessary information but, the designer had chosen a startlingly garish style. The oblong signs were divided diagonally: one half a dazzling orange the other an equally strong and complementary blue. Complementary colours are used when designers want vibrancy in design, usually when they need to draw the viewers' attention to something. So, from a purely design perspective, where the intention was to attract attention and convey information, the signs met the design brief. But, in the setting of a wild and wonderful Atlantic coastline, the proposed signs would have shone out like glaring beacons. They may have been informative, but they would also have assaulted the senses and sensitivities of visitors.

What if the managers had shared their management ethic with the designers? It could, for example, have been a much simplified version of Leopold's (1949) land ethic: 'A thing is right when it preserves integrity and beauty, wrong when it tends otherwise'. Ethics will, at the very least, give guidance and provide some consistency when making decisions. Each individual decision might be minor and seemingly of little consequence, but a sequence of misguided and inconsistent decisions will have a much greater impact.

> A land ethic, then, reflects the existence of an ecological conscience, and this in turn reflects a conviction of individual responsibility for the health of the land. Health is the capacity of the land for self-renewal. Conservation is our effort to understand and preserve this capacity. (Leopold 1949)

8.2 What Does Nature Conservation Mean?

This chapter deals with values and begins by specifically asking the question, why conserve wildlife? Before we can answer this, we need to consider 'nature conservation' or, at least, what I think we mean by nature conservation. We all use the words, but how often, if at all, do we think about the meaning? The following examples illustrate the diverse range of definitions:

> Conservation biology strives to maintain the diversity of genes, populations, species, habitats, ecosystems and landscapes, and the processes normally carried out by them, such as natural selection, biogeochemical cycling, photosynthesis, energy transfer, and hydrologic cycles. It is a dynamic play, with players and action on many different spatial and temporal scales, with old actors disappearing and new ones arriving, but the play ultimately comes down to one thing: dynamic evolutionary processes in a changing ecological background.

> Conservation biology attempts to keep those normally evolutionary processes working within a functioning setting. (Meffe et al. 1997)

> A personal relationship, a matter of decisions about human actions in the light of their implications for non human nature. I believe that perhaps the most critical element in conservation is the form of human engagement with non-human nature. (Adams 2003)

> Nature conservation can be thought of as a social movement working to develop or reassert certain values in society concerning the human-nature relationship. (Jepson and Canney 2003)

> Nature conservation is a highly organised system for expressing environmental preferences. (Cole-King 2005)

My belief is that nature conservation is about the future. It is about developing, or enabling the development of, populations, habitats and ecosystems that are resilient and robust enough to survive environmental change, whatever the reason for that change. It is about enabling natural processes and accepting the outcomes of these processes even when the outcomes are unpredictable. Nature conservation is not necessarily about preventing any change to the current situation or recreating conditions that may have existed at some time in the past. It is not preservation, but preservation can be a useful conservation tool. We will need to preserve, or hold reservoirs of, wildlife until such time as natural processes can be reinstated. Nature conservation is an expression of a relationship between people and nature: it is something that we do, a human or anthropogenic activity. We are the only species that engages in this activity. No other species presumes to make decisions about our environment or the fate of the other species that inhabit this earth.

Given this unique and very privileged position, I am reminded of the Welsh motto of my old school: '*Ymhob braint y mae dyletswydd*'. Loosely translated into English, this means, 'With every privilege there is an obligation'. Do we have a moral obligation to protect the natural world? Does wildlife have intrinsic value regardless of its value to humanity? Some people will not support these beliefs but will recognise that nature provides resources to satisfy both our physical and emotional needs. They may also realise that our survival on this planet is inextricably linked to the survival of nature. I believe that nature conservation is not only something that we do but something that we must do.

8.3 Values – Introduction

> Like winds and sunsets, wild things were taken for granted until progress began to do away with them. Now we face the question whether a still higher 'standard of living' is worth its cost in things natural, wild and free. (Leopold 1949)

Fig. 8.3 Sunset, Freshwater West

People value wildlife/wild places/nature for so many different reasons. For some it provides, or contributes to, their livelihood. Others value the opportunities for leisure activities, which may range from the hunting and taking of game to passive bird watching. Some will recognise spiritual or religious values.

Legislation or statutory law is, in any democracy, an expression of society's values. Wildlife legislation, both domestic and international, is a powerful driver for many organisations, so much so that the law, or an interpretation of the law, can replace deeper reasoning. But, setting aside any negative issues, anyone engaged in making decisions about wildlife management must have a detailed understanding, and take account, of all relevant legislation.

In recent decades, science has tended to dominate nature conservation. Some hold the view that nature conservation is, or should be, a scientific process. Others believe that there needs to be a balance between scientific and intuitive approaches to nature conservation: they do not see conservation as a science but recognise that science is an essential tool which helps us to describe, understand and manage wildlife.

Religious beliefs can have very significant and profound implications for our relationship with nature. Religious attitudes cover a broad spectrum, ranging from those that regard nature as something that has no value other than to provide for human need or greed, to those that profess extreme reverence and respect for the natural environment. In many parts of the world, religious beliefs have been set aside, and society is governed entirely by secular law. In others, there is no distinction between religious and secular law. While this is an extremely complicated, and for many a confusing, area, it is essential that we respect the religious beliefs held by others. I am not a theologian, and religion is far too important a subject to be dealt with superficially here, but I believe it is imperative that individuals responsible for preparing management plans understand how the prevalent religious beliefs and values should influence their decisions.

This wide range of beliefs is not restricted to religion. Many organisations and individuals engaged in nature conservation express an almost entirely utilitarian view: conservation is about ensuring that nature continues to provide a variety of services for people. There are also fundamentally different environmental philosophies, including the belief that wildlife has an intrinsic value, regardless of its usefulness to people. (I will return to this debate later in the chapter.)

Monetary values are a vital component of the discussion, but, since the scope of this chapter gives time for only a cursory glance at human values, the subject will have to be skimmed very briefly. Some economists suggest that monetary values should provide a basis for making decisions about land use, for example, comparing the value of wildlife to potential agricultural yield. In contrast, many conservationists argue that nature should never be valued in monetary terms: what price do we give to the song of a skylark? Monetary value can have some rather obvious negative impacts, for example, the commercial trading, legal and illegal, in endangered species or their body parts, such as elephant tusks, rhinoceros horn and tiger pelts. I am not, however, suggesting that monetary values should be dismissed. They are very important, and they can be used to demonstrate the contribution that nature, or a particular site, makes to a local or global economy. This can help, for example, by providing additional justification for site acquisition, or by making a contribution to defending a threatened site, or justifying expenditure on a site. However, with the exception of the need in some protected areas to physically protect endangered species from illegal poaching or exploitation, monetary values have very limited, if any, relevance to management planning. By that I mean they will have little direct impact on the way in which sites are managed, unless, of course, the key function of a protected area shifts from nature conservation to the provision of ecosystem services.

This chapter is not about challenging human values: it is an attempt to highlight the relationship that should exist between values and actions. The essential point, of course, is that we find some reason to value the natural world. Precisely what those reasons are is less important. The greatest threat of all is complacency or, worse, a failure to recognise any value in nature.

When planning conservation management we must give appropriate attention to the values and beliefs of others. This chapter is not the place for moral or religious guidance, but we all need to recognise and be aware of our motivations and inspirations. We need to understand why we are personally committed to conserving nature, and we need the confidence and conviction to ensure that our beliefs, as far as possible, govern our decisions and actions.

8.4 Scientific Values

> Wildlife management is not a science. Wildlife managers apply the science of biology. They use methods of science. But management is an art. Science is any body of organised, tested, and accepted knowledge; or it is research: the process of finding, testing, organising, and communicating knowledge. As an art wildlife management is the application of knowledge to achieve goals. Wildlife management is primarily application of biology, especially ecology. Wildlife managers use scientific methods to obtain information about populations and habitats. They require the objectivity of scientists. They also require manual and communication skills achieved through experience. They use judgement and form compromises especially when decisions must be based on limited information. In selecting goals, they compare and judge values. Science does not deal in compromises and value judgements. (Bailey 1982)

Conservation managers and scientists have become so reliant on science as the principal tool for identifying, justifying, and even defending, nature conservation priorities and actions that they often forget that science is just one of the many ways in which humanity can express values and preferences. Perhaps we should, just occasionally, remind ourselves that conservation did not begin with science. In Britain, as an example, we cannot ignore the role of the romantic poets. The British concept of a 'National Park' began with the publication in 1810 of William Wordsworth's *Guide to the Lakes*. Wordsworth recognised the need for conservation and clearly understood the difficulty of preserving something that must change (an insight not always shared by scientists). North American conservation efforts were strongly influenced by the writings in the mid 1800s of Ralph Waldo Emerson, writer, poet and philosopher, and of Henry David Thoreau, author, transcendentalist, naturalist and philosopher. They were among the first to argue that nature has values other than economic benefits to mankind. They both used aesthetic and philosophical arguments for the preservation of wilderness.

However, since the 1940s science has been the main, and sometimes only, legal basis for justifying nature conservation. There is a suggestion that because many countries were so reliant on science during the Second World War scientists and science gained extremely high status. The status of science was such that 'scientific interest' alone was sufficient to justify legislation, public spending and conservation management (Adams 2003). In Britain, The National Parks and Access to the Countryside Act 1949 made provision for the notification of areas of land of special interest because of the 'flora, fauna, geological or physiological features' that they contained. These sites became the Sites of Special Scientific Interest (SSSIs) (Fig. 8.4). They represent the best examples of the British natural heritage of wildlife habitats, geological features and landforms. SSSIs are currently areas that have been notified as being of special interest under the Wildlife and Countryside Act 1981 (the 1981 Act was further strengthened by the Countryside and Rights of Way Act 2000).

Fig. 8.4 Coed Ganllwyd SSSI

Perhaps the most remarkable issue is that in Britain the most important domestic legislation aimed at protecting wildlife implies that the flora and fauna, geology and landforms are important only because they are of *interest* to science. Adams (2003) reminds us of a legal case in the 1960s when the construction of a major reservoir was proposed for a site which had been notified as an SSSI. The supporters of the development were able to defeat the conservation argument by demonstrating that very few scientists had actually visited or shown any interest in the site until it was threatened. The case was lost because the defenders relied entirely on one value, 'scientific interest', and this could not be demonstrated. The most significant point is that the defenders did not offer any other values to support their case. Adams goes on to suggest, 'that SSSIs have always reflected a broader suite of values other than the narrowly scientific'. His reasoning is that the individuals at all levels engaged in the selection of the sites took account of wider values. He says that: 'Science is necessary to select (and manage) SSSIs, but it no longer figures prominently as a reason for their selection. SSSIs are places for nature, not for scientists.' Adams is, of course, implying that human society and the way that we justify conservation has moved on. We value nature, in addition to its 'scientific interest', for many other reasons, including, for some people, intrinsic values.

Scientific value: an oxymoron, a contradiction? Can science have values? These are not easy questions to answer. When I looked up 'science' in the *Oxford Concise Science Dictionary* I found no entry for that word. Presumably anyone referring to a science dictionary, except me, knows what 'science' means. Elsewhere, a typical dictionary definition is: 'The systematic study of the structure and behaviour of the physical and natural world through observation and experimentation.' (*Oxford English Dictionary*) If this is science, what are scientific values or ethics? This is where it gets difficult: some scientists completely reject the idea that it is possible to include ethics or values as part of science (Kaiser 2000; Nielsen 2001). Science is regarded by many as being, of necessity, objective and value free: that is, it can have no imposed human values. Focusing on conservation science, some scientists, Van Houtan (2006) for example, suggest that there is a view that pure science is a fiction and that no scientific observation is value free. Once again, as is so often the case when attempting to discuss values, there are many different, often contradictory, beliefs. The best I can offer to demonstrate that these contradictions exist is a brief account of one example of how 'scientific' values have been applied to nature conservation in Britain:

The UK Nature Conservation Review (NCR) criteria (Ratcliffe 1977) were developed for, 'the selection of biological sites of national importance to nature conservation in Britain'. 'A number of different criteria have, by general agreement and established practice, become accepted as a means of judging the nature conservation value of a defined area of land' (Ratcliffe 1977). (As an aside, it is difficult to locate any references to 'established practice' or any use of these criteria prior to the publication of the NCR.) The criteria were designed to assess comparative site quality. That is, each site was compared with others of a similar type. Ratcliffe makes the following statement: 'This is the most difficult part of the whole process to rationalise satisfactorily, since it is essentially subjective, even when based on a consensus view.' The critical point here is that the criteria were developed by a group of the most competent and highly respected conservation scientists of the day to enable them to identify the most important conservation sites in the UK. They made it abundantly clear that theirs was not an objective view, though without doubt, given their standing, it was a scientific view, i.e. a view held by scientists. Despite these assertions, many individuals and organisations have come to believe that the use of the criteria represents a scientific or objective method for identifying the important features. For example, the Royal Society of Edinburgh published a document that contained the following: 'The Nature Conservation Review (NCR) provides scientific criteria for selection of the majority of SSSIs and continues to be a sound and objective basis on which to designate SSSIs.' (RSE 1998)

Some of the criteria have received attention in a different context, i.e. they are recognised as useful conservation values. For example, Margules and Usher (1981) suggest that 'naturalness', 'diversity', 'rarity' and 'area' are the only criteria that have a scientific basis. Angermeier (2000) goes a little further and considers 'naturalness' to be an imperative value in nature conservation. Naturalness and diversity have probably attracted more attention than all other conservation values. Some scientists have worked towards developing an objective or scientific methodology for describing 'naturalness' (Machado 2004). However, I have not yet found anything to persuade me that I should question the assertion made by the author of the NCR, Ratcliffe (1977), that the use of the criteria 'is essentially subjective'. I believe that they represent human and not scientific values.

In 1984, E O Wilson argued that a scientific understanding of life's variety increases the ethical significance of protecting it. He clearly made a distinction between science and ethics. Later, Van Houtan (2006) posed a rather fundamental question: 'Is nature conservation a virtue or is it just good science?' If conservation is simply science then 'empiricism reigns'. However, if conservation is a virtue, scientific arguments alone are insufficient; they must be joined by ethics and cultural values.

> To succeed as a social cause, conservation needs a hope that academic science itself cannot provide. Conservation needs a cultural legitimacy that inspires enthusiasm, allegiance, and personal sacrifice – in other words, actual changes in human behaviour. Such a vision does not provide a straight path to easy answers; rather, it offers a description of ethics currently estranged from conservation science. (Van Houtan 2006)

8.5 Conservation (Environmental) Ethics

> A storm had stranded thousands of starfish on a beach. A boy began to pick up the starfish and throw them back into the water. A man came along and asked the boy "Why are you doing that? It won't make a difference." The boy picked up another starfish, and replied, "It will to this one." (Anon)

Nature conservation only exists because some people value nature. It would be difficult, if not impossible, to justify any conservation effort in the absence of human values or ethics, and so it would not make any sense to write about management planning without spending some time considering why we conserve nature. We will only conserve what we value; this is why values are so central to conservation planning and management. It is only by understanding 'why', even if at a very superficial level, that we can begin to make decisions about what we want. This section may have less relevance to the management of statutory conservation sites where there is a legal obligation to protect the conservation features, but it still has relevance and will guide the identification of purpose and priorities on non statutory sites, and the management of non-qualifying features on statutory sites.

Ethic: a set of moral principles.

Ethical: (a) relating to moral principles.
(b) morally correct.

Ethics: (a) the moral principles that govern a person's behaviour or how an activity is conducted.
(b) the branch of knowledge concerned with moral principles.

Moral: (a) concerning the principles of right and wrong behaviour.
(b) following accepted standards of behaviour.

(Soanes et al. 2006)

Many people claim an ethical basis for supporting or defending their particular cause. For example: 'There is a moral responsibility to restore species that we as humans made extinct.'; 'We have a moral obligation to conserve wildlife.' Other people will identify rather more closely with a moral obligation to combat human suffering and starvation. Consider, then, an area of land that is undeniably important for wildlife but which could also feed a starving village. Can ethics help us to make the right decision? Is there a right decision?

Does, or can, a single environmental ethic or set of moral principles exist; something that all reasonable people could subscribe to; something that can guide our activities as conservation managers; some kind of ethics rule book? Most conservation managers will, sooner or later, be faced with difficult moral decisions, for example, to cull or not to cull a troublesome species. This may not be such a difficult decision when dealing with aggressive, alien, invasive species, but is not so easy when one native species threatens another; although it would seem to be easier when the species are plants and more difficult when they are mammals.

Clare Palmer's *Bibliographic Essay on Environmental Ethics* (1994) provides a very accessible and easily assimilated account of environmental ethics. The first part of her essay identifies some of the central questions in the environment ethics debate. She begins with 'value':

> What is considered to be valuable, and from where does its value come? Is value subjective or objective? Is all value the creation of human subjectivity, or are values already 'out there' and to be discovered rather than created?

Some people believe in 'ethical monism': that a single, governing, ethical principle or set of consistent principles, which can be applied to all ethical problems, can exist. This has, until recently, dominated environmental ethics, and indeed ethics as a whole. More recently 'ethical pluralism' has gained support. This is the belief that no one ethical principle, or set of principles, can possibly be relevant to all situations.

Ratcliffe (1976) examined the issues underlying the philosophy of nature conservation. He made a distinction between economic and cultural reasons and dealt mainly with the latter. 'The question of whether we conserve wildlife and habitat because of its right to existence poses one of the ultimate philosophical problems, and can only be answered in terms of personal credo.' Ratcliffe's conclusion was clear: he believed that there is no universal ethical position; conservation ethics are a matter of personal belief.

Perhaps the most significant point is that there are many issues or questions that divide environmental ethicists. They do not share a common belief or beliefs; there is a wide and diverse range of opinion. So, would it be anything other than naïve to suggest that there exists somewhere a universal answer to all our ethical problems?

8.5.1 Intrinsic and Instrumental Values

Intrinsic values (deep ecology): All life has intrinsic value regardless of its value to people. Humans are not separate from non-human life. The world is a network of phenomena that are fundamentally interconnected and interdependent.

Instrumental values (ecosystem services): These are anthropocentric or human centred values. Humans are outside, separate and not part of nature. All values are human values, i.e. the value that only we can ascribe to something. Nature can only have value when it is of use to people.

The instrumental/intrinsic debate is nothing new but is possibly one of the most important debates. For well over a century it has quietly rumbled in the background, interspersed with occasional eruptions as each new generation recognises the tension between human and non-human life. It is so unfortunate that, at each successive eruption, those new to the debate, those who have not had an opportunity to acquaint themselves with the historic perspective, reinvent superficial arguments. They can be so impressed by their 'inventions' that they rush to promote these new ideas with a fervour and zeal that is only available to the uninformed. Over time, different words have been used to describe and debate the same things. In the 1930s people spoke of 'instrumental' values and others used 'anthropocentric' values. Until recently, many talked about 'environmental goods and services'. We now seem to favour the term 'ecosystem services'.

We cannot be entirely certain when this debate began, but it was probably sometime around the beginning of the twentieth century. It can easily be traced back to the USA in the 1900s. Gifford Pinchot, the first Chief of the United States Forest Service (1905–1910), is remembered for reforming the management and development of forests in the United States and for advocating the conservation of the nation's reserves by planned use and renewal. He called it 'the art of producing from the forest whatever it can yield for the service of man'. His approach to nature was utilitarian. He championed the economic benefits of nature, the application of scientific principles to the use of forests and rivers for the good of mankind: 'the greatest good for the greatest number for the longest time'. This set him in opposition to his one-time friend and camping companion John Muir, the Scottish-born American author and early advocate of preservation of in the United States.

John Muir was one of the most influential early naturalists and conservationists. His ideas continue to inspire those who care for wildlife and wild places. Both men were opposed to the reckless exploitation of natural resources, but, where Pinchot saw conservation as a means of managing the nation's natural resources for long-term sustainable commercial use, Muir valued nature for its spiritual and transcendental qualities. A famous early conservation debate (1906) in the USA was in response to plans by the City of San Francisco to construct a dam at the mouth of Hetch Hetchy Valley in the Yosemite National Park. The Hetch Hetchy Valley was described by Muir as 'one of nature's rarest and most precious mountain temples. The pristine Tuolumne River flowed along the valley floor, surrounded by flowered meadows and ancient forests which teemed with bears and bobcats, rushes and eagles.' Muir summed up the basic arguments against the dam: 'These temple destroyers, devotees of ravaging commercialism, seem to have a perfect contempt for Nature, and, instead of lifting their eyes to the God of the mountains, lift them to the Almighty Dollar.' Pinchot, opposed to conservation of wilderness or wildlife for its own sake, was one of the most influential supporters of the proposed dam. Enabled by an act of 1913, the dam was completed in 1923.

One of the most significant participants in the intrinsic/instrumental debate was Aldo Leopold. He began his career as a supporter of Pinchot and his ideas, but later, as his ecological understanding developed, as he gained experience and confidence, he gradually moved over to support Muir. Eventually, recognising that people need a spiritual relationship with the land, he published his famous land ethic. Throughout this chapter, and indeed the rest of this book, I rely heavily on Aldo Leopold for sources of reference and for personal inspiration. It is remarkable that today, over 60 years since Leopold's *A Sand County Almanac* (1949) was first published, so many books and other publications turn to him as a principle source of guidance on environmental ethics. It may be because his writing is so accessible, so lyrical and such a joy to read. Or it might be that anyone who cares so passionately for nature and wildlife that they choose to make it their profession can readily identify with someone who, as a forester, ecologist, teacher, writer and philosopher, took some of the first steps along the trail that we feel compelled to follow. Or, is it because his writings are as fresh and relevant today as they ever were? As a book reviewer once wrote, 'if there are cracks in time, Aldo Leopold fell through one'.

A Sand County Almanac is essential reading for anyone who wishes to gain an insight into conservation ethics. Leopold describes an 'ethical sequence'; ethics, he suggests, is a philosophical evolution. To illustrate the sequence he begins with an account of Odysseus who, on his return from the wars in Troy, hanged a dozen servant girls because he suspected that they had been misbehaving. There was no impropriety in this because the girls were his property to be disposed of as a matter of expediency, not of right or wrong. Leopold describes Odysseus' fidelity to his wife during his long years of absence to demonstrate that concepts of right and wrong existed in ancient Greece. The ethics of that time covered wives but not human chattels: 'During the 3,000 years which have since elapsed, ethical criteria have been extended to so many fields of conduct, with corresponding shrinkages in those judged by expediency only.'

Leopold believed that a 'land' ethic is both an evolutionary possibility and an ecological necessity:

> It is inconceivable to me that an ethical relation to land can exist without love, respect, and admiration for land, and a high regard for its value. By value, I of course mean something far broader than mere economic value; I mean value in the philosophical sense. (Leopold 1949)

It is not possible to read *A Sand County Almanac* and not be aware that Leopold's land ethic is derived from his deep love and respect of 'land'. He regarded a land ethic as a simple enlargement of the ethical boundaries of the human community to include 'soils, waters, plants and animals, or collectively: the land.' It is so important to note that when Leopold extends ethics to include 'land' he does not displace other human ethics.

> In short a land ethic changes the role of Homo Sapiens from conqueror of the land-community to plain member and citizen of it. It implies respect for his fellow members, and also respect for the community as such. A thing is right when it tends to preserve the integrity, stability and beauty of the biotic community. It is wrong when it tends otherwise. (Leopold 1949)

Today we might label this as a 'biocentric' ethic. Leopold's land ethic is grounded in evolutionary and ecological biology, and this may be the reason why so many conservationists who subscribe to a biocentric ethic gravitate towards Leopold, regardless of their religion or cultural background. In the 60 years since the Almanac, are we any closer; have we yet evolved sufficiently to make significant improvements to his land ethic?

Meffe et al. (1997) draw our attention to the fact that in Leopold's time scientists held the belief that habitats progressed towards a static equilibrium. This has since been replaced by the non-equilibrium paradigm which recognises that systems or habitats do not exist in a single, internally-regulated, stable state. They are dynamic and continually changing in response to the influence of a range of natural factors: 'Though ecology now acknowledges the normalcy of change and disturbance in nature, the Leopold Land Ethic, appropriately revised in light of these recent developments in science, remains the guiding environmental ethic for conservation biology.' Consequently, Meffe et al. offer a revised version of the land ethic, though I have to confess that I find Leopold's version much more inspiring:

> A thing is right when it tends to disturb the biotic community only at normal spatial and temporal scales. It is wrong when it tends otherwise.

8.5.1.1 Intrinsic Value

> Uniquely in us, nature opens her eyes and sees that she exists. (Friedrich Wilhelm Joseph Schelling) (1775–1854)

Fig. 8.5 Nature opens her eyes

What does 'intrinsic value' mean? The simplest definition is: all life has value in itself, independent of its usefulness to humans. The most coherent and widely applied definitions of intrinsic value come from deep ecology. This is a biocentric environmental movement which began in the early 1970s. It was inspired by the

Norwegian philosopher Arne Naess. He was not the first, or only, philosopher to recognise the need for a radical change to humanity's relationship to nature, but in 1973 he was the first to use the term 'deep ecology'. Deep ecology is founded on the principles that all systems and all life on earth are interrelated, and that values which are human-centred (anthropocentric) are inadequate. Humans are integral components of the fabric of life. Arne Naess and George Sessions devised a deep ecology platform which comprises the eight points of the deep ecology movement. The platform is not intended as a rigid and inflexible set of principles but as a series of ideas which are open for discussion (Harding 2005).

The deep ecology platform:

1. All life has value in itself, independent of its usefulness to humans.
2. Richness and diversity contribute to life's well-being and have value in themselves.
3. Humans have no right to reduce this richness and diversity except to satisfy vital needs in a responsible way.
4. The impact of humans in the world is excessive and rapidly getting worse.
5. Human lifestyles and population are key elements of this impact.
6. The diversity of life, including cultures, can flourish only with reduced human impact.
7. Basic ideological, political, economic and technological structures must therefore change.
8. Those who accept the foregoing points have an obligation to participate in implementing the necessary changes and to do so peacefully and democratically.

> 'Blind faith' is a category we usually associate with religion, where it has been exercised to the point of affirming the absurd – and doing so proudly as a measure of one's devotion. But believing – whether in general principles, guiding values, or the statements of respected authorities – is one of the chief ways in which all of us find meaning in the world. Beliefs are frames of reference that help all of us, the pious and the sceptical alike, to sort through conflicting information and arrange what we take to be valid in a meaningful pattern. (Roszak 2001)

Only a tiny minority of people give intrinsic value any consideration and within that minority there are a range of quite different and often contradictory opinions. Some hold that it is simply a belief that comes from somewhere deep within ourselves, something inexplicable, and something that we do not need to justify. Others believe that we have a moral responsibility to protect all life; nature has no voice, only we can speak for her. This entirely altruistic perspective is extremely rare, as most of us have long lost the essential connection and harmony that may have once existed between people and their natural world. We have created barriers so impenetrable that nature, now isolated, has become almost irrelevant. For most people nature and wildlife are little more than just another form of television entertainment. Distant and remote, the players in the nature soap operas are no more real, relevant or precious than the characters and stories followed so avidly during nightly visits to the fictitious streets or villages portrayed on primetime television.

Other people believe that the logical conclusion of our understanding of ecology, and the interrelationships and interdependence of species, is that we have no choice: we must recognise the intrinsic value of all life because human survival is so inextricably reliant on the wellbeing and survival of non-human life. This view is perhaps best regarded as a very weak version of intrinsic value in that it implies that nature is only important because it provides for us. (I will return to this discussion later in this chapter when I introduce Norton's convergence theory.)Whatever people believe, it is the effect of their beliefs on their actions that perhaps matters most.

> Their oft-stated objections to the rape of the forests is that they might include within them some rare plant which bears the cure for cancer, or that the trees fix carbon dioxide, and that if they are cut down we may no longer enjoy our privilege of private transport. None of this is bad, only stupid. We are failing to recognise the true value of the forest, as a self regulating system that keeps the climate of the regions comfortable for life. Without the trees there is no rain, and without rain there are no trees. (Lovelock 1991)

Of the minority who recognise intrinsic value, most will adopt a middle road version, perhaps because they are outwardly uncomfortable with expressing simple beliefs or convictions. So, although they may hold deep personal beliefs, they seek rational arguments to support these beliefs. Early in his career, Aldo Leopold (1949) wrote in an unpublished essay:

> Possibly, in our intuitive perceptions, which may be truer than our science and less impeded by words than our philosophers, we realise the indivisibility of the earth – its soil, mountains, rivers, forests, climate, plants and animals, and respect it collectively not only as a useful servant but as a living being, vastly less active than ourselves in degree, but vastly greater than ourselves in time and space – a being that was old when the morning stars sang together, and, when the last of us has been gathered unto his fathers, will still be young.

Unfortunately despite recognising 'our intuitive perceptions' Leopold, as a scientist, was possibly uncomfortable with this statement. The essay was unpublished, and he never repeated the idea in print. However, much later in his life, he wrote:

> To sum up: a system of conservation based solely on economic self-interest is hopelessly lopsided. It tends to ignore and thus eventually to eliminate, many elements in the land community that lack commercial value, but that are (as far as we know) essential to its healthy functioning. It assumes falsely, I think, that the economic parts of the biotic clock will function without the uneconomic parts. (Leopold 1949)

The two statements have similar meanings: 'indivisibility of the earth' in his 1923 statement is not very different to 'will function without the uneconomic parts' in his 1949 text.

I own a rather special watch – a Swiss railways pocket watch – and, as you might expect, it is extremely accurate. I am very fond of this watch and would be very upset if it stopped working. Imagine, then, that I decide one day to dismantle and clean all the components. I very carefully pull everything apart and lay the pieces neatly on the table. I clean each individual part with an expensive horological degreaser, and then begin to reassemble my watch. After an hour or so the task is completed . . . well almost. There are three parts – three very small insignificant parts – left over. Since I have no idea what they are or what they do, I decide that they must be so insignificant that neither I, nor my watch, need them. And so I discard them. I wind up the watch, set it to the correct time and … nothing happens. The second hand sits motionless; the watch is dead. It will never work again because three tiny, insignificant parts have been discarded, lost forever. This is precisely what we are doing to our world, the global ecosystem. We discard insignificant parts: species and habitats. They are insignificant because we do not understand their function and because we have not learned to value them. And yet, we expect the ecosystem to continue to function, to provide for us, to sustain life.

Gaia

In 1979 James Lovelock published *'Gaia – A new look at life on earth'*. Fifty six years had passed since Aldo Leopold described earth as a 'living being, vastly less active than ourselves in degree, but vastly greater than ourselves in time and space'. The Gaia hypothesis or theory, which implies that the Earth may be considered as a vast living system, was originally proposed in the late 1960s and early 1970s by Dr James Lovelock, a British atmospheric scientist, in collaboration with Dr Lynn Margulis, an American microbiologist.

Gaia theory recognises that:

- Earth's physical and biological processes are inextricably bound together to form a self-regulating system.
- Life (living organisms) and the inorganic environment have evolved together as a single living system.
- The Gaia system self-regulates global temperature, atmospheric content, ocean salinity, and other factors in an automatic manner.
- Earth's living system has shaped and then maintained an optimal environment for life on earth.

In James Lovelock's own words:

> …the physical and chemical condition of the surface of the Earth, of the atmosphere, and of the oceans has been and is actively made fit and comfortable by the presence of life itself. This is in contrast to the conventional wisdom which held that life adapted to the planetary conditions as it and they evolved their separate ways. The entire range of living matter on Earth from whales to viruses and from oaks to algae could be regarded as constituting a single living entity capable of maintaining the Earth's atmosphere to suit its overall needs and endowed with faculties and powers far beyond those of its constituent parts…[Gaia can be defined] as a complex entity involving the Earth's biosphere, atmosphere, oceans, and soil; the totality constituting a feedback of cybernetic systems which seeks an optimal physical and chemical environment for life on this planet. To what extent is our collective intelligence also a part of Gaia? Do we as a species constitute a Gaian nervous system and a brain which can consciously anticipate environmental changes? (Lovelock 1979)

> The name of the living planet, Gaia, is not a synonym for the biosphere – that part of the Earth where living things are seen normally to exist. Still less is Gaia the same as the biota, which is simply the collection of all individual living organisms. The biota and the biosphere taken together form a part but not all of Gaia. Just as the shell is part of the snail, so the rocks, the air, and the oceans are part of Gaia. Gaia, as we shall see, has continuity with the past back to the origins of life, and in the future as long as life persists. Gaia, as a total planetary being, has properties that are not necessarily discernible by just

> knowing individual species or populations of organisms living together . . . Specifically, the Gaia hypothesis says that the temperature, oxidation, state, acidity, and certain aspects of the rocks and waters are kept constant, and that this homeostasis is maintained by active feedback processes operated automatically and unconsciously by the biota. (Lovelock 1988)

Gaia is such a huge theory. In its early years it was often dismissed as being unscientific, even fantasy. Many saw overwhelming teleological implications, i.e. Gaia implied that natural processes are being shaped for some *purpose*. This was strenuously denied by both Lovelock and Margulis. The language of Gaia is often anthropomorphic. It can also be poetic. This probably fuelled the attitudes of sceptics and cynics. Today, almost everything described in the theory has some level of support from the wider scientific community. An exception, that some people find difficult to accept, is the idea that 'the biota manipulates its environment for the purpose of creating biologically favourable conditions for itself'. Conventional science places a considerable emphasis on rejecting the idea that there is any purpose in nature. Whatever our beliefs, it would be dangerous to ignore Gaia: human life can only exist in the presence of non-human life.

Intrinsic Value – Gaining Support

Intrinsic value, as a concept, may remain contentious, but it is gaining in credibility, popularity and application. One of the most significant early examples of formal recognition of this concept was in the World Charter for Nature adopted by the General Assembly of the United Nations in 1988. The definition used was: 'All life warrants respect regardless of its usefulness to man.'

Fig. 8.6 All life warrants respect

More recently, the Welsh Assembly Government document *Environment Strategy for Wales* (2006) contained the following statements:

> We recognise that our environment has an intrinsic value.
>
> Biodiversity is 'the variety of life' on Earth. It includes all plants, animals and micro-organisms (species) and the places where they live (habitats). We value it for itself, as well as for the role it plays in many natural processes and its direct and indirect economic, social, aesthetic, cultural and spiritual benefits.

The official policy of the European Union appears to favour an anthropocentric justification for maintaining biodiversity, but it also mentions intrinsic value:

> Biological diversity (biodiversity) is essential to maintain life on earth and has important social, economic, scientific, educational, cultural, recreational and aesthetic values. *In addition to its intrinsic value* biodiversity determines our resilience to changing circumstances. Without adequate biodiversity, events such as climate change and pest infestations are more likely to have catastrophic effects. It is essential for maintaining the long term viability of agriculture and fisheries for food production. Biodiversity constitutes the basis for the development of many industrial processes and the production of new medicines. Finally, biodiversity often provides solutions to existing problems of pollution and disease.

It is important to recognise that whenever one set of people hold a particular belief, especially when that belief is extreme or simply new, there will be opposition and ridicule. There will invariably be another faction who quite genuinely have an opposing view. The deep ecology movement has many more antagonists than supporters. Opposition from industrialists and developers is easily understood. People driven by the pursuit of economic growth and personal wealth (apparently at any cost) will inevitably regard anything that stands in the way of progress as unreasonable and unacceptable. Numerous examples of the extreme opposition to intrinsic value can be found in the writings of George Reisman, a professor of Economics at Graziadio School of Business and Management, Pepperdine University. I offer the following without comment:

> The idea of nature's intrinsic value inexorably implies a desire to destroy man and his works because it implies a perception of man as the systematic destroyer of the good, and thus as the systematic doer of evil.
>
> The doctrine of intrinsic value is itself only a rationalization for a pre-existing hatred of man. It is invoked not because one attaches any actual value to what is alleged to have intrinsic value, but simply to serve as a pretext for denying values to man.
>
> If an alleged scientific expert believes in the intrinsic value of nature, then to seek his advice is equivalent to seeking the advice of a medical doctor who was on the side of the germs rather than of the patient, if such a thing can be imagined.
>
> It is important to realize that when the environmentalists talk about destruction of the "environment" as the result of economic activity, their claims are permeated by the doctrine of intrinsic value. Thus, what they actually mean to a very great extent is merely the destruction of alleged intrinsic values in nature such as jungles, deserts, rock formations, and animal species which are either of no value to man or hostile to man. (Reisman 1996)

Surprisingly, the more significant and articulate opposition comes from people with apparently strong anthropocentric views, people who put people first. One of the most outspoken, and most often quoted, critics is Ramachandra Guha, a historian and anthropologist who lives in Bangalore. He is extremely critical of deep ecology, regarding it as shallow and arrogant (believing itself to be the leading edge of environmental ethics). He clearly opposes what he sees as a focus on the preservation of unspoiled wilderness and the restoration of natural areas. From his perspective, the concept of National Parks is simply an antidote to modern civilization, and wilderness is an amenity that can only exist in an already affluent consumer society. The two most frequently encountered of his quotes are:

> The Northern wilderness lover has largely been insensitive to the needs and aspirations of human communities that live in or around habitats they wish to "preserve for posterity."

> At the same time, he or she has also been insensitive to the deep asymmetries in global consumption, to the fact that it is precisely the self-confessed environmentalist who practices a lifestyle that lays an unbearable burden on the finite natural resources of the earth.
>
> Wilderness lovers like to speak of the equal rights of all species to exist. This ethical cloaking cannot hide the truth that green missionaries are possibly more dangerous, and certainly more hypocritical, than their economic or religious counterparts. (Guha 2000)

Guha's writings are sometimes quite difficult to understand, and there are often contradictions. He sometimes gives the impression that he has only a superficial grasp of conservation science. For example, to support his arguments he often claims that biodiversity is best represented in areas which have been modified by people. He also suggests that a series of small nature reserves would be as good, if not better, than much larger sites. He is clearly a champion of people: he places their rights above all else. Perhaps this is reasonable, or certainly no more unreasonable than environmentalists who believe that there is no place for people in nature. Some of those who talk about re-wilding or recreating wilderness have a distorted view of what natural might be. They regard the presence of people, indigenous or otherwise, and all human activities as destructive. The deliberate exclusion of people in order to protect wildlife can diminish the bond between people and nature. A failure to recognise this paradox could threaten the future of wildlife.

8.5.1.2 Ecosystem Services (Anthropocentric Values – Instrumental Values)

> Penguins are only important because people enjoy seeing them walk about on rocks .... In short, my observations about environmental problems will be people-oriented, as are my criteria. I have no interest in preserving penguins for their own sake. (Baxter 1974)

Today, there remains a significant division between people and organisations that apply anthropocentric (instrumental) values when attempting to promote or defend nature and those who recognise and use intrinsic or biocentric values (Benson 2000; Jepson and Canney 2003). The debate is slightly complicated because if we subscribe to biocentric values we can, without significant contradiction, also embrace and use *some* of the values that are more generally associated with the anthropocentric view. That is, we can, and should, value nature both for what it is and what it provides for us.

The anthropocentric view could be expressed as: wildlife has no intrinsic value independent of human interest or values. John Benson (2000) provides three divisions within the green spectrum. The first is light green:

> The world of non-human beings, living and non-living, has no independent moral status, but only matters insofar as it matters to human beings (which do have independent moral status).

This would appear to be the position adopted by some professionals involved in nature conservation. At least, this is the position that they hold publicly. For example, in 2005 the heads of The Convention on Biological Diversity, CITES, The Convention on Migratory Species of Wild Animals, Ramsar and The World Heritage Centre wrote to the world leaders who were meeting at the UN to discuss progress towards achieving the Millennium Development Goals. The letter called on the leaders to recognise that biodiversity needs to be used in a sustainable way and its benefits shared more equitably.

The letter offered the following justification for the conservation of biodiversity in which every single justification represents an anthropocentric view. They have provided an extremely useful list of the various anthropocentric arguments that can be used to support nature conservation:

> Biodiversity is the variety of life on earth: genes, species, ecosystems. The services we use from ecosystems, such as clean water, food, fuel and fibre, medicines, and climate control, cannot be provided without biodiversity.
>
> Failure to conserve and use biological diversity in a sustainable way will perpetuate inequitable and unsustainable growth, deeper poverty, new and more rampant illnesses, continued loss of species, and a world with ever-more degraded environments which are less healthy for people.
>
> A wide range of crop and livestock genetic diversity is essential to ensure that our agro-systems can adapt to new challenges from climate, pests and diseases.
>
> The biological wealth in marine environments will be needed to feed growing populations and provide livelihoods for coastal communities around the world.
>
> Wetlands are needed as water regulators to protect us from floods and storm surges, to help in moderating climatic change with other ecosystems such as forests, and to act as living filters for pollutants and excess fertilizers.
>
> We must not forget that biodiversity is central to many of the world's cultures, the source of legend and myth, the inspiration for art and music. It is the basis for medicinal knowledge, drawing on the property of a variety of plants and animals for healing. Provision of these services across all these ecosystems depends on maintaining biological diversity.
>
> Biodiversity can indeed help alleviate hunger and poverty, can promote good human health, and be the basis for ensuring freedom and equity for all. All of us rely on biodiversity, directly or indirectly for our health and welfare.

An example of the division between anthropocentric and intrinsic beliefs can be taken from a debate in 2005 between members of The World Convention on Protected Areas (WCPA). The issue in question was the proposed de-gazetting of the Amboseli National Park in Kenya. Most participants in the debate were concerned that the proposed diminished status of the park could lead to a loss of wildlife. The relevance to our debate is that while many WCPA members felt that the protection of such an obviously important international park should take precedence over local issues there was an opposing view. One member accused others of being 'radical, misanthropic, biocentric conservationists'. He, in reply, was labelled 'a radical, anthropocentric, biophobic developmentalist'.

I use this example simply as evidence that such divisions do occur in the conservation profession. Anthropocentric values are clearly recognised as being important to many organisations and individuals central to the nature conservation movement. But why do they hold these values? Jepson and Canney (2003) suggest that the conservation movement is characterised by six distinct values. I have summarised them as follows:

1. Where people exploit nature they also have a moral responsibility to protect it.
2. The unnecessary destruction of wildlife is unacceptable (uncivilised).

3. The aesthetic and intellectual values of nature are an important part of the cultural inheritance of many people and as such they must be protected.
4. Healthy ecosystems are essential to economic growth, quality of life and social stability.
5. It is important to preserve the gene pool which has provided the basis for our agriculture and much of our medicine.
6. Society has a moral duty to permit traditional peoples inhabiting natural landscapes to choose their own destiny in a way that is appropriate to their history and culture.

They suggest that it is the combination of these values that defines conservation, but that the first three, which reflect intrinsic values, have been forgotten. They give three reasons for this:

1. 'Development' has become the overriding international cultural ideal.
2. There is a recognition that government support for conservation is best justified by using economic development terms.
3. Scientists have recognised that their ability to represent nature in units (species, habitats, etc.) creates an opportunity to integrate ecological theory with economics. This is because dividing nature into parts creates discrete units that can be assigned a monetary value, thereby creating the possibility of treating units of nature as commodities. The alignment of nature with economic costs and benefits is attractive to politicians because it allows them to make decisions that ignore the problems which arise as a consequence of different intrinsic and cultural values.

Ecosystem Services

> By at least trying to put money values on some aspects of environmental quality we are underlining the fact that environmental services are not free. They do have values in the same sense as marketed goods and services have values. (David Pearce 1992)

Ecosystem services, the latest manifestation of anthropocentric values, are defined as services provided by the natural environment that benefit humans. The millennium ecosystem assessment framework is a widely accepted method that has categorised ecosystem services into four broad categories:

- Supporting Services, such as soil formation, nutrient cycling, water cycling and primary production. These underpin the provision of the other 'service' categories.
- Provisioning Services, such as food, fibre, fuel, bio-materials and water.
- Regulating Services, such as climate regulation, flood protection, pollination and air/soil/water quality.
- Cultural Services, such as education, cultural heritage/sense of place, health, recreation, tourism and aesthetic value.

We should welcome the recognition of 'ecosystem services' as a means of demonstrating our dependence on the natural world. It is important that we value our ecosystems for what they provide, for example, a carbon store, agriculture, forestry, fisheries and leisure. However, if the value of these services becomes the only way in which we can measure or value ecosystems then all our good intentions will fail.

We may develop an ability to assess the cost of environmental damage in terms of our losses. It is also relatively easy to understand the impact of our actions when the consequential environmental losses are described as our losses. But, once again, we must be extremely cautious: we can only deal with the known benefits of an ecosystem. There will certainly be many more that we do not yet understand or recognise.

We can place monetary value on extractive industries, such as forestry or fisheries. We may be able to measure tertiary services such as tourism. The intangible benefits of intrinsic *appeal* will always be extremely difficult, perhaps impossible, to calculate, but it will never be possible to quantify the intrinsic *value* of an ecosystem. There is a significant danger that in the final calculation, given that we tend to measure gains within the constraints of a human lifetime, intrinsic value will always remain less important than human gains. If we rely on ecosystem services as the main justification for conserving biodiversity there is a risk, possibly a small risk, that if it is shown that there are alternative artificial means of obtaining some of these services that justification will be discredited.

Fig. 8.7 Lowland beech woodland

There is a common assertion that if ecosystems are delivering the desired goods and services biodiversity will be conserved. A logical, but seriously erroneous, consequence could be the replacement of policies and objectives concerned with conserving biodiversity with those concerned with maintaining ecosystem services. It is wrong to believe that if ecosystems are delivering the desired services then biodiversity will be conserved. In reality, a seriously depleted ecosystem, that fails to meet its biodiversity potential, and with many rare and endangered species absent, can provide valuable, but not necessarily sustainable, ecosystem services. Many ecosystem services are provided by individual, or groups of, species, and not by intact, functional ecosystems. Some of these species may be robust and resilient to adverse factors, both anthropogenic and natural. Some species can be replaced by others which provide the same, or similar, services. An obvious example is woodland, which can provide many services, including carbon sequestration, the supply of water and erosion control. The presence of *specific* tree species, along with their associated and often dependent species, is far less important in this context than the presence of any tree species, including alien species. A commercial coniferous woodland plantation, because it can provide a carbon store, prevent soil loss, contribute to flood management and provide leisure opportunities, could easily be perceived as being more valuable than a native oak woodland.

Fig. 8.8 Commercial forest plantation

Ecosystem services can be used to support biodiversity conservation, but if the value of these services is the only measure that we apply to ecosystems the risk would certainly contradict the precautionary principle. Biodiversity conservation would become a sophisticated version of cherry picking. We should not direct our attention to whatever we happen to perceive as being valuable in any place at any particular time. We should understand that values will change with time; something that has no recognisable value today may be extremely important tomorrow. The value of ecosystem services should never become a surrogate measure of biodiversity. The only measure of biodiversity is life itself: the variety of species, communities and habitats.

Some authors (Norton 1991) suggest that the division between instrumental and intrinsic values has done much more harm than good. Norton offers a ‘convergence hypothesis’, suggesting that a sufficiently broad instrumental view would provide an adequate basis for conservation policies. He does not believe that there is a need to use intrinsic values to justify conservation: even if we believe that our natural environment has no value other than its value to human beings, logic might lead us to conclude that, given our very existence is dependent on the condition of our environment, we must protect it for our, if not its, own sake.

Meffe et al. (1997) dismiss Norton's view. They point to the difference of the 'burden of proof' between instrumental and intrinsic values. If biodiversity is only valued in terms of the environment services that it delivers to people, the burden of proof in defending biodiversity, habitats or species is placed on the conservationists: they will have to prove that the habitats or species have value to people. If, on the other hand, we recognise that biodiversity has intrinsic value the burden of proof shifts to society at large, who would have to justify their intended actions whenever they place biodiversity at risk. The simplicity of Norton's view may be attractive, but it is also risky. Perhaps these divisions need to be blurred. We might, as individuals, be motivated or driven by intrinsic values but recognise that there is no contradiction in applying instrumental values to support our position.

Intrinsic value is rarely, if ever, accepted by governments, developers, industrialists and others in the commercial world. However, in trying to give the impression that they are genuinely concerned about our environment, they will often advocate the use of ecosystem services, both as a means of justifying nature conservation and also as a measure of successful conservation. In reality, this is a rather transparent form of 'sleight of hand', something the Americans call 'green-washing'. This is where, for example, an organisation claims green environmental policies but where their activities are anything but green. A recent example placed considerable emphasis on the importance and protection of biodiversity and yet relegated biodiversity as just one of many ecosystem services: services that can be traded, exchanged, devalued or discarded. The inevitable and inescapable consequence of trading in biodiversity is that somewhere something important will be lost.

> If a price can be put on something, that something can be devalued, sold, and discarded. It is also possible for some to dream that people will go on living comfortably in a biologically impoverished world. They suppose that a prosthetic environment is within the power of technology, that human life can still flourish in a completely humanised world . . . (Edward O Wilson 1994)

Both the biocentric and anthropocentric ethical positions represent different, and legitimate, values that we give to wildlife. Without any doubt, the most serious problems arise when people fail to recognise *any* value in wildlife or in their environment. It is important that professional conservationists are aware of the debate and recognise that there are implications for nature conservation. It will influence:

- The way in which we understand nature conservation and, consequently, how we define our obligations (objectives) and manage our sites (outcome or process).
- The ways in which we communicate (see Chapter 3).

Ideally, all countryside professionals would develop personal philosophies of nature conservation. We all need to recognise and be aware of our motivation and inspirations. We need to understand why we are personally committed to conserving nature. We need the confidence and conviction to ensure that our beliefs govern our decisions and actions. We should also be able to share our beliefs and feelings with colleagues and others, since this sort of discussion will help to develop or reaffirm group, organisational and corporate ethics. It is only through understanding why we conserve nature that we will recognise what we are trying to achieve and what we need to do.

We need to discuss our values with others, because a love of nature and the wild is not shared by society at large. This may be an overly cynical view. Certainly, there is an increasing awareness of environmental issues and a general move towards 'greener' approaches to living, but, for now, this is peripheral. As nature or 'natural values' guide conservation, conservationists must appreciate that they have a key role in guiding society towards valuing nature. There can be no values without understanding, no understanding without awareness and no awareness without experience.

As I mentioned earlier, society will only treasure and conserve what it values: without the support of society at large we will fail, even on the most secure sites. We must never forget that the sites we mange will not survive as islands in a desolate, sterile or artificial sea. The sites themselves, in isolation, will never be sufficient to ensure sustainable conservation. The distinction between protected and non-protected areas must diminish. (Incidentally, these distinctions may also contribute to the divisions in society between the values held by conservationists and those of the majority.)

But we must be careful: even a sense of value does not necessarily engender care or respect. Excessive consumerism is not restricted to the high street or supermarket. Its influence can extend to wild places, including nature reserves, national parks and the wider countryside. Our demand-led society, with unfettered ideas of human superiority and lacking a conservation ethic, will, if uncontrolled, destroy the very places that it values so highly as recreational resources. So, if we recognise an obligation to encourage others to share and enjoy these special places, then surely it is even more important that we also attempt to share our values, passions, care and respect for nature. If we do this, we might inspire others to adopt these values, and there is a hope that they, in turn, will learn to care.

> The ecological ego matures towards a sense of ethical responsibility with the planet that is as vividly experienced as our ethical responsibility to other people. It seeks to weave that responsibility into the fabric of social relations and political decisions. (Roszak 2001)

Fig. 8.9 The value of nature

Recommended Further Reading

Benson, J. (2000). *Environmental Ethics – An Introduction with Readings*. Routledge, London.

Harding, S. (2006). *Animate Earth – Science, Intuition and Gaia.* Green Books Ltd, Dartington, UK.

Leopold, A. (1949). *A Sand County Almanac, and Sketches Here and There*. Oxford University Press, New York.

Lovelock, J. E. (1991). *Gaia The Practical Science of Planetary Medicine.* Gaia Books Ltd, London, UK.

Naess, A. (2002). *Life's Philosophy: Reason and Feelings in a Deeper World*. The University of Georgia Press, Athens and London.

Roszak, T. (2001). *The Voice of the Earth, an Exploration of Ecopsychology.* Phanes Press INC, Grand Rapids, MI, USA.

Wilson, E. O. (1994). *The Diversity of Life.* Penguin, Harmondsworth.

References

Adams, W. M. (2003). *Future Nature a Vision for Conservation*, revised edition. Earthscan, London.

Angermeier, P. L. (2000). *The natural imperative for biological conservation*. Conservation Biology, 14, 373–381.

Bailey, J. A. (1982). *Implications of "Muddling Through" for Wildlife Management*. Wildlife Society Bulletin, 10(4), 363–369.

Baxter, W. F. (1974). *People or Penguins: The Case for Optimal Pollution*. Columbia University Press, New York, USA.

Benson, J. (2000). *Environmental Ethics – An Introduction with Readings*. Routledge, London.

Cole-King, A. (2005). Personal communication.

Guha, R. (2000). *The paradox of global environmentalism*. Current History, 99(640).

Harding, S. (2005). *What is Deep Ecology*? Resurgence, issue 185.

His Holiness the Dalai Lama (1987). *An ethical approach to environmental protection.* In *Tree of Life: Buddhism and Protection of Nature.* Davies, Shann (ed.) (1987). Buddhist Perception of Nature. Geneva.

Jepson, P. and Canney, S. (2003). *Values-led conservation.* Global Ecology and Biogeography, 12, 271–274.

Kaiser, J. (2000). *Taking a stand: ecologists on a mission to save the world.* Science, 287. 1188–1192.

Leopold, A. (1949). *A Sand County Almanac, and Sketches Here and There*. Oxford University Press, New York.

Lovelock, J. E. (1979). *Gaia – A New Look at Life on Earth.* Oxford University Press, Oxford, UK.

Lovelock, J. E. (1988). *The Ages of Gaia – a biography of our living earth.* Oxford University Press, Oxford, UK.

Lovelock, J. E. (1991). *Gaia The Practical Science of Planetary Medicine.* Gaia books Ltd, London, UK.

Machado, A. (2004). *An index of naturalness*. Journal for Nature Conservation, 12, 95–110.

Margules, C. and Usher, M. B. (1981). *Criteria used in assessing wildlife conservation potential: a review.* Biological Conservation, 21, 79–109.

Meff G, K., Carroll, C. R. and contributors (1997). *Principles of Conservation Biology, Second Edition*. Sinauer associates, INC Sunderland Massachusetts, USA.

Nielsen, L. A. (2001). *Science and advocacy are different – and we need to keep them that way.* Human Dimensions of Wildlife, 6, 39–47

Norton, B. G. (1991). *Towards Unity Among Environmentalists*. Oxford University Press, New York, USA.
Palmer, C. (1994). *A Bibliographic Essay on Environmental Ethics*, Studies in Christian Ethics. Edinburgh.
Ratcliffe, D. A. (1976). *Thoughts Towards a Philosophy of Nature Conservation*. Biology and Conservation, (9).
Ratcliffe, D. A. (ed.) (1977). *A Nature Conservation Review*. Cambridge University Press, Cambridge.
Reisman, G. (1996). *Capitalism*. Jameson Books, Ottawa, Illinois, USA.
Roszak, T. (2001). *The Voice of the Earth, An Exploration of Ecopsychology*. Phanes Press INC, Grand Rapids, MI, USA.
Royal Society of Edinburgh (1998). *People and nature: a new approach to SSSI designations in Scotland*. Response to Scottish office consultation on SSSIs, Edinburgh.
Soanes, C. Hawker, S. and Elliot, E. (2006). *Paperback Oxford English Dictionary*. Oxford University Press, Oxford, UK.
Van Houtan, K. S. (2006). *Conservation as a virtue: a scientific and social process for conservation ethics*. Conservation Biology, 20(5), 1367–1372.
Welsh Assembly Government (2006). *Environment Strategy for Wales*. Cardiff, Wales, UK.
Wilson, E. O. (1994). *The Diversity of Life*. Penguin, Harmondsworth.

Chapter 9
What Do We Value?

Humanity is but a part of the fabric of life – dependent on the whole fabric for our very existence. As the most highly developed tool-using animal, we must recognize that the unknown evolutionary destinies of other life forms are to be respected, and act as gentle stewards of the earth's community of being. (Gary Snyder, *Four Changes*, 1974)

Abstract This chapter explores some of the potential outcomes for nature conservation. It begins with the Rio Convention on Biodiversity which, despite providing a common global purpose, is unfortunately not always understood and is sometimes misrepresented. For example, one of the more common misconceptions is that there is an obligation, in all circumstances, to maximise diversity. There are no universally accepted definitions of natural or of wilderness but, whatever definition we use, taking a global perspective, the protection of the last surviving tracts of wilderness must be the highest nature conservation priority. The value of cultural landscapes and their semi-natural habitats is explored. Many parts of the world have few wilderness areas: in place of wilderness we have a glorious landscape that has been shaped over thousands of years as the, mainly unintentional, by-product of generations of people toiling to provide a living for their families. Many conservation managers and scientists suggest that 'natural values' are a priority and should guide all nature conservation outcomes. Wildlife management is often about moving from a less natural to a more natural state, or about attempting to maintain the natural elements within systems and populations. The implications of the three varieties of naturalness (Peterken 1993) are discussed: original naturalness, the state that existed before man became a significant ecological factor; present naturalness, the state that would prevail now if man had not become a significant ecological factor; and future naturalness, the state which would develop if man's influence were completely and permanently removed. These definitions provide an extremely useful basis for academic debate: they may not represent something that we want for the future, but they can help us explore the options. It should be possible to create at least some sustainable places where, as far as possible, we rely on, or

M. Alexander, *Management Planning for Nature Conservation: A Theoretical Basis & Practical Guide*, DOI 10.1007/978-94-007-5116-3_9,

enable, natural processes, where opportunities for wildlife are optimal, where human interaction is not exploitation, and where our mutual dependence is recognised. In other words, could these be tomorrow's natural places?

9.1 Biodiversity – The Rio Convention

In Chapter 7 I discussed conservation values and ethics, or *why* we conserve nature. In this chapter I will explore *what* our obligations might be. Taking a global perspective, it seems reasonable to begin with the Rio Convention on Biological Diversity, although this is not to suggest that conservation began with the convention.

The convention has three main goals:

- The preservation of biological diversity
- The sustainable use of resources
- The equitable distribution of the benefits arising from the use of genetic resources.

The first goal is the most relevant to this discussion, and so it would be timely to remind ourselves how the convention defines biological diversity:

> Biodiversity is the term given to the variety of life on Earth and the natural patterns it forms. The biodiversity we see today is the fruit of billions of years of evolution, shaped by natural processes and, increasingly, by the influence of humans. It forms the web of life of which we are an integral part and upon which we so fully depend.
>
> This diversity is often understood in terms of the wide variety of plants, animals and micro-organisms. Biodiversity also includes genetic differences within each species. Yet another aspect of biodiversity is the variety of ecosystems such as those that occur in deserts, forests, wetlands, mountains, lakes, rivers, and agricultural landscapes. In each ecosystem, living creatures, including humans, form a community, interacting with one another and with the air, water, and soil around them.

We might also reflect on the fact that the convention was signed by 150 government leaders at the 1992 Rio Earth Summit. The scale of the international commitment is such that we have been provided with a tangible global purpose. Unfortunately, the intention of the convention and the definition of biodiversity are not always understood and are sometimes misrepresented. For example, one of the more common misconceptions is that there is an obligation, in all circumstances, to maximise diversity (everything everywhere, with no room for anything).

9.2 Cultural Landscapes and Biodiversity

Fig. 9.1 A landscape shaped by people

In many parts of the developed world our ecosystems, habitats and populations are at best semi-natural. In UK, the Nature Conservancy Council (1989) defined a semi-natural habitat as: 'modified by human activity from its original state but with a vegetation composed of native species, similar in structure to natural types, and with native animal species'. The Rio Convention unequivocally recognises that biodiversity can be the product of both natural processes and the combination of natural and human influences, and so implies that semi-natural is important. Taking a European perspective, Natura 2000 does not differentiate between natural and semi-natural habitats or communities: they have equal value. This means that there is a legal obligation in Europe to safeguard the important semi-natural habitats. In addition to the legal imperative, semi-natural communities represent some of the most highly valued components of our cultural landscape. Some people believe that the 'pre-industrial agricultural landscape' represents an ideal state, or view, of nature and provides a reference point for restoration. Others take a very different view and suggest that this idyll, so lavishly portrayed by the Romantic British painters, has provided a distorted vision of nature: a vision that has driven us to put far too great an emphasis on the maintenance, or restoration, of something that is a poor substitute for wild nature.

This is an extract from *Human Relationships with the Natural World: an historical perspective*, ECOS 24 (1) 2003:

> The natural world and its landscapes tell sometimes painful stories and our relationships with nature may hold the potential to include and heal. In Shakespeare's Henry the Fifth, after the bloody battle of Agincourt, the English court visits the French to shape a

prospect of peace. From the lips of the Duke of Burgundy comes an extraordinary plea for reconciliation – extraordinary because of its metaphoric use of landscape as a symbol of human conflict, in particular the way in which the meadows now lie abandoned because the men who mowed have gone to war:

"The even mead, that erst brought sweetly forth
The freckled cowslip, burnet and green clover,
Wanting the scythe, all uncorrected, rank,
Conceives by idleness, and nothing teems
But hateful docks, rough thistles, kecksies, burs,
Losing both beauty and utility…"

My commentary on the play says how realistically Shakespeare depicts the battlefield but he certainly also understood the ecology of the flood meadows of the Avon, for here in the middle of this speech is a precise description of the effect of neglect on what we call the Alopecurus-Sanguisorba meadow which, without mowing, shifts to what we know as the Arrhenatherum grassland, Primula veris, Sanguisorba officinalis and Trifolium pratense fading away and Rumex crispus, Rumex obtusifolius, Cirsium arvense, Cirsium vulgare, Heracleum sphondylium and Anthriscus sylvestris growing up among the rank grasses.

This is perhaps one of the most sophisticated ways of relating to the natural world that we could conceive – that we see how we have helped make nature what it is, that we read back from its changes how we are ourselves unmade and yet may glimpse the prospect of ourselves once again gaining both beauty and utility. Finding ourselves with new enemies these days, we had better learn again who cares for the natural world and how to live in it together. (Rodwell 2003)

If we apply international definitions of wilderness (see Chapter 10), in many parts of the world there are few wilderness areas, probably none in much of the old world. In place of wilderness we have a glorious landscape that has been shaped over thousands of years as the, mainly unintentional, by-product of generations of people toiling to provide a living for their families. Our cultural landscape is special and precious; its values should be celebrated and not diminished through comparison with something that happened at an earlier time or which happens elsewhere. Our cultures are enhanced by landscapes that represent our place within an ecosystem, as opposed to landscapes where everything apart from humanity is removed, or landscapes where everything human is removed.

Regrettably, some of those who so strongly and rightly defend the cultural landscape sometimes forget that landscape is often a patchwork of semi-natural communities. They overlook the 'natural' component. They also forget the connection between culture and the original wild. This connection is clearly expounded by Aldo Leopold (1949):

Wilderness is the raw material out of which man has hammered the artefact called civilisation. Wilderness was never a homogeneous raw material. It was very diverse, and the resulting artefacts are very diverse. These differences in the end-product are known as cultures. The rich diversity of the world's cultures reflects a corresponding diversity in the wilds that gave them birth.

Each step away from our connection to the wild diminishes or dilutes our culture. Thus, a countryside where everything is an artificial, bright, emerald green, a landscape that is entirely the product of human endeavour, will no longer provide a link with, or maintain, our cultural roots. We seem to understand the consequences of setting aside the vernacular style in building, and would mourn the passing of an old wall replaced by a wire fence, or the stone barn substituted for something out of a packet from the local builders' merchant. We must value the semi-natural components of landscape as vigorously as we defend the cultural component, for these are two inseparable and interdependent parts of a single entity. There can be no doubt that we must continue to safeguard these areas, though we will certainly have to accept some compromises; cultural landscapes can only survive as evolving or 'living' landscapes which include people. It is essential to realise that environmental change has the potential to overwhelm our ability to retain every precious vestige of original, natural habitat. The future may have to be very different.

> Pam, Arglwydd, y gwnaethost Cwm Pennant mor dlws,
> A bywyd hen fygail mor fyr?
>
> Why, Lord, did you make Cwm Pennant so beautiful,
> And the life of an old shepherd so short?
>
> Eifion Wyn (1867–1926)

Fig. 9.2 Cwm Pennant, Snowdonia, North Wales

9.3 Wilderness

There are no universally accepted, or comfortable, definitions of natural or of wilderness. Some definitions exclude all human activity and this can be extended to include indigenous peoples, even when they follow a lifestyle which is in apparent harmony with their environment. Others believe that people must be included, and that we should not create artificial barriers between human and non-human life. Whatever definition we use, taking a global perspective, the protection of the last surviving tracts of wilderness must be *the highest nature conservation priority*. Once destroyed wilderness is lost forever: nothing can replace the diverse products of aeons of unfettered nature.

9.4 Outcomes Delivered by Natural Processes

Fig. 9.3 Freshwater West sand dune system

There is an increasingly popular view that there is a need to move away from an approach to nature conservation that is almost entirely based on achieving defined outcomes (the semi-natural state). Although it will not be possible to recreate wilderness, it may be possible, in some circumstances, to adopt an approach that enables natural processes and accepts the resultant outcomes. These outcomes may not always be predictable, at least in terms of the detail. The relationship between plant communities and environmental factors, for example, altitude, slope, aspect, geology, soil and climate (Rodwell 2006, Personal communication), will ensure that everything cannot happen everywhere. There will always be a limited range of potential outcomes. This approach is appropriate for habitats at the natural end of the spectrum, for example, unmodified sand dunes (Fig. 9.3). In these sites, instead of defining what communities we want within a habitat and where we want them, we would have to be content with whatever nature delivered. Protagonists of this approach would like to see its application on a landscape scale, the implication being that this would deliver a 'more natural' countryside.

9.5 Natural/Naturalness

The discussion in each of the preceding three sections has relied heavily on the use of 'natural'. It is impossible to consider 'diversity' as a value without immediately straying towards a second and equally important value, 'naturalness'. So, before progressing any further, there is a need to discuss what 'natural' might mean.

'Natural' can be used in two different contexts (Machado 2004):

- *To describe* the state or condition of a habitat or population (a feature). For example, a heath land can be described as 'semi-natural'.
- *As a conservation value.* For example, 'naturalness' was one of ten different criteria used in the UK Nature Conservation Review (Ratcliffe 1977) to select the key areas for nature conservation in the UK.

These quite different contexts are often confused or combined, and the use of the word 'natural' has become, rightly or wrongly, value laden. A failure or reluctance to recognise the difference between 'natural' as a descriptor and 'natural' as a value can lead to an assertion that 'natural' is the only, or at least the prime, nature conservation value, with all other values being subservient. The consequence could be, for example, that a very common, robust, species-poor, but 'natural', habitat would always be regarded as more important than an exceptionally rare, fragile, species-rich meadow which is the product of human intervention.

Many, if not most, nature conservationists, managers and scientists regard 'naturalness' as the most important value that can be applied to an ecosystem, habitat or population. This is not to suggest that naturalness is an all-or-nothing quality (Angermeier 2000) or that cultural landscapes with plagioclimatic communities lack natural, or any other, values.

Some scientists have worked towards developing objective or scientific methodologies for describing 'naturalness'. Naturalness, the degree to which something is natural, is a continuous gradient from completely natural to completely unnatural or artificial. For example, Machado (2004) has developed ten 'naturalness' categories, with the following representing the extremes at either end of the scale:

10. Natural virgin system; only natural elements and processes. Possible anecdotal presence of negligible or hardly noticeable anthropic elements, or totally insignificant physical-chemical pollution coming from exterior anthropic sources.

1. Artificial system; high closure; without self-maintained macroscopic life; microscopic life absent or in containers.

These, and all other, categories are very contrived. They represent one view of how the progression from natural to artificial might be punctuated. The definitions of 10 and 1 are debatable. There will be many other, equally valid but very different, versions, all expressing a matter of opinion and not fact. Angermeier (2000) suggests that the extremes are only abstractions: 'entirely natural areas no longer exist'. If the definition of the extreme values which define a range can be questioned so can all the intermediate values.

An important issue, the relationship between humanity and all other species on this planet, is largely avoided by Machado. When applied to nature conservation some authors suggest that 'natural' defines a state that has not been made or influenced by humans (Hunter 1996). Thus, 'natural' could simply be the absence of anthropogenic factors, past, present and future. In contrast, in his book *The Diversity of Life*, Edward O Wilson (1994) makes the essential connection between humanity and nature: 'We did not arrive on this planet as aliens. Humanity is part of nature, a species that evolved among other species.' If we are part of nature, how can any human influence be considered unnatural? The debate is further complicated because some people believe that human actions which rely on technology might be considered 'unnatural', while those actions that do not rely on technology (particularly those by indigenous peoples), even though they modify or transform natural ecosystems, are 'natural'.

Aldo Leopold differentiates between mechanised and non-mechanised man. He suggests that the actions of non-mechanised man can harmonise with, and even complement, nature: 'There were once men capable of inhabiting a river without disrupting the harmony of its life' (Leopold 1949). When did we drift apart, when did we become mechanised man, and when did our actions become something unnatural? This is, in essence, the long-running debate. Can it ever be resolved?

> Gone are the days when ecologists (and conservationists) could conceive of 'nature' in equilibrium, and hence portray human-induced changes in those ecosystems as somehow 'unnatural'. (Adams 2003)

I have not yet found anything that seriously challenges the assertion made by Ratcliffe (1977), the author of one of the first publications to apply 'naturalness' as a conservation value. When introducing the NCR criteria for site selection (naturalness is one of the criteria) Ratcliffe describes their use as being 'essentially subjective'. It is important to recognise that 'naturalness' is a subjective value, but equally important that our inability to apply objectivity to 'naturalness' does not diminish its value.

Whatever our definition of 'natural' may be, many conservation managers and scientists believe it is something that nature conservation should deliver. Wildlife management is often about moving from a less natural to a more natural state, or about attempting to maintain the natural elements within systems and populations. I am not setting aside the idea that humanity is part of nature but, for now, I will turn to George Peterken's (1993) concepts of 'natural'.

George Peterken identifies three varieties of naturalness:

Original naturalness:	The state that existed before man became a significant ecological factor (not a single state but a succession).
Present naturalness:	The state that would prevail now if man had not become a significant ecological factor.
Future naturalness:	The state which would develop if man's influence were completely and permanently removed.

He also identifies a fourth variety, *potential naturalness*, which he describes as 'the state that would develop if human influence were completely removed and the resultant succession was completed in a single instant '. However, given that, according to Peterken, this variety cannot actually exist, it is probably not relevant to this debate.

9.5.1 Original Naturalness

Is 'original naturalness' something that we should attempt to recreate? There are many people who appear to hold this view. Some of the most often used words in the nature conservation vocabulary are: reintroduce, restore, recreate and (more recently) re-wild. There is a North American plan, 'Pleistocene re-wilding', developed by Cornell University. Their intention is to re-establish mega-fauna (lion, cheetah, elephant and camel populations) in North America. These animals would 'stand in' for the original North American Pleistocene animals. They claim that, 'this would help maintain ecosystems and boost biodiversity'. Peter Taylor (2005), in his book *Beyond Conservation*, presents a case for the restoration of the natural processes of wild nature in parts of Britain. Wilding in his context requires the introduction, or reintroduction, of herbivores, including large ungulates, and carnivores, such as lynx and wolves:

> However, in large area initiatives, an important element of wilding is to leave the herds subject to natural selection, including the periodic scarcity of feed, winter temperatures, disease and wounding.
>
> British wild-land proponents must begin to consider the re-introduction of lynx, wolf, and possibly bear and hence the reinstatement of predator-prey dynamics. (Taylor 2005)

Can we assume that the North American approach represents an attempt to recreate Peterken's original naturalness? Taylor's version is slightly more ambiguous in its meaning, but he does talk about 'bringing back the carnivores', which suggests that he is talking about some version of original naturalness. Is the intention to recreate something that happened at sometime (although when is not defined)? This 'something' was inhabited by many species now long since extinct. The presence of a very important predator, pre-historic humans, and particularly their role as a *natural* component of the past wilderness, seems to be discounted in this vision for the future.

There is a considerable volume of literature concerned with the large-scale and widespread extinction of animals in the Holocene. Some suggest that man was the most significant agent.

> On present evidence, this dreadful syncopation – humans arrive, animals disappear – seems to have occurred to a greater or lesser degree on every land mass except the continents of earliest human evolution, Africa and Eurasia. (MacPhee1998)

Ross MacPhee goes on to suggest that the main reason for these extinctions may have been diseases introduced by man. He asks why in Asia and Africa during the relevant time period there were so few extinctions. One hypothesis proposes that the Afro-Asian mega-fauna developed behavioural mechanisms to deal with the evolving human predator.

Peter Taylor suggests that vegetation changed as a consequence of the combination of natural climate change and human factors. He also suggests that the extinction of the larger grazing animals was the consequence of human intervention, with the implication that this was not an act of nature. (If it was a 'natural extinction' then the argument for the reintroduction of large grazing animals and predators might be diminished.)

John G. Evans (1975), in *The Environment of Early Man in the British Isles*, takes a different view: 'Where man was involved, he acted simply to increase the rate of change of processes already in operation.' Another author, A.J. Stuart (1993), makes a similar point. He suggests that extinctions follow from human predation, but only at times of fundamental changes in the environment. Can we, or should we, even attempt to differentiate between natural and anthropogenic factors? It is safe to assume that most people will, at least grudgingly, accept that, at some time or in some state, people can be regarded as a natural component of a wilderness. If this is the case, should the concept 'original naturalness' exclude people? The key questions could be: when was yesterday, or, which yesterday do we want for tomorrow?

> The natural world is dynamic. Successions wax and wane: natural disturbances punctuate the lives of organisms and the structures of communities, ecosystems, and landscapes: and climate change and migrations move organisms and change the rate of transforming matter and energy in ecosystems. (Pickett 1998)

George Peterken's concept of 'original naturalness' recognises that it was not a single state but a succession. Peter Taylor also clearly recognises the variety of original naturalness. So, should tomorrow be all our yesterdays?

9.5.2 Present Naturalness

Fig. 9.4 Beech in an upland oak wood

The value of a reference point for future management which is based at some time, or times, in the 'natural' past must be challenged. We need to take the discussion further and move to 'present naturalness' (the state that would prevail today if man had not been a significant ecological factor). Present natural is an extremely important concept because it reminds us that, even if anthropogenic influences were completely absent, ecosystems would have changed with time. A false concept of 'natural' can lead to inappropriate conservation objectives and management. For example, until recently, beech *Fagus sylvatica* was not considered to be a native species in north-western Britain. As a consequence, it was regarded as an alien invasive species and attempts were made to eradicate it.[1] There is evidence to suggest that it very rarely occurred naturally in Britain beyond a line drawn roughly from the Wash to the Severn Estuary. However, we know that beech can survive in northern Britain and that, had man not destroyed the wildwood, thus preventing colonisation by beech, we may now have to regard beech as a natural component of northern woods. Clearly, the 'present natural state' is an extremely important concept as it helps us avoid making inappropriate decisions. However, the actual conditions that might now prevail are a matter of conjecture, debate and interpretation. There will be many different visions of what might have been. Can we rely on any of these visions to provide a reference for whatever we are trying to achieve for the future?

9.5.3 Future Naturalness

Should nature conservation be rather more concerned with what we require of the future, and less about recreating the past or fossilising some intermediate state? Perhaps we should complement the sites managed to preserve wildlife by creating some large areas of Peterken's future natural. There is considerable and growing support for the creation of large wilderness areas (Adams 2003; Taylor 2005). However, I am not entirely sure that they actually mean 'future natural' as defined by Peterken. The key issue is: will it ever be possible to exclude man's influence, and particularly the impact of past intervention? For example, we will not be able to eliminate the impact of atmospheric pollution or invasive alien species. Thus, human influence, at some level, is likely to prevail.

Another, slightly obscure, argument might be that any version of future natural will be the product of human intervention in that we will have to make a number of decisions and take many actions to facilitate the process. These interventions could, for example, include the introduction of large herbivores and carnivores (Taylor 2005). Once we embark on an interventionist course, however minimal that may be, we end up with relative degrees of naturalness. In other words, something that is the product, at least in part, of human design.

We might modify Peterken's definition by considering what part people could play in the future. Will it ever be possible to exclude human influence entirely, and if we could, should we remove human influence completely and permanently? Should we regard ourselves as part of the natural world? How can we be anything else?

---

[1] We continue to control beech in some sites, not because it is regarded as an alien invasive, but because it does not provide for the variety of species that we would expect to find in an oak wood.

Can we create at least some sustainable places where, as far as possible, we rely on, or enable, natural processes, where opportunities for wildlife are optimal, where human interaction is not exploitation, and where our mutual dependence is recognised? In other words, could these be tomorrow's natural places?

Tomorrows natural places in the making:

Fig. 9.5 Cwm Idwal – an experiment in habitat management

Fig. 9.6 Stackpole Head, Pembrokeshire – a coast for tomorrow

References

Adams, W. M. (2003). *Future Nature a Vision for Conservation*, revised edition. Earthscan, London.

Angermeier, P. L. (2000). *The natural imperative for biological conservation*. Conservation Biology, 14, 373–381.

Evans J. G. (1975). *The Environment of Early Man in the British Isles*. Elke Books Ltd, London, UK.

Hunter, M. L., Jr. (1996). *Benchmarks for managing ecosystems: are human activities natural.* Conservation Biology, 10, 695–697.

Leopold, A. (1949). *A Sand County Almanac, and Sketches Here and There*. Oxford University Press, New York.

Machado, A. (2004). *An index of naturalness*. Journal for Nature Conservation, 12, 95–110.

NCC (1989). *Guidelines for the Selection of Biological SSSIs.* Nature Conservancy Council, Peterborough, UK.

Peterken, G. F. (1993). *Woodland Conservation and Management,* 2nd edn. Chapman and Hall, London.

Pickett, S. T. A. (1998). *Natural processes-biological diversity and heterogeneity*. In *Status and Trends of the Nation's Biological Resources*, Vols. 1 and 2, Mac, M. J., Opler P. A., Puckett Haeker C. E. and Doran, P. D. U.S. Department of the Interior, U.S. Geological Survey, Reston, VA, USA.

Rodwell, J. (2003). *Human Relationships with the Natural World: an historical perspective*, ECOS 24 (1).

Ratcliffe, D. A. (ed.) (1977). *A Nature Conservation Review.* Cambridge University Press. Cambridge.

Stuart, A. J. (1993). *The Failure of Evolution: Late Quaternary Mammalian Extinctions in the Holarctic*. Quaternary International 19, 101–107.

Snyder, G. (1974). *Turtle Island with Four Changes*. New Directions Publishing Corporation, New York.

Taylor, P. (2005). *Beyond Conservation, A Wildland Strategy.* Earthscan, London.

Wilson, E. O. (1994). *The Diversity of Life.* Penguin, Harmondsworth.

Chapter 10
Approaches to Conservation Management

Abstract We need to employ a wide variety of different approaches to managing wildlife. No single management solution can suit all circumstances. In broad terms, nature conservation planning and management can be divided into three, possibly four, main approaches. *Management planning by prescription* is when a plan prescribes or describes management actions, and although particular outcomes are required these are not specified. Management by prescription is rarely a valid approach, although it may occasionally be an inadequate, but necessary, compromise. *Management by defining conservation outcomes* was developed in recognition of the legal requirement to protect specified features on statutory, and other, sites. In outline, this approach is based on identifying the most important features on a site. The desired outcome for each feature is defined: these are the management objectives. Although a features approach is imperfect, it is the most appropriate approach for many sites. *Wilderness management* usually implies an acceptance that natural processes will maintain a wilderness ecosystem providing anthropogenic threats or factors are removed or kept under control. There is an immediate issue with defining what is meant by control, but the real problem lies in the idea that 'natural' excludes humanity. Regardless of how we choose to define wilderness, with or without people, and given that the only thing that we can be sure of is that these places will change, how will we know that the changes that we observe are acceptable? Will we ever be able to differentiate between changes that are a consequence of anthropogenic or natural factors? The purpose of *experimental management* is to test ideas and practices, and the outcomes will not necessarily be of benefit to wildlife. We know that, if we are to progress and become more effective and efficient at conserving wildlife, we need large-scale experimental management, and we need to explore new ideas and directions.

M. Alexander, *Management Planning for Nature Conservation: A Theoretical Basis & Practical Guide*, DOI 10.1007/978-94-007-5116-3_10,

10.1 Introduction

Fig. 10.1 Woodland with wild garlic

The single most important stage in any management plan is deciding what we want: these are the desired outcomes or objectives of management. It is extremely difficult to decide what we want to achieve if we have little or no idea of what can be achieved. This is not to suggest that almost anything can happen anywhere, but there will, in many situations, be some choices.

We need to employ a wide variety of different approaches to managing wildlife. No single management solution can suit all circumstances since the condition of natural and semi-natural habitats varies from almost pristine to completely derelict. Populations of species can range from being robust and viable to fragile and close to extinction. Since the starting point for management (i.e. the condition of a feature) varies so too will the outcomes. We will often have to make choices (though these will be limited by natural factors) guided by legislation and influenced by corporate or personal ethics.

Usually, the management of protected natural areas is about trying to prevent people from doing things, or about attempting to undo (or at least minimise) the impact of things that they have done, are doing or will do. Occasionally, we need to persuade or encourage people (sometimes with payments) to continue doing something that they may no longer wish to do. Or, in their absence, we do whatever they once did.

Whatever we do, we need to start doing it on a much larger scale. The need for a landscape-scale approach has become widely recognised and supported. As with all new ideas, this concept has a cohort of enthusiasts, and a few seem to take the view that we should abandon other approaches, i.e. the traditional, small nature reserve or protected area. There are, in fact, many very good reasons for maintaining smaller sites, certainly until we have sufficient large sites (if that is ever achievable). Small sites can grow and provide the foundations for large sites. We will also need to retain some smaller secluded areas because of their importance for the preservation of the local genetic variation found in naturally isolated populations. Small sites also provide the stepping stones that allow species to move.

We should have no doubt that climate change is happening. Conservation managers must be prepared to take whatever actions are necessary to ensure that wildlife has opportunities to cope with change. In Europe, we already have evidence that species are moving northwards. For example, many butterflies with a southerly distribution in Britain have expanded in the northern part of their range. More significantly, many species have failed to respond to recent climate changes because of the lack of suitable habitat (Hill et al. 2002). Consequently, in addition to establishing landscape-scale sites, there is also an urgent need to improve connectivity. Connectivity is the creation of corridors, or connections, between sites that allow species to move from place to place.

> An ecosystem is a tapestry of species and relationships. Chop away a section, isolate that section, and there arises the problem of unravelling. (Quammen 1996)

Quammen very eloquently makes the point that fragmented ecosystems or habitats are neither robust nor sustainable. Thus, connectivity not only provides opportunities for movement but also creates larger areas in which wildlife is less likely to become extinct and more likely to survive global change. Connectivity might be regarded as the process of creating large sites from smaller ones.

Conservation management, and indeed nature conservation, will benefit from a pluralistic approach. We should recognise that there are a variety of equally valid, but sometimes contradictory, theories, approaches and actions. In broad terms, nature conservation planning and management can be divided into three, possibly four, main approaches:

1. Management planning by prescription
2. Management planning by defining outcomes (features approach)
3. Management planning by enabling process
4. Experimental management (This does not strictly stand alone but could represent a scientific investigation of any of the above.)

10.2 Management Planning by Prescription

All management plans must identify and describe the actions necessary for the successful management of the site. Management planning by prescription is when a plan prescribes or describes management actions, and although particular outcomes are required these are not specified.

Management by prescription is sometimes used when organisations are obliged to adopt a bureaucratic approach to nature conservation. In these circumstances, there are some advantages to this method: it is easy to cost activities and relatively easy to monitor compliance. The worst examples can be found in some agri-environment schemes where farmers are paid for doing things, with an expectation that their actions will deliver wildlife, but nobody has told the farmers what those wildlife outcomes should be. Some years ago, a very large gathering of European nature conservationists in Ireland was challenged by a farmer from the Western Isles. He said, 'I am sick and tired of you conservationists. You tell me what to do but never what you want. Tell me what you want, and I will probably know what needs to be done.'

Some management plans contain statements that are called objectives but which are, in fact, a prescription or an activity. For example: 'To maintain the characteristic flora and fauna by grazing and/or mowing' (Eurosite 1999). This means that the most important section in any plan, the decisions and expressions of what we want to achieve, is omitted. In the place of outcomes, we find lists of activities (things to do) even though their purpose is unclear. Management by prescription is often the default position encountered in many inadequate management plans. Unfortunately, this approach to planning, which was commonplace before the mid 1990s, is now reappearing. For example, a Eurosite publication, *Management Planning for Protected Areas*, contains the following example of an objective: 'To ensure that there is sufficient grazing to prevent invasion by scrub' (Idle and Bines 2005). Plans based on this and many similar approaches identify prescriptions that are presumably intended to deliver something of value, though that 'something' is not described.

One of the more recent contributions to nature conservation management has been the introduction of evidence-based conservation management (Pullin and Knight 2001; Sutherland 2000; Sutherland et al. 2004). This suggests that that management decisions based on personal experience should be supplemented, or even replaced, by decisions based on evidence. This evidence is, in fact, the collective experiences of others, access to which will allow conservation managers to learn from others and not repeat their mistakes.

Many management techniques can be widely applied with little modification, for example, methods for controlling invasive alien species, such as rhododendron *Rhododendron ponticum* in western Britain, or for eradicating rats *Rattus rattus* on offshore islands. The desired outcome will always be the same, i.e. the effective and efficient eradication of the introduced species. There can be no doubt that the collation and dissemination of examples of these kinds of conservation management techniques, which are supported by sound evidence of success, will provide obvious and very significant benefits.

As the volume and quality of evidence increases, there may be a tendency for some managers to rely too heavily on prescription and to place less emphasis on defining their objectives. There is the risk that a technique which is successful in some places may not be appropriate for all sites. The only test of a management procedure is whether or not it produces the required outcomes. This might not be an issue when managing a non-statutory site, where the plant communities are not defined and protected by legislation. Plagioclimatic communities are the consequence of past management. Hay meadows, for example, were originally managed to provide a crop of hay; wildlife was an unintentional by-product. There may, therefore, be a sound argument for managing these places to produce hay, regardless of any particular outcome in terms of the diversity or abundance of individual species. However, if this approach is adopted, it would be best regarded as experimental management, as there is always a possibility that it will not deliver wildlife. Clearly, this type of management is not appropriate for sites where outcomes *must be* the maintenance or enhancement of nature conservation values, and this will include most statutory nature conservation sites.

There are some additional reasons for being cautious and for not assuming that management techniques will be applicable everywhere simply because there is evidence that they have been successful somewhere:

Management Will Vary from Place to Place
Management, although entirely appropriate in one location, will not necessarily be useful elsewhere. With the exception of some very specific techniques, management must vary because factors will vary from place to place. Not only will the range of factors vary but also the level of impact will vary. Subtle changes in climate, aspect and edaphic factors can have a significant impact on management intensity. Similarly, the presence of invasive species or wild herbivores can influence management.

Management Should Vary from Place to Place
Sites must be managed in different ways to maintain the local distinctiveness which is often a reflection of our cultural diversity. That is, different people will do things differently in different places. A diversity of management approaches will provide a variety of different opportunities for wildlife.

Management Will Change with Time
Regardless of how appropriate and successful management may be at any particular time, the factors that influence the feature (habitat or species) will change. For example, a pasture will have a significantly lower tolerance to grazing pressure during periods of drought. When features are unfavourable, the recovery management (management required to move a feature to a favourable state) will usually be different in intensity or type to the maintenance management (management required to hold a feature in the required state). For example, a neglected pasture which has begun changing to scrub will often require cutting, clearing and heavy grazing to return it to a favourable condition, and this is followed by much lighter grazing for maintenance. There is a further complication, because factors will have both an individual and a collective influence on features. Often, an individual factor will have little or no impact until it is combined with another factor. For example, the

population of feral goats in the mountains of North Wales is a factor that can influence tree regeneration, but, usually, it has little discernible impact because the population is low and dispersed. However, a second factor, a gradual and measurable increase in winter temperatures, has led to increased goat productivity and/or survival. The consequence is that a management action is required to protect the woodlands, i.e. the goats are culled.

Fig. 10.2 Grazing management

There is another problem associated with management by prescription and most other management approaches. Management is often a compromise. For example, one of the biggest problems when attempting to maintain pastures, meadows and heath is agricultural abandonment. These highly valued habitats were often created as a by-product of now unsustainable, and sometimes archaic, subsistence management. Traditional management, once abandoned, cannot be easily replaced. For example, traditional shepherding is no longer possible, and market pressures dictate both the breed and age of the grazing animals that are generally available. The consequence is that managers are often obliged to use whatever management tools are available, even when there is no certainty that these are appropriate. This is all the more reason for moving from management by prescription to management by defining outcomes. If an adaptive management process is applied, it may be possible to develop alternative management techniques that will deliver broadly similar and acceptable outcomes.

In conclusion, many of the issues mentioned above will not be a problem if conservation management evidence is used as intended: intelligently and in a way that builds on other tried and successful approaches. Management by prescription is rarely a valid approach, although it may occasionally be an inadequate, but necessary, compromise, and, even in these circumstances, there is a great deal to be gained from making some attempt to describe the intended outcome.

> A fixed pattern of management cannot be applied to a forest as it can to a collection of machinery in a factory, nor, because a certain forest has been managed in one way with successful results, does it follow that another forest, though apparently similar, will respond in the same way to the same system. Ready-made systems of management cannot be taken off a peg and applied to various classes of forest. (Brasnett 1953)

Fig. 10.3 Roundton Hill NNR

10.3 Management by Defining Conservation Outcomes

Management by defining outcomes, sometimes called 'a features approach to management planning', has become the favoured approach for planning and management on most statutory conservation sites in Europe, and is also encountered in many other parts of the world. The International Union for the Conservation of Nature, IUCN, maintains a worldwide list of protected areas. IUCN protected area management categories classify protected areas according to their management objectives. Areas are divided into six categories. Category IV is described as the 'habitat/species management area'.

> Category IV protected areas aim to protect particular species or habitats and management reflects this priority. Many category IV protected areas will need regular, active interventions to address the requirements of particular species or to maintain habitats, but this is not a requirement of the category.

IUCN identify a number of approaches that are suitable for managing this category. They are:

Protection of particular species: to protect particular target species, which will usually be under threat (e.g., one of the last remaining populations);
Protection of habitats: to maintain or restore habitats, which will often be fragments of ecosystems;
Active management to maintain target species: to maintain viable populations of particular species, which might include for example artificial habitat creation or maintenance (such as artificial reef creation), supplementary feeding or other active management systems;
Active management of natural or semi-natural ecosystems: to maintain natural or semi-natural habitats that are either too small or too profoundly altered to be self-sustaining, e.g., if natural herbivores are absent they may need to be replaced by livestock or manual cutting; or if hydrology has been altered this may necessitate artificial drainage or irrigation;
Active management of culturally-defined ecosystems: to maintain cultural management systems where these have a unique associated biodiversity. Continual intervention is needed because the ecosystem has been created or at least substantially modified by management. The *primary aim* of management is maintenance of associated biodiversity.

Active management means that the overall functioning of the ecosystem is being modified by e.g., halting natural succession, providing supplementary food or artificially creating habitat.

The emphasis on a need for active management clearly implies an intention that management will maintain specific populations of species or habitat conditions and that this will be achieved through human intervention. The inference must be that category IV areas will require defined outcomes. In other words, in order to actively manage anything we need to know what we are trying to achieve.

Management by defining conservation outcomes was developed in recognition of the legal requirement to protect specified features on statutory, and other, sites. In outline, this approach is based on identifying the most important features on a site. (Biological features can be habitats, communities or populations of species.) The desired condition (outcome) for each feature is defined, and these are the management objectives.

When dealing with statutory sites, where there is a legal obligation to protect specified features, we may have no choice but to apply an outcome-driven approach. However, we should be aware that there are potential shortcomings. Most of what we manage in our cultural landscape is the incidental by-product of past human endeavours to provide a living, food and shelter. Today, legislation is sometimes interpreted in a way that suggests an obligation to fossilise these remains, to preserve them as in aspic.

A features approach to habitat management divides or reduces a site or ecosystem into easily recognised and manageable components (features). Unfortunately, the whole is usually greater than the sum of the parts and, consequently, there are some problems. A general, but often unspoken, assumption is that if all the important features on a site are in a favourable condition then the site as a whole is also favourable. Clearly, this is a potentially dangerous assumption, but, unfortunately, in the practical, resource-deficient world of nature conservation management, there is usually no affordable alternative. As already mentioned, on statutory sites there is usually a legal obligation to protect the features that form the basis of the designation. Regrettably, this is often all that we can afford to do, so legislation can sometimes restrict choices at site level.

It is sometimes possible, particularly when dealing with more natural habitats, to comply with the legal requirements (particularly under European laws) and yet make significant allowance for process. A flexible approach can be obtained by applying an appropriate level of definition for a feature. For example, on a statutory sand-dune site, if the features are defined as particular plant communities, such as humid dune slacks, we are compelled to specify how much we want, where we want it, and give some indication of quality. However, in reality, we know that, over time, these slacks will increase or decrease and move around the site. We also recognise that this mobility is highly desirable. A more appropriate alternative would be to regard the entire dune system as the feature. In other words, the level of definition has changed. With this approach, management can focus on ensuring that, as far as possible, natural processes are enabled. The consequence is that individual communities can appear, disappear and move without giving any cause for concern.

I do not wish to give the impression that statute and resources alone are responsible for the features-based outcome planning. There are many extremely good reasons for adopting this approach. In many parts of the world, sites are managed to maintain semi-natural or plagioclimatic vegetation. Hay meadows and heath are good examples. A features approach, which defines the outcome that we require for these plant communities, is, without doubt, the most appropriate for use on these sites. The preservation of existing habitats and species, at the very least, provides reservoirs of wildlife to repopulate the wider countryside. These places could also provide routes for species that may have to move in order to survive a rapidly changing climate.

There is also a very strong case for preserving much of our cultural heritage as expressed in the variety of important semi-natural communities, such as hay meadows and coppice woodland. One final point: we should recognise that this approach to management, particularly in circumstances where we suppress natural processes, can be very expensive.

To summarise, although a features approach is imperfect, it is the most appropriate approach for most sites. Currently, there are no alternatives for sites where the features have legal status and were the basis for site selection. It is always applicable in situations where the features are plagioclimatic plant communities or populations of a species.

Management by defined outcomes will always be necessary when managing species, semi-natural habitats and statutory sites where the features are legally defined.

Fig. 10.4 Razorbills, a protected species

Fig. 10.5 Semi-natural coastal heath, Ramsey Island

10.4 Management by Enabling Process

Management by enabling process is about allowing natural processes to deliver a succession of outcomes, some of which may be unpredictable but all of which should be considered acceptable. At face value, management by enabling natural processes should be the obvious and ideal approach to managing very large tracts of countryside, natural habitats and wilderness.

10.4.1 Wilderness Categories

IUCN Categories Ia and Ib define wilderness areas. The main differences are:

- Ib protected areas will generally be larger and less strictly protected from human visitation than category Ia.
- Ia protected areas are strictly protected areas, generally with only limited human visitation.
- There would usually not be human inhabitants in category Ia, but use by indigenous and local communities takes place in many Ib protected areas.

Category Ia Strict Nature Reserve
Category Ia are strictly protected areas set aside to protect biodiversity and also possibly geological/geomorphological features, where human visitation, use and impacts are strictly controlled and limited to ensure protection of the conservation values. Such protected areas can serve as indispensable reference areas for scientific research and monitoring.

Primary Objective

To conserve regionally, nationally or globally outstanding ecosystems, species (occurrences or aggregations) and/or geodiversity features: these attributes will have been formed mostly or entirely by non-human forces and will be degraded or destroyed when subjected to all but very light human impact.

The area should generally:

- Have a largely complete set of expected native species in ecologically significant densities or be capable of returning them to such densities through natural processes or time-limited interventions;
- Have a full set of expected native ecosystems, largely intact with intact ecological processes, or processes capable of being restored with minimal management intervention;
- Be free of significant direct intervention by <u>modern</u> humans that would compromise the specified conservation objectives for the area, which usually implies limiting access by people and excluding settlement;
- Not require substantial and on-going intervention to achieve its conservation objectives;
- Be surrounded when feasible by land uses that contribute to the achievement of the area's specified conservation objectives;

– Be suitable as a baseline monitoring site for monitoring the relative impact of human activities;
– Be managed for relatively low visitation by humans;
– Be capable of being managed to ensure minimal disturbance (especially relevant to marine environments).

The area could be of religious or spiritual significance (such as a sacred natural site) so long as biodiversity conservation is identified as a primary objective. In this case the area might contain sites that could be visited by a limited number of people engaged in faith activities consistent with the area's management objectives.

Category Ib Wilderness Area

Category Ib protected areas are usually large unmodified or slightly modified areas, retaining their natural character and influence, without permanent or significant human habitation, which are protected and managed so as to preserve their natural condition.

Primary Objective

To protect the long-term ecological integrity of natural areas that are undisturbed by significant human activity, free of modern infrastructure and where natural forces and processes predominate, so that current and future generations have the opportunity to experience such areas.

The area should generally:

– Be free of modern infrastructure, development and industrial extractive activity, including but not limited to roads, pipelines, power lines, cellphone towers, oil and gas platforms, offshore liquefied natural gas terminals, other permanent structures, mining, hydropower development, oil and gas extraction, agriculture including intensive livestock grazing, commercial fishing, low-flying aircraft etc., preferably with highly restricted or no motorized access.
– Be characterized by a high degree of intactness: containing a large percentage of the original extent of the ecosystem, complete or near-complete native faunal and floral assemblages, retaining intact predator–prey systems, and including large mammals.
– Be of sufficient size to protect biodiversity; to maintain ecological processes and ecosystem services; to maintain ecological refugia; to buffer against the impacts of climate change; and to maintain evolutionary processes.
– Offer outstanding opportunities for solitude, enjoyed once the area has been reached, by simple, quiet and non-intrusive means of travel (i.e., non-motorized or highly regulated motorized access where strictly necessary and consistent with the biological objectives listed above).
– Be free of inappropriate or excessive human use or presence, which will decrease wilderness values and ultimately prevent an area from meeting the biological and cultural criteria listed above. However, human presence should not be the determining factor in deciding whether to establish a category Ib area. The key objectives are biological intactness and the absence of permanent infrastructure, extractive industries, agriculture, motorized use, and other indicators of modern or lasting technology.

However, in addition they can include:

Somewhat disturbed areas that are capable of restoration to a wilderness state, and smaller areas that might be expanded or could play an important role in a larger wilderness protection strategy as part of a system of protected areas that includes wilderness, if the management objectives for those somewhat disturbed or smaller areas are otherwise consistent with the objectives set out above.

Where the biological integrity of a wilderness area is secure and the primary objective listed above is met, the management focus of the wilderness area may shift to other objectives such as protecting cultural values or recreation, but only so long as the primary objective continues to be secure.

10.4.2 Wilderness Management

The following examples are typical of the way in which objectives are expressed in management plans for wilderness sites:

Maintain and enhance the wilderness values of naturalness, outstanding opportunities for solitude and primitive recreation, and protect special features of the mountain wilderness by:

- Rehabilitating the impacts of three closed vehicle tracks….
- Notifying the State of federal Water Rights for any available water…
- Eliminating unauthorised motor vehicle access….
- Improving opportunities for recreation while preserving naturalness…

There is nothing in the objective or, indeed, the plan that helps the reader to understand what is meant by 'the wilderness values of naturalness'. When will they know that they have enhanced the naturalness? What, in any case, does naturalness mean? It is very easy to be critical, but this seems to be the best that can currently be achieved on wilderness sites. Most objectives convey even less meaning. For example:

The objectives of conservation areas are:

- to conserve natural biological diversity
- to conserve geological diversity
- to preserve the quality of water and protect catchments

(Parks and Wildlife Service Tasmania 2003)

What do 'conserve' and 'preserve' mean? The problem is, of course, that we cannot, and perhaps should not, specify what we want to achieve in these places. This is because natural and semi-natural habitats depend on natural processes and factors for their existence and survival. Over time, they will change, either following some natural, catastrophic event or in response to the changing influence of natural factors. The consequence is that we cannot be sure of the precise outcome or direction of change. The only thing that we can be certain about is that these places will be occupied by a succession of different conditions.

In the USA publication *Principles of Conservation Biology*, Meff et al. (1997) clearly endorse the idea that conservation management ('biology' in their text) is concerned with enabling this succession of different conditions. They believe that there are three principles which are so basic to conservation practice that they should permeate all aspects of conservation and should be part of any endeavour in the field. Their three guiding principles of conservation biology are:

Principle 1: 'Evolution is the basic axiom that unites all of biology.'

This is perhaps an obvious principle, but can it be applied to the management of sites that are established, sometimes as a consequence of legislation, to maintain semi-natural or plagioclimatic vegetation? If we accept a need to 'preserve' some semi-natural communities (see previous section) then clearly not. However, perhaps 'preservation', even if it is possible, should be the exception and not the rule. All communities, natural or otherwise, have changed and will continue to change. Their first principle is, however, certainly relevant to wilderness management and management by enabling natural processes to deliver a succession of outcomes.

Principle 2: 'The ecological world is dynamic and largely non-equilibrial.'

The second principle is important because it represents the move away from the equilibrium paradigm: the belief that habitats and ecosystems evolve to a balanced or stable state which would be maintained indefinitely. The equilibrium paradigm has been replaced by the non-equilibrium paradigm which recognises that systems or habitats do not exist in a single, internally-regulated, stable state. They are dynamic and continually changing in response to the influence of a range of natural factors, for example, flood, fire, storms, volcanic activity, disease, etc. Peterken (1996) describes the importance of natural disturbances in northern woodlands: he mentions, wind, fire, drought, and biotic factors (Dutch elm disease). Sprugel (1990) argues that vegetation would not be stable over long periods of time even without human influence: 'One must recognise that there are often several communities that could be the 'natural' vegetation for any given time'. He uses African savannas, the Big Woods of Minnesota and the lodge pole pine forests of Yellowstone National Park as examples.

Meff et al. (1997) also make the point that it is important to understand that 'nonequlibrial processes' does not imply that species interactions (communities) are ephemeral or unpredictable. Communities are not chaotic assemblages of species, but community structure and species composition have a long evolutionary history, and they will continue to change over time.

Principle 3: 'The human presence must be included in conservation planning.'

The third principle needs some explanation. Meff & Carroll include this principle because they believe that:

- Humans are and will continue to be a part of both natural and degraded ecosystems.

- There is no way to 'protect' nature from human influences, and those influences must be taken into account in planning efforts.
- Native cultures are a historical part of the ecological landscape and have ethical rights to the areas where they exist. (Furthermore, they themselves add other types of diversity – cultural and linguistic diversity – which the earth is rapidly losing.)
- There are benefits to be gained by explicitly integrating humans into the equation for conservation. (For example; indigenous or local people will possess knowledge and skills that can help with planning and management.)

Regrettably, except for possibly the final point, the third principle is not widely recognised. When some people talk or write about wilderness management or enabling processes, they usually mean that anthropogenic threats or factors are removed or kept under control. There is an immediate issue with defining what is meant by control, but the real problem lies in the idea that 'natural' excludes humanity. Can, or should, we differentiate between anthropogenic and natural factors? The idea that 'natural' excludes human influence, and that nature conservation should be mainly concerned with producing habitats that are the direct result of natural processes, has influenced nature conservation management for a considerable period.

The ethical issues concerning the relationship between humanity and the rest of the natural world are unresolved. Consequently, there are no universally accepted, or comfortable, definitions of natural or of wilderness. For example, there are two IUCN definitions of wilderness (Chape et al. 2003):

> Large area of unmodified or slightly modified land, and/or sea, retaining its natural character and influence, without permanent or significant habitation, which is protected and managed so as to preserve its natural condition.

And

> Ecosystems where since the industrial revolution (1750) human impact (a) has been no greater than that of any other native species, and (b) has not affected the ecosystem's structure. Climate change is excluded from this definition.

The second definition could be taken to imply that people were part of nature until the industrial revolution which began in 1750. Many other sources suggest that wilderness is nature or natural processes uninterrupted by humans. In Chapter 9, I briefly mentioned the North American plan, *Pleistocene re-wilding,* developed by Cornell University (Donlan 2005). Their intention is to re-establish megafauna (for example, lion, cheetah, elephant and camel populations) in North America. These animals would 'stand in' for the original North American Pleistocene animals. The claim is that, 'this would help maintain ecosystems and boost biodiversity'. In other words, this view of wilderness excludes even the past influence of people. The megafauna of the North American Pleistocene (mastodons, camels, tapirs, giant ground sloths and many others) may have been driven to extinction by people. The extinctions occurred shortly after people arrived, about 11,000 years ago (Martin and Klein 1984). Whatever the reason, the extinctions were complete long before the industrial revolution. Humanity has been able to control, damage, exploit and destroy natural ecosystems for a very long time.

Regardless of how we choose to define wilderness, with or without people, and given that the only thing that we can be sure of is that these places will change, how will we know that the changes that we observe are acceptable? What might acceptable mean? Will we ever be able to differentiate between changes that are a consequence of anthropogenic or natural factors? The impact of humanity is all-pervasive: there is no corner of this world that has escaped our influence.

10.4.3 Management Options

The concept of 'enabling process' when managing nature reserves in cultural settings has been recognised in Britain for at least a quarter of a century, and probably for much longer. Conservationists talk about a 'non-intervention' approach to management as shorthand for implying that they want something natural or something delivered through natural processes. 'Non-intervention' is one of three management options described in *A Handbook for the Preparation of Management Plans* (NCC 1983). The options are:

- Non-intervention
- Limited-intervention
- Active-management

Non-intervention, as defined by NCC, is a 'climax or natural vegetation concept'. It is clear that 'natural' excludes, as far as possible, the influence of man. In the absence of natural (primeval) habitats in Britain, the aim is to acquire semi-natural habitats and allow them to develop free from as much adverse human influence, direct and indirect, as possible.

> It has sometimes been considered that the aim should be to achieve the type of climax vegetation that is postulated to have existed in the past, but as this cannot be known with certainty and may be difficult to achieve in present day conditions, the selection of this non-intervention option is considered to be the next best alternative. (NCC 1983)

If 'non-intervention' is to remain true to its original definition, that is, the influence of people is excluded, the results would approximate to Peterken's (1993) future naturalness: 'the state that would develop if man's influence were completely and permanently removed'. Is there any justification for non-intervention that may, in reality, mean abandonment? Perhaps we should take the risk and see what happens. There are plenty of examples where abandoned areas (even post-industrial sites) have, without any conservation management, become spectacular refuges for wildlife.

Non-intervention management in a cultural landscape might best be regarded as experimentation rather than nature conservation, and we should probably not take the risk on our most precious sites. Options are, in fact, shorthand for describing the way in which a site is managed. Management requirements will change with time, even where there is an intention to apply a non-intervention option. Nearly all sites would need some, if not considerable, periods of active-management before they could be left to nature. For example, on many sites invasive species would have to be removed.

The original definition of *limited-intervention* is rather woolly and suggests that this is an ephemeral option, employed when there is doubt that non-intervention will deliver. This option might be applied to process management if the definition is modified, hence the recent appearance of *minimal-intervention*. A minimal-intervention option would enable natural processes but would control, or remove, the influence of undesirable anthropogenic factors (although some people would argue that all anthropogenic factors are undesirable). In a tropical forest, for example, management aims to prevent illegal logging, poaching, squatters, invasive species, etc. (Alexander et al. 1992). But this is not necessarily consistent with an obvious obligation to protect the rights of indigenous peoples. Even if we should and could control anthropogenic factors, there will be some factors that we are not aware of, and there will always be the potential for new factors to appear.

I have mentioned that, although the outcomes of both non-intervention and minimal-intervention would be unpredictable, since they are more or less the product of natural processes any outcome could be regarded as acceptable. However, if nature reserves are to have any purpose they should make a contribution towards preventing, or at least reducing, the rate of extinction of species and habitats. This would mean that process management should be about working with, or enabling, natural processes to deliver something that at least optimises opportunities for nature. When working in cultural landscapes, and with the exception of experimental sites, ideally we should have some means of defining and measuring nature conservation benefits or, at least, of obtaining evidence to suggest that conditions on a site are moving in an acceptable direction and certainly not declining.

Process management, of a sort, can be applied to semi-natural habitats such as coppice woodland and hay meadows. In these examples, the management process, originally designed to produce woodland products or a crop of hay, incidentally creates semi-natural habitats that have become highly valued by scientists and conservationists. As an alternative to defining an outcome precisely in terms of species composition and structure, perhaps we should concentrate on producing a good crop of woodland products or hay, although, given that we cannot be certain that this will deliver anything of value, this should probably be treated as experimental management.

10.5 Experimental Management

Experimental management can be any of the above: management by prescription, by defining outcomes and by enabling process. Its purpose is to test ideas and practices: the outcomes will not necessarily be of benefit to wildlife. Some people argue that all conservation management should be experimental. However, if experiments are to have scientific validity, there is a need to apply appropriate scientific methods and protocols. In reality, we cannot afford that luxury; conservation managers never claim to have adequate resources.

There are very few examples of habitat-scale experiments. The most famous is Oostvaardersplassen in The Netherlands (Fig. 10.6). Oostvaardersplassen represents a version of process management allied to a minimal-intervention option. There are, without doubt, very significant nature conservation gains: the list of breeding and visiting birds is very impressive, though these may be regarded as the serendipitous by-product of a grand experiment.

Oostvaardersplassen may be contentious: there are animal welfare considerations, and many people claim that the approach has limited application elsewhere. However, Oostvaardersplassen has demonstrated that there are real alternatives to outcome-driven conservation management. We now know that if we are to progress and become more effective and efficient at conserving wildlife, we need large-scale experimental management, and we need to explore new ideas and directions.

Fig. 10.6 Konik ponies graze the Oostvaardersplassen

References

Alexander, M., Oliver, D. M. and Perrins, J. M. (1992). *Monteverde Cloud Forest Management Plan*. Tropical Science Centre, San Jose, Costa Rica.

Brasnett, N. V. (1953). *Planned Management of Forests*. George Allen and Unwin, London

Chape, S., Blyth, S., Fish, L., Fox, P. and Spalding, M. (compilers) (2003). 2003 *United Nations List of Protected Areas*. IUCN, Gland, Switzerland and Cambridge, UK and UNEP-WCMC, Cambridge, UK.

Donlan, J. (2005). *Re-wilding North America*. Nature, 436/18

Eurosite (1999). *Toolkit for Management Planning*. Eurosite, Tilburg, The Netherlands.

Hill, J. K., Thomas, C. D., Fox, R., Telfer, M. G., Willis, S. G., Asher, J. and Huntley, B. (2002). *Responses of butterflies to twentieth century climate warming: implications for future ranges*. Proceedings of the Royal Society of London Series b-Biological Sciences, 269, 2163–2171

Idle, E. T. and Bines, T. J. H. (2005). *Management Planning for Protected Areas, A Guide for Practitioners and Their Bosses*. Eurosite, English Nature, Peterborough, UK.

Martin, P. S. and Klein, R. G. (eds.) (1984). *Quaternary Extinctions: A Prehistoric Revolution.* University of Arizona Press, Tuscon, Arizona, USA.
Meff, G. K., Carroll, C. R. and contributors (1997). *Principles of Conservation Biology,* 2nd edn. Sinauer associates, INC Sunderland Massachusetts, USA.
NCC (1983). *A Handbook for the Preparation of Management Plans.* Nature Conservancy Council, Peterborough, UK.
Parks and Wildlife Service Tasmania (2003). *Moulting Lagoon Game Reserve (Ramsar Site) Management Plan.* Parks and Wildlife service, Department of Tourism, Parks, Heritage and the Arts, Tasmania, Australia.
Peterken, G. F. (1993). *Woodland Conservation and Management,* 2nd edn. Chapman and Hall, London.
Peterken, G. F. (1996). *Natural Woodland, Ecology and Conservation in Northern Temperate Regions.* Cambridge University Press, Cambridge, UK.
Pullin, A. S. and Knight, T. M. (2001). *Effectiveness in conservation practice: pointers from medicine and public health.* Conservation Biology, 15, 50–54.
Quammen, D. (1996). *The Song of the Dodo.* Pimlico, London, UK.
Sprugel, G. S. (1990). *Disturbance, Equilibrium, and Environmental Variability: What Is 'Natural' Vegetation in a Changing Environment?* Biological Conservation, 58, 1–18.
Sutherland, W. J. (2000). *The Conservation Handbook: Research, Management and Policy.* Blackwell, Oxford.
Sutherland, W. J., Pullin, A. S., Dolman, P. M. and Knight, T. M. (2004). *The need for evidence-based conservation.* Trends in Ecology and Evolution, 19, 305–308.

Chapter 11
Legislation and Policy

Abstract All management plans must contain a section on legislation and policy: together they provide the foundations that support the plan and act as a guide to the direction that the process should follow. This chapter outlines the significance of legislation to planning and uses examples to demonstrate some of the main areas of influence. The management of all sites will be influenced to some extent by legislation. On statutory conservation sites management may be governed almost entirely by legislation. Non-statutory sites do not escape the implications of legislation: there is a, sometimes bewildering, range of national and local laws, all requiring compliance. There is little purpose in spending any time discussing the failings, or otherwise, of legislation in a management plan. If there are issues these are probably best resolved elsewhere. Policies, or more specifically organisational policies, are a high-level statement of the purposes of an organisation (why it exists). These policies will lead to an expression of their intentions, or a course of actions that they have adopted or proposed. The policies may have been adopted voluntarily, imposed by legislation, or they may be a combination of both. At best, they provide an operational guide for an organisation, although sometimes, usually in less enlightened organisations, they are imposed as a set of incontrovertible rules.

11.1 Legislation

It is not the place of this book to document or discuss the implications of all the legislation that may influence conservation management. This section will outline the significance of legislation to planning and will use examples to demonstrate some of the main areas of influence.

The management of all sites will be influenced to some extent by legislation. On statutory conservation sites management may be governed almost entirely by legislation. In these circumstances, managers can sometimes find that their

M. Alexander, *Management Planning for Nature Conservation: A Theoretical Basis & Practical Guide*, DOI 10.1007/978-94-007-5116-3_11,

freedom is severely limited. Even non-statutory sites do not escape the implications of legislation: there will be health and safety legislation, access legislation and a sometimes bewildering range of other national and local laws, all requiring compliance.

Occasionally, legislation can be controversial, for example, the legislation that provides for designation and conservation of European habitats and species (Natura 2000). The main issue is that some people suggest that legislation does not favour dynamic, changing processes on sites. There is a belief that legislation is rather more about preservation than conservation. It may be the case that legislators have not given adequate consideration to the dynamic nature of our natural environment. However, it is clear that, in some cases, it is the interpretation of legislation that may be the real issue. There is also a view that this problem is not insurmountable: legislation, and the interpretation of legislation, will change with time, and, in any case, the legislation in question is intended mainly for sites in a cultural landscape and not for wilderness areas. There is a general consensus that the management of habitats in cultural landscapes is best achieved through defining outcomes for the important site features (see Chapter 15). There is little purpose in spending any time discussing the failings, or otherwise, of legislation in a management plan. If there are issues these are probably best resolved elsewhere.

> Many conservationists, and those affected by nature conservation laws, see wildlife legislation as too blunt an instrument, as an inflexible set of rules that stifle innovation and flexibility, create bureaucracy, respond too slowly to social change and even hinder rather than help the conservation cause. In this respect nature conservation is no different to any other field of social activity where people feel constrained or threatened by the law rather than protected by it. But the law is simply a set of rules by which society organises itself. In democracies, legislation is made – and can be changed – by our elected representatives. Meanwhile common law develops over time through social custom and the work of the courts. Despite its failings the law of nature conservation, like any other, is ultimately an expression of social values and choices. And because law takes time to change – typically years – it tends to enshrine values and priorities that evolve slowly over time, rather than responding to whatever is making the headlines today. Law is indeed an imperfect tool, but since so much of it inadvertently enshrines rights to destroy biodiversity, is there really a viable alternative to a statutory basis for nature conservation?
>
> Much of the antipathy towards conservation law, and environmental legislation generally, is because of its complexity, although compared to the law of, say, financial services or international trade nature conservation law is relatively uncomplicated. But if conservation law is simplified too much, it becomes either draconian or toothless and fails to reflect the huge diversity of people's relationships with nature. Most of the complexity in environmental legislation is because the legislators somehow have to express society's demand for 'balance' between a wide range of environmental objectives and an even wider range of cultural, social and economic aspirations. And on top of that, we demand fair and transparent rules of engagement. The only alternatives to a legal framework for nature conservation capable of balancing conflicting interests, monitoring outcomes and accounting for itself, are to permit anything or ban everything. So maybe those turgid directives and regulations are worth having.
>
> (Cole-King 2007)

11.1.1 *Examples of Legislation*

The following section contains a few examples of legislation. They are typical of the types of legislation that anyone involved in preparing a management plan will need to consider. It is essential that this section in any management plan is taken seriously and given adequate attention. Legislation is, in most circumstances, intended to protect wildlife or people (both managers and visitors). In general, legislation will have a very positive impact. Indeed, it is usually the most important mechanism for protecting wildlife. There will also be a wide range of obligations where, occasionally, compliance can be expensive, sometimes prohibitively so. For example, in Britain the need to comply with Health and Safety law, particularly regarding the use of safety equipment and the provision of certificated training, can mean that some management activities requiring the use of dangerous power tools become too expensive. Whatever the benefits or costs, all operations on a site must be legal: the cost of compliance will be significantly less than the cost of being prosecuted.

The legislation section in a plan can be conveniently divided into two main sub-sections: wildlife legislation and general legislation

11.1.1.1 Wildlife Legislation

Site Designation

The statutory status of a site, along with the obligations that this imposes, is one of the most important considerations in any plan. When statutory sites are established to protect specified features, species and/or habitats then the management plan must place the obligation to protect these features above all other considerations. Sometimes, when organisations have limited resources they will concentrate on the statutory features and give little, if any, attention to other features.

This section should also take account of the various international designations, for example, Ramsar, Biosphere and World Heritage, which will also influence the management of the site features.

Examples of legislation that leads to site designation:

International – Ramsar

The Convention on Wetlands, signed in Ramsar, Iran, in 1971, is an intergovernmental treaty which provides the framework for national action and international cooperation for the conservation and wise use of wetlands and their resources. There are 154 Contracting Parties to the Convention, with 1,636 wetland sites, totalling 145.7 million hectares, designated for inclusion in the Ramsar List of Wetlands of International Importance. Contracting Parties are expected to designate sites for the List 'on account of their international significance in terms of ecology, botany, zoology, limnology or hydrology' (Article 2.2), *and* to 'formulate and implement their planning so as to promote the conservation of the wetlands included in the List, and as far as possible the wise use of wetlands in their territory' (Article 3.1).

Fig. 11.1 Dyfi, a Ramsar site

European – Natura 2000

In the European Union, nature conservation policy is based on two main pieces of legislation: the Birds Directive and the Habitats Directive. This has led to the development of the network of Natura 2000 sites. These are statutory nature conservation sites where the designation identifies specific species and habitats (features) which will be protected. Article 6 of the 'Habitats' Directive (92/43/EEC) sets out provisions which govern the conservation and management of Natura 2000 sites. One of the key requirements for the Natura Network is that the habitats or species must be at Favourable Conservation Status across their natural range within the network. However, because the network will depend on the contribution of each individual site, it will always be necessary (although there is no legal requirement) to assess Favourable Conservation Status at site level. The implications of Favourable Conservation Status for the preparation of conservation management objectives are discussed in detail in Chapter 15.

Fig. 11.2 Rhinog SAC, North Wales

Fig. 11.3 Pembrokeshire seabird cliffs SPA

National (Britain) – Sites of Special Scientific Interest (SSSI)

Sites of Special Scientific Interest are the best examples of Britain's natural heritage of wildlife habitats, geological features and landforms. An SSSI is an area that has been notified as being of special interest under the provisions of the Wildlife and Countryside Act 1981. This was strengthened by the Countryside and Rights of Way Act 2000, which amends the 1981 Act and improves protection for SSSIs in England and Wales. The improvements include: enabling the conservation agencies (Natural England and the Countryside Council for Wales) to refuse consent for damaging activities; providing new powers to combat neglect; increasing penalties for deliberate damage and a new court power to order restoration; improving powers to act against cases of third party damage; and placing a duty on public bodies to further the conservation and enhancement of SSSIs. The sites are selected because they contain special or 'interest' habitat or species features. The UK Joint Nature Conservation Committee published *A Statement of Common Standards Monitoring* in 1998. This sets out a commitment to prepare conservation objectives which define the favourable condition for each 'interest' feature.

National (USA) – Wilderness Act

Federal legislators in the USA have taken a different approach. The Wilderness Act protects more than 105 million acres of wild land throughout the United States. The act allows research and recreation (such as hiking, canoeing and camping) but prohibits mechanized vehicles and all development within the designated wild areas.

The 'Wilderness Act' (1964) defines a wilderness as:

> A wilderness, in contrast with those areas where man and his own works dominate the landscape, is hereby recognized as an area where the earth and its community of life are untrammelled by man, where man himself is a visitor who does not remain. An area of wilderness is further defined to mean in this chapter an area of undeveloped Federal land retaining its primeval character and influence, without permanent improvements or human habitation, which is protected and managed so as to preserve its natural conditions and which (1) generally appears to have been affected primarily by the forces of nature, with the imprint of man's work substantially unnoticeable; (2) has outstanding opportunities for solitude or a primitive and unconfined type of recreation; (3) has at least 5,000 acres of land or is of sufficient size as to make practicable its preservation and use in an unimpaired condition; and (4) may also contain ecological, geological, or other features of scientific, educational, scenic, or historical value.

Species Legislation – The Following Are Examples of Legislation Intended to Protect Species:

International – CITES

The Convention on International Trade in Endangered Species (CITES) is an agreement between governments. Its aim is to ensure that international trade in specimens of wild animals and plants does not threaten their survival. CITES operates by controlling the international trade in specimens of selected species. These are listed in three appendices according to the degree of protection they require. All the imports, exports and introductions of species covered by the Convention have to be authorized through a licensing system. Currently, 169 countries have joined the Convention. CITES control is obtained through the establishment of regional and domestic management and scientific authorities. (CITES, specifically the species listed in the appendices, can make an extremely useful contribution to the process of identifying the important site features.)

European Legislation – Article 5 of Council Directive 79/409/EEC of 2 April 1979 on the Conservation of Wild Birds

This is the wild birds directive: it requires member states to establish a system for conserving wild birds. This is a general system for protecting wild birds by prohibiting the deliberate capture, killing, damage to nests, taking of eggs, and disturbance, particularly during the breeding season.

National Legislation – The UK Wildlife and Countryside Act 1981

Part 1 of the Act contains provisions for the protection of wild birds: any bird of a kind which is ordinarily resident in, or a visitor to, Great Britain in a wild state. The Act prohibits the deliberate capture or killing, protects nests when in use, and prohibits the collection or destruction of eggs. Protection from disturbance only includes birds listed in Schedule 1 (Cook 2004).

11.1.1.2 General Legislation

This can be extremely complex and difficult to deal with, because the management of protected areas is always subject to a vast array of regional, domestic and local legislation. This aspect of planning cannot be treated lightly since legislation will influence the management of all sites. It would be impossible to produce a representative list of typical legislation because this will vary so enormously from country to country and from site to site. Instead, the following are a few examples of legislation which are relevant to most UK sites. They are included to illustrate the diversity of legislation that has to be considered.

Health and Safety at Work Act 1974

The main principle of the Health and Safety at Work Act 1974 is that those who create risk as a result of a work activity are responsible for the protection of workers and members of the public from any consequences.

The Act places specific duties on employers, the self-employed, employees, designers, manufacturers, importers and suppliers.

Legislation associated with the Act places additional duties on owners, licensees, managers and people in charge of premises.

The main points of the Act place general duties on people. Examples are:

- Employers must maintain a safe workplace
- Anyone who undertakes a work activity must protect the public
- Goods must be designed so as to be safe and without risks to health
- Employees are required to co-operate with their employer in taking care

The Control of Substances Hazardous to Health Regulations 2002 (COSHH)
This requires employers to control exposure to hazardous substances to prevent ill health. They have to protect both employees and others who may be exposed.

Hazardous substances include:

- Substances used directly in work activities (e.g. adhesives, paints, cleaning agents)
- Substances generated during work activities (e.g. fumes from soldering and welding)
- Naturally occurring substances (e.g. grain dust)
- Biological agents such as bacteria and other micro-organisms

Disability Discrimination Act 1995
The DDA defines a disabled person as: 'a person who has, or has had in the past, a physical or mental impairment which has a substantial and long-term adverse effect on his or her ability to carry out normal day-to-day activities.'

There are numerous sections under this Act. Perhaps most relevant to the management of protected sites is the section dealing with, 'Discrimination in relation to goods, facilities and services'. Here, Section 19 of the Act reads:

> It is unlawful for a provider of services to discriminate against a disabled person-
>
> - in refusing to provide, or deliberately not providing, to the disabled person any service which he provides, or is prepared to provide, to members of the public;
> - in failing to comply with any duty imposed on him by section 21 in circumstances in which the effect of that failure is to make it impossible or unreasonably difficult for the disabled person to make use of any such service;
> - in the standard of service which he provides to the disabled person or the manner in which he provides it to him; or
> - in the terms on which he provides a service to the disabled person.

The Provision and Use of Work Equipment Regulations 1998 (PUWER)
The Regulations require risks to people's health and safety from equipment that they use at work to be prevented or controlled. In addition to the requirements of PUWER, lifting equipment is also subject to the requirements of the Lifting Operations and Lifting Equipment Regulations 1998.

In general terms, the Regulations require that equipment provided for use at work is:

- suitable for the intended use;
- safe for use, maintained in a safe condition and, in certain circumstances, inspected to ensure this remains the case;
- used only by people who have received adequate information, instruction and training; and
- accompanied by suitable safety measures, e.g. protective devices, markings, warnings (HSE 2004).

Occupiers' Liability Acts of 1957 and 1984

The Occupiers' Liability Act 1957 sets out the duty of care to visitors – i.e. people invited or permitted to use land, whether expressly or by implication. There is an obligation to take reasonable care that visitors will be safe doing whatever it is that they have been invited or permitted to do on a site.

The Occupiers' Liability Act 1984 sets out the duty of care to people that have not been invited or permitted to be on a site, i.e. trespassers. In most circumstances, there is a duty of care where a site manager is aware of, or believes that there is, a danger and knows that people may be in, or can come into, the vicinity of the danger. There is an expectation that the manager will provide some protection.

Where there is a risk, as outlined above, the manager will have to take reasonable care that people do not suffer injury on the site. It may be possible to discharge this duty of care by warning people about a danger, but this may not always be sufficient. For example, some extra precautions may be needed if there is reason to believe that unsupervised children are likely to use the site.

The duty of care does not apply to risks that adults willingly accept on behalf of themselves or those immediately in their care (NE 2005).

11.2 Policies

Policies, or more specifically organisational policies, are a high-level statement of the purposes of an organisation (why it exists). These policies will lead to an expression of their intentions, or a course of actions that they have adopted or proposed. The policies may have been adopted voluntarily, imposed by legislation, or they may be a combination of both. At best, they provide an operational guide for an organisation, although sometimes, usually in less enlightened organisations, they are imposed as a set of incontrovertible rules.

Most organisations will have general policy statements which cover their entire operation. In addition to managing sites they can be involved in a very much wider range of activities. This can be anything from funding to enforcement, from lobbying to providing advice. It is rarely necessary for a plan to include details of all the policy statements of an organisation, but a reference to their existence may be appropriate. The policies, and in some cases the remit, of an organisation will determine how it manages its sites and for what purpose. Ideally, organisations should prepare policies which are specific to site management: they should be unequivocal and concise. Broad-based, general policies will be open to interpretation and are often difficult to apply.

The policy section should begin with the inclusion of all relevant organisational policies. This should be followed, if necessary, by an assessment of the extent to which organisational policies can be met on individual sites. Local conditions can significantly influence the ability to meet policies. For example, although an organisation may have a policy to encourage stakeholders, particularly local communities, to take an active role in the management of sites, in some circumstances (the site may be very remote) this will not be possible. As another example, it can be very difficult to meet access policies on sites which contain dangerous features, fragile wildlife, or where the site is inaccessible.

The following is an example of a policy developed by a government agency in the UK. A policy statement for the management of a suite of National Nature Reserves – Countryside Council for Wales (This is an unpublished document for internal use.)

The purpose of NNRs as guided in legislation is as follows:

- They are statutory sites where the primary land use is management for nature or earth science conservation. This sets them apart from **other** SSSIs, Natura 2000, Ramsar and Biosphere sites, where, for example, the primary land use could be agriculture.
- They are sites where CCW's duty is to maintain or restore the nature conservation and geological features to Favourable Conservation Status.
- They provide special opportunities for research relating to flora, fauna and earth science.
- They have to be of national importance for wildlife or earth science.

Having acknowledged that our prime responsibility is to safeguard the nationally important conservation features, the Welsh Assembly Government expects CCW to consider its wider remit for landscape, recreation/access and public understanding. In this regard, the suite of Welsh NNRs will:

- be managed to the highest possible standard.
- be one of many tools at CCW's disposal for habitat, species and earth science protection.
- include adequate representation of all key Welsh biodiversity habitats.
- be managed within a context of sustainable development i.e. optimising opportunities for NNRs to contribute social benefits such as access, enjoyment and education, and consequently making a contribution to local economies.

Furthermore, providing that all activities or uses are sustainable and compatible with the conservation management of the site features the NNRs will:

- provide CCW with direct practical experience of a wide range of land management practices for wildlife.

- provide opportunities for study and research.
- be exemplars of conservation management with a geographical distribution that will provide opportunities for demonstration, training and education at all levels for most people in Wales.
- provide opportunities for the development and trial of new and innovative management techniques.
- be sites with a geographical distribution (to ensure that they are available to people from most areas of Wales) where public access, appreciation and enjoyment, including provisions for people of all abilities, will be encouraged.
- become sites where stakeholders, in particular local communities, will be encouraged to take an active role in management, and adopt a sense of value and shared ownership.
- be sites which contribute to local economies through visitor spend on local services.

CCW will encourage the sustainable public use of National Nature Reserves in Wales in so far as such use:

- Is consistent with CCW's duty to maintain or restore the nature conservation and geological features to Favourable Conservation Status.
- Does not expose visitors or staff, including contractors, to any significant hazards

  All legitimate and lawful activities will be permitted in so far as these activities:

- are consistent with CCW's duty to maintain or restore the nature conservation and geological features to Favourable Conservation Status.
- do not expose visitors or staff, including contractors, to any significant hazards.
- do not diminish the enjoyment of other visitors to the site.

Recommended Further Reading

Cook, K. (2004). *Wildlife Law: Conservation and Biodiversity*. Cameron May, London, UK.

European Communities (2000). *Managing NATURA 2000 Sites, The Provisions of Article 6 of the 'Habitats' Directive 92/43/CEE*. European Communities, Luxembourg.

Reid, C. T. (2002). *Nature Conservation Law*. W. Green & Son Ltd, Edinburgh, UK.

Van Heijnsbergen, P. (1996). *International Legal Protection of Wild Fauna and Flora*. IOS Press, Fairfax, VA, USA.

Reference

Cole-King, A. (2007). Personal communication.

Chapter 12
Description

Abstract The description is fundamentally a collation exercise. All relevant data are located and arranged under various headings. The order in which the headings are organised is of no particular significance and, initially, the headings should be regarded as having equal value. The description should only include statements of fact which are collated and recorded: at a later stage, they will provide the basis for evaluation and decision-making. All the deficits in the information are recorded, and whenever it is considered that a shortfall will impede decision-making this is noted. This can then be discussed at the appropriate place in the evaluation sections. The full description outlined in this chapter will not be appropriate for many sites. The various subsections should be completed only if the information has relevance to site management or the planning process.

12.1 Introduction

The description, usually the third section in a plan, is fundamentally a collation exercise. All relevant data are located and arranged under various headings. The order in which the headings are organised is of no particular significance and, initially, the headings should be regarded as having equal value.

The description should only include statements of fact. This is not the place for judgements. The facts are collated and recorded, and, at a later stage, they will provide the basis for evaluation and decision-making. All the deficits in the information are recorded, and whenever it is considered that a shortfall will impede decision-making this is noted. This can then be discussed at the appropriate place in the evaluation sections.

Although planning guides do not, in general, place an excessive emphasis on the preparation of a description, many, if not most, plans contain disproportionately large descriptions. It is not uncommon for the size of the description to exceed the remainder of the plan, and examples where the description is in excess of 75% of the plan are not uncommon. An over-emphasis on the description is a particular

M. Alexander, *Management Planning for Nature Conservation: A Theoretical Basis & Practical Guide*, DOI 10.1007/978-94-007-5116-3_12,

problem when resources are scarce. Planners can become preoccupied with the idea that they have to prepare an exhaustive and definitive site description, which includes sections on all conceivable information, regardless of its relevance to planning or managing the site. This is perhaps not such an issue when planning has been in place for a long period and the description has grown over time. But even here we should not forget that management plans are about communication; they should be as succinct as possible.

It is very unlikely that any individual would be able to complete all the sections without assistance. The author should consider his/her position as editorial, and should seek help and guidance from others. For example, there is no point tackling the climate section when a full climatological description of the site can be bought more cheaply than the raw meteorological data. Often, specialists will already have prepared accurate, detailed descriptions of site features. Where these are reasonably concise, there is little point in rewriting them for the plan, but they must obviously be checked for accuracy and relevance. Where a description is acceptable, it should be incorporated in the plan and attributed to the original author. If a report is too large to be incorporated in a plan, a summary should be prepared. In these cases, provide a reference including the location of the original document. Sites and populations of species are dynamic and continually changing in response to natural and man-induced trends. The description must accommodate these changes. The various sections will require review and update as additional information becomes available. It is generally a good idea to give the name of the author and date of each update.

The full description outlined below will not be appropriate for many sites. The various subsections should be completed only if the information has relevance to site management or the planning process. For example, on a coastal site, where the solid geology is totally obscured by a great depth of blown sand, there would be little point in producing a detailed geological description. However, if the section was omitted, readers may wrongly assume that the planner had forgotten to include, or even consider, geology. A simple statement, 'Geology is not believed to be a significant consideration in this plan', would remove any ambiguity.

There is no purpose in using all the subheadings for small or uncomplicated sites, or when resources for planning are in short supply. The headings can be used as prompts to guide the process of preparing a simple description. The minimal description should at least contain enough information for readers to understand the later sections in the plan.

Any system designed to hold information in some sort of logical or structured form is likely to give rise to one of the two most frequent errors:

- Insufficient categories or headings: The consequence is that far too much data is held under a single heading and, as a result, locating any particular item can be time-consuming and tedious.
- Too many categories or headings: The consequence is that little or no information is contained under each heading. Another problem associated with descriptions that contain a plethora of headings is that people feel compelled to make an entry under each, regardless of its relevance to the plan.

Both errors can be avoided by using a structure that grows by dividing and subdividing in relation to the volume of information. For example, on a small, simple site all the biological information can be held under one heading. As the quantity of information grows, 'biological' can be subdivided into flora, fungi and fauna. Each of these can be further divided, if there is a need, following standard taxonomic classification.

12.2 Description – Contents

3 Description (Note: The description is usually part 3 of a plan)

- **3.1 General Information**
  - 3.1.1 Location & Site Boundaries
  - 3.1.2 Zones (Compartments)
  - 3.1.3 Tenure
  - 3.1.4 Past Status of the Site
  - 3.1.5 Relationships with Any Other Plans or Strategies
  - 3.1.6 Management/Organisational Infrastructure
  - 3.1.7 Site Infrastructure
  - 3.1.8 Map Coverage
  - 3.1.9 Photographic Coverage
- **3.2 Environmental Information**
  - 3.2.1 Physical
    - 3.2.1.1 Climate
    - 3.2.1.2 Geology & Geomorphology
    - 3.2.1.3 Soils/Substrates
    - 3.2.1.4 Hydrology/Drainage
  - 3.2.2 Biological
    - 3.2.2.1 Flora
      - 3.2.2.1.1 Flora – Habitats/Communities
      - 3.2.2.1.2 Flora – Species
        - Vascular Plants
        - Bryophytes
        - Lichens
    - 3.2.2.2 Fungi
    - 3.2.2.3 Fauna
      - 3.2.2.3.1 Mammals
      - 3.2.2.3.2 Birds
      - 3.2.2.3.3 Reptiles
      - 3.2.2.3.4 Amphibian
      - 3.2.2.3.5 Fish
      - 3.2.2.2.6 Invertebrates
    - 3.2.2.4 Alien Invasive/Pest Species

Important: This section follows the numbering system used in the preceding contents list and not the general numbering sequence used elsewhere in this book.

3.1 General Information

Contents:

Note: Information on the legal basis for establishing the site and all the legislation which will influence the management of the site and its features are not included in the description. This information will be presented and discussed in the earlier section, 'Legislation and policy' (Chapter 11).

3.1.1 Location & Site Boundaries

This section should provide the information that will enable the site to be easily located. A map showing the location of the site is often sufficient. Any additional information that may help people locate or gain access to the site is also provided. This could include, for example, the main routes to the site and the name of the nearest town or village, country, state, region and county/department boundaries.

The national grid reference or latitude and longitude can be given, but this must be accompanied by an indication of what the reference relates to, for example, the centre of the site, a car park or gateway.

The location of the site boundaries is obviously essential information. The only sensible way to present this information is on a map. The map can be annotated or supplemented with information on how the boundaries can be located on the ground, for example, how they are marked. All obligations or responsibilities for maintaining the boundaries should be included.

3.1.2 Zones (Compartments)

In Britain, the word 'compartment' is often used in place of zones. This originally came from forest management plans (Brasnett 1953). There is no difference between compartments and zones.

Important: Deciding on the best time to tackle the issue of zoning in a plan is not easy. The establishment of meaningful zones requires an analysis based on information derived from the management objectives and their associated rationales. However, objectives cannot, and must not, be completed until much later in the planning process. This leaves the planner with two alternatives: either prepare a provisional zonation, which may need to be amended at a later stage, or wait until the objectives and rationale have been completed before attempting anything. Whichever option is adopted, the zone map should be placed at this early stage in the final plan, as one of the key functions of the zonation map is to help describe the site and particularly the management activities. Zones should be developed as the plan progresses and regarded as provisional until the plan is complete.

Sites may be divided into zones to meet a wide variety of purposes, for example, to describe management actions or to guide or control a number of activities. It is often very difficult to describe, or even consider, the management of large or complicated sites unless they are divided into a series of zones.

If zones are established, the following guidelines apply:

- The basis or justification for their selection should be outlined.

- A concise description of the function, including any restrictions that apply within each zone, should be included.
- They should be clearly shown on a map.
- Maps must be made available to all interested parties.
- The boundaries of zones must be easily recognised and located on the ground. Physical features, such as rivers, walls or roads, can be used as boundaries, but some of these may move over time. On large, homogeneous sites, where there are no obvious landscape features, it may be necessary to install some form of permanent marker or use GPS. A commonly used, and useful, approach is to mark the map grid intersections on the ground.
- Zones should not generally be defined by the location and boundaries of habitats or communities, particularly when the vegetation is dynamic and changing.
- Zones should be identified with a unique and, if possible, meaningful code. In some cases, a simple numerical code will be quite adequate.

It is important that zonation systems are regarded as flexible management tools that can be introduced, removed or modified according to need. They can be used for a very wide range of different purposes.

3.1.3 Tenure

This section must be completed. It is essential that the individuals preparing the plan have a full understanding of the land tenure and legal status of the site. Tenure documents are usually over-complex and written in a style that makes them difficult to understand. The role of the planner is to translate the document into everyday language, but it is important that the translated documents are not used for legal purposes. The first sentences in all cases should be: 'This is not a legal document. Please refer to the original tenure documents before taking any decision or any action which may have legal implications.' The location of all legal documents should be noted.

Where tenure is complicated by the presence of more than one owner/occupier, land holding or status, each separate area should be individually described. A map showing the different areas of tenure, rights of way, etc. should be included.

This subsection should include, for each tenure area, all the information that is relevant to planning and managing a site. The following headings may be used. This is not a complete list, but it represents the minimum requirement.

- Type of holding (for example, owned, leased or agreement)
- The names and contact details of owners and occupiers
- Date of acquisition or agreement
- Length of lease/agreement
- The area of each individual lease, holding, etc.
- Conditions and reservations: describe all the conditions imposed in respect of ownership, lease, tenancy, agreement, etc.
- Legal rights of access

- Legal rights held by others (e.g. collection of shellfish, peat cutting or hunting)
- Obligations and legal responsibilities arising from tenure

For sites, or the parts of a site, that are owned by the organisation responsible for management, some background information which describes the reason for, and process of, acquisition should be included. For example, some sites are purchased because they have been identified in a formal acquisition strategy, others simply because an opportunity arose and, occasionally, a site is acquired through legal compulsory purchase.

The location of title deeds should be confirmed and recorded. It may, in some cases, be necessary to include copies of deeds in an annex, but, in practice, these are rarely consulted.

All service routes entering, crossing or lying immediately adjacent to a site should be described. Examples include roads, water pipes, gas pipes, electricity pylons and cables, drainage ditches and canals. All arrangements for maintenance, the rights of access and the normal frequency of activity should be noted.

3.1.4 Past Status of the Site

This section provides a brief historic review of the interest shown in the site. Although this will usually refer to the attentions of scientists, it could also include naturalists, artists, writers and others. This should be followed by details of any past legal conservation status. This information is effectively an assessment or evaluation of the site made at an earlier time by others. It will often indicate the prime reasons for site acquisition, and can prepare the way for the discussion in the evaluation section of the plan.

3.1.5 Relationships with Any Other Plans or Strategies

The management of a site will often be influenced, or even regulated, as a consequence of other plans, for example, National, Regional and Local Biodiversity Action Plans and Regional Structural Plans. Situations also arise where a site contains features where the responsibility for management and planning is held by a different organisation, for example, archaeological, historic and geological features. It is important that all legitimate plans are recognised in the site plan.

3.1.6 Management/Organisational Infrastructure

This section should contain a brief outline of the organisational structure and the staff deployed in managing the site. This can include details of staff responsibilities.

This statement should be in respect of present staffing levels. Later sections in the plan may identify a need to revise the staffing structure on a site.

3.1.7 Site Infrastructure

A description of all significant buildings and any other structures should be included, along with any relevant information, for example, their purpose, suitability, condition, etc. Examples will include visitor centres, hides, workshops and toilets. Maps showing locations will enhance this section.

3.1.8 Map Coverage

Record any relevant contemporary maps and any useful historic maps. Include maps showing topography, geology, soil, land use, vegetation, etc. Give the date, scale and location of the maps. There is little or no purpose in attempting to locate every map ever produced which covers the site. Many historical maps are very inaccurate and have little more than curiosity value. The degree to which the site is believed to have changed or developed in recent times should influence the amount of effort put into locating historical maps. For example, on dynamic coastal sites early maps, particularly admiralty charts, can provide useful information about past conditions and trends.

3.1.9 Photographic Coverage

The record of photographic coverage can contain sections on aerial and ground photographs. Historical photographs can be a useful source of information on past land use and management. Where individual photographs are of special interest they should be listed and described. Any reference to an individual photograph should give a location and include comments on the contents and quality. It is often sufficient to make general comments on the availability, or otherwise, of photographs. For example:

'The site records contain over 500 colour transparencies depicting a wide range of views, species and activities. The collection has not been sorted or catalogued, and many of the photographs are of poor quality. It is essential, given the need to maintain a photographic record and also to provide material for talks, displays, etc., that the collection is improved.'

3.2 Environmental Information

When planning the management of small, uncomplicated sites this subheading, without any subdivisions, may be adequate. As the size or complexity of a site increases, further tiers of subdivisions may be introduced:

3.2.1 *Physical*

Contents

3.2.1.1 Climate

Climate is an extremely important factor. However, a simple outline or summary will be adequate for most sites. Even when detailed records are available, there is usually no point in including them in the plan. A brief description of the data, along with its location, will be sufficient.

Recent changes should be mentioned, and if any trends have become apparent these should be included.

Microclimate can be important on some sites. For example, when managing mosses and liverworts in deep woodland gorges humidity is one of the most important factors. High humidity levels are maintained, in part, because air movement is suppressed by the trees.

3.2.1.2 Geology and Geomorphology

A simple, general description will be sufficient in most cases. There are occasional exceptions where there is justification for detailed accounts. These are:

- When the site contains important geological or geomorphological features which require protection and/or management.
- When active geomorphological processes are a feature or when other important site features are dependent on these processes, for example, river shingle banks and active sand dune systems.

3.2.1.3 Soils/Substrates

Describe the major soils or substrate types using a map whenever possible. Note anything that may be relevant to site condition or management.

3.2.1.4 Hydrology/Drainage

The relevance of hydrology and drainage to the management of the site features will once again determine the level of attention given to this subject. Obviously, when planning the management of rivers, catchments, bogs, fens and, in fact, any wetland habitat, hydrology will be a significant factor. It will also be important in less obvious circumstances. For example, the survival of humid slack communities in a sand dune system is entirely dependent on the height of the water table.

Any significant human intervention, past and present, should be described, for example, past land management, including drainage, peat cutting and river canalisation, and current off-site land use, including drainage or water extraction within the catchment area.

3.2.2 Biological

When planning the management of small, uncomplicated sites this subheading, without any subdivisions, may be adequate. As the size or complexity of a site increases, further tiers of subdivisions may be introduced. The obvious first tier is to divide the data between flora and fauna. Fungi are usually included under flora, but it might be more appropriate to keep them separate. Flora can be subdivided into habitats, communities and species. Further divisions of flora and fauna can be based on taxonomic classification. A subheading for alien, and in particular invasive or pest, species can also be included.

Contents:

3.2.2 Biological
- 3.2.2.1 Flora
  - 3.2.2.1.1 Flora – Habitats/Communities
  - 3.2.2.1.2 Flora – Species
    - 3.2.2.3.1.1 Vascular Plants
    - 3.2.2.3.1.2 Bryophytes
    - 3.2.2.3.1.3 Lichens
- 3.2.2.2 Fungi
- 3.2.2.3 Fauna
  - 3.2.2.3.1 Mammals
  - 3.2.2.3.2 Birds
  - 3.2.2.3.3 Reptiles
  - 3.2.2.3.4 Amphibians

3.2.2.1 Flora

3.2.2.1.1 Flora – Habitats/Communities

The habitat/communities subsection is used to describe the habitats and plant communities. Whenever possible, a standard approach should be adopted. Where a standard classification system has been used to identify communities, it will be sufficient to record the system by name and give a location for any documentation that provides methodology, along with a description of the individual communities. If a non-standard approach is used, a description of methodology and communities should be included in the plan, either in this section or possibly the appendices. Clearly, the most appropriate way of presenting this information is by producing a vegetation map.

In Britain, for example, there are many different classification systems, with two in common use:

Phase 1 Habitat Classification

This classification was developed in the 1980s for mapping terrestrial and freshwater habitats within SSSIs and nature reserves, and for larger scale strategic surveys. The classification has subsequently been used extensively for major surveys. It was originally published by NCC (reprinted by JNCC) and is supported by a field manual (JNCC 2004b).

The National Vegetation Classification (NVC)

This is a phytosociological classification of terrestrial and freshwater vegetation. It is employed as the main classification for terrestrial habitats in *Guidelines for the Selection of Biological SSSIs* (NCC 1989) and has been used to interpret Annex I of the EC Habitats Directive, where relevant. The UK conservation agencies and others have extensive data holdings coded using the NVC.

The NVC is published as a five-volume series entitled *British Plant Communities: Woodland & Scrub* (Rodwell 1991a); *Mires and Heaths* (Rodwell 1991b); *Grasslands and Montane Communities* (Rodwell 1992); *Aquatic Communities, Swamps and Tall-herb Fens* (Rodwell 1995); *Maritime Communities and Vegetation of Open Habitats* (Rodwell 2000).

If any habitats or communities are qualifying features (features that formally led to the legal site designation) they must be identified and described in this section. Occasionally, there can be differences in the way the various national and international designations describe more or less the same community. This is the place in the management plan to resolve these differences for the purposes of site management.

Species – (3.2.2.1 Flora and 3.2.2.3 Fauna)

Although many managers recognise a need to complete and maintain species lists for sites these lists have no place in the main body of the management plan. If, for any reason, their inclusion is considered necessary, they should be attached to the plan as appendices. Species lists can be misleading: the size or accuracy of a list will often be a reflection of the effort that has been put into recording on the site. In many situations, a great diversity of species is an indication of the health or general good condition of the site, for example, a forest wilderness. In other circumstances, high diversity may be an indication that a site is in extremely poor condition. For example, disturbed raised bogs, where the peat has been cut, will usually contain many more species than pristine, or uncut, bogs.

It is important that all notable or endangered species, along with any other species that may have specific management requirements, are recorded. This must include all species that are given specific legal status or protection and, most importantly, species which are qualifying features (i.e. features which formally led to the legal site designation).

Any significant surveys, or other projects that may have relevance to the data presented in this section, should be described. It is also essential, as with all other sections in the description, that any shortfall of data is recorded. It may be that species recording is so incomplete that subsequent management decisions will be difficult or impossible.

3.2.2.4 Alien Invasive/Pest Species

The presence of alien invasive species is possibly, with the exception of global climate change, the single most frequently encountered and serious problem that conservation managers will face. The problem is global and increasing.

For now, all alien invasive plant and animal species that are present on a site or found close to the site should be recorded. Many of the problem plant species will be obvious and include (in Britain) rhododendron *Rhododendron ponticum,* Japanese knotweed *Reynoutria japonica* and Himalayan balsam *Impatiens glandulifera.* Other less aggressive species may gain an advantage as climate changes. (This could become a significant ethical and practical issue associated with climate change. Alien species could become the only species that will survive in a modified climate). There is little purpose in crystal ball gazing: as changes take place, the description, and all other sections in the plan, will be amended. Alien animal introductions that are generally regarded as pest species, for example, American mink *Mustela vision*, grey squirrel *Sciurus carolinensis* and feral goats, should also be recorded.

The description of any past management used to control these species can be included in this subsection.

3.3 Cultural

This section deals with the impact of man and with human values.

Contents:

3.3.1 Archaeology

The presence of any archaeological or historical remains on the site should be recorded, along with any implications for management (Fig. 12.1). Ancient monuments are often legally protected, and the site manager may be responsible for ensuring their safeguard. Even when there is no need to provide active management, it is essential that other management operations do not in any way threaten these remains. It is important, therefore, that all recorded remains, particularly all legally protected monuments, are noted, and shown on a map whenever possible. Where nothing is known, this may indicate the need for future surveys.

Fig. 12.1 Newborough Warren NNR

Archaeological remains, along with a recorded history of past land use, can provide valuable guidance for future management. This is particularly important when dealing with semi-natural or artificial habitats.

3.3.2 Past Land Use

An appreciation of past land use will often provide the planner with an essential guide to understanding the current condition of the features on a site. This is particularly important when dealing with damaged or semi-natural features. Although of academic interest, there is generally little purpose in looking too far into the past. Consider the period that is most likely to have affected the present condition.

3.3.3 Present Land Use

Record present land use, but exclude management for nature conservation. Record all aspects of land use, i.e. forestry, agriculture, water extraction, etc. Note the impact that any of these activities are known to have on the site.

3.3.4 Past Management for Nature Conservation

This should be the easiest section to complete for all managed sites. In reality, records have usually not been adequately maintained and, consequently, this essential section is often difficult, or impossible, to write. When records have been kept, the best way to present the information is to follow the structure used in the plan. The projects should be grouped in relation to the objective that they serve.

3.4 People – Stakeholders, Access, Etc.

This section is used to describe all aspects of current (at the time of plan preparation) public use and interest in the site. 'Public' is taken to mean anyone with an interest in the site, and will include local people, tourists and special interest groups.

Contents:

3.4 People – Stakeholders, Access, Etc.

- 3.4.1 Stakeholders
- 3.4.2 Access and Tourism
- 3.4.3 Interpretive Provisions
- 3.4.4 Educational Use

3.4.1 Stakeholders/Stakeholder Analysis

A stakeholder is any individual, group or community living within the influence of the site or likely to be affected by a management decision or action, and any individual, group or community likely to influence the management of the site.

This section in the description is slightly different to others as it includes some analysis, as opposed to simply presenting factual information. Please see Chapter 4 for a full explanation.

3.4.2 Access and Tourism

For very small sites where access is not a significant issue, a simple statement (a paragraph or two) under a broad heading may suffice. The following list of headings can be used for larger sites and particularly for sites where the provision of access is important. The headings are neither exclusive nor exhaustive; additional headings should be added when necessary. Maps can be used where these will help to convey the information. This could include maps to show the area surrounding the site, provision by others in the area, where visitors come from, etc.

Contents:

3.4.2 Access and Tourism
- 3.4.2.1 Visitor Numbers
- 3.4.2.2 Visitor Characteristics
- 3.4.2.3 Visit Characteristics
- 3.4.2.4 Access to the Site
- 3.4.2.5 Access Within the Site
- 3.4.2.6 Visitor Facilities and Infrastructure
- 3.4.2.7 The Reasons Why People Visit the Site
  - 3.5.2.7.1 Wildlife Attractions
  - 3.5.2.7.2 Other Features That Attract People
- 3.4.2.8 Recreational Activities
- 3.4.2.9 Current and Past Concessions
  - 3.4.2.10 Stakeholder Interests
  - 3.4.2.11 The Site in a Wider Context

3.4.2.1 Visitor Numbers

This subsection could include the annual total number of visitors, the average number or a range over a specified number of years, trends up or down over a specified period, the proportional use of different parts of the site/access points, etc. An account of how this information was collected should be included. For example, are the data based on estimates or accurate information from data-loggers?

3.4.2.2 Visitor Characteristics

This is a subsection on visitor demographics. The most commonly relevant sub-headings are age, gender, ethnicity, nationality, occupation, education or language profile. Their origin (for example, local, tourist, foreign tourist, even country of origin) can be included, but only if this is relevant to the later stages of the management plan. Other important information can include: the size of groups or parties; do people visit singly or in small family groups; are the groups formal, with a leader, or informal?

3.4.2.3 Visit Characteristics

This can include, for example, the amount of time that visitors spend at the site and the timing and seasonality of visits. Different types of visitor may spend different periods of time on a site; it is not uncommon for local people to make frequent but very short visits, while people from further away often spend considerably longer on the site. The length of time that people spend on a site can have significant implications for the type of facilities that they will require.

3.4.2.4 Access to the Site

The ways in which people gain access to the site, or the modes of transport used, are described. This can include departure points, availability of public transport and any other relevant information about their journey.

3.4.2.5 Access Within the Site

This is a description of how people currently gain access within a site. An account of all routes, for example, entrance points, car parks, paths, trails, tracks, roads, boardwalks, bridges, etc., is included. An annotated map is obviously one of the most appropriate ways of presenting this information. Attention should be given to the condition of the various routes, for example, the difficulty of the routes or their suitability for people with mobility problems. If vehicles, bicycles, horses, etc. are permitted on the site this should be included in the description.

3.4.2.6 Visitor Facilities and Infrastructure

This subsection contains a description of the current visitor facilities and infrastructure; it includes both quantitative and qualitative information. All buildings, such as visitor centres, hides, shelters, toilet facilities, etc., along with an account of their condition, should be included. Interpretive and education facilities can be described under this general heading, unless the provision is substantial. In these cases, separate headings may be more appropriate. All information signs, including direction signs, safety signs and location signs, are mentioned.

3.4.2.7 The Reasons Why People Visit the Site

There is always a temptation to treat this section as an evaluation or discussion, but it is simply a description of the features that attract people to a site. They are listed along with an indication of the level of attention that they attract. The following subheadings are sometimes useful:

Wildlife Attractions

These could be the wildlife features that led to the site designation, although it is dangerous to assume that visitors will find these attractive. With the exception of a few specialists, slime moulds and obscure invertebrates rarely attract visitors. Of course, it is much easier when the wildlife is spectacular, furry or feathered. Sometimes people visit a site to see wildlife features that are not strictly part of the reason for site designation. An example of this is the profusion of bluebells on a Welsh island designated for its seabird populations.

Fig. 12.2 Bluebells on Skomer Island

Other Features That Attract People

This could be the landscape or wilderness quality of a site. Some sites have spectacular geological or geomorphological features, waterfalls, caves and other karst features. Archaeological, historical and other cultural features are often important. There may be other reasons, for example, the search for solitude or spiritual experiences.

3.4.2.8 Recreational Activities

All activities should be included because, depending on intensity, all have the potential to have some impact on the features or on the quality of experience enjoyed by other visitors. Even people walking quietly around a site can have a significant impact if there are too many of them. For example, breeding birds can be disturbed, footpaths can become erosion zones or unsightly scars on the landscape, and visitors, particularly those seeking solitude, may be distracted by the presence of others. The description of activities can be conveniently divided into three categories:

- Activities that are, or have been, encouraged
- Activities that are, or have been, tolerated
- Activities that are, or have been, excluded

(When a rationale or discussion which supports the decisions that led to the inclusion or exclusion of any activity is available a reference should be provided, or, if it is reasonably succinct, the text may be included in the description.)

3.4.2.9 Current and Past Concessions

This will include information on current concessions, if there are any. This is particularly important if the period specified for any of the concessions extends beyond the implementation date of the plan in preparation.

3.4.2.10 Stakeholder Interests

Stakeholder interests and involvement in the site will have been described in the general description. If there is anything that is specifically relevant to access it should be included here, for example, the use of the site by tour operators, local transport providers, local shops, etc.

3.4.2.11 The Site in a Wider Context

This will include a description of all other relevant or similar provisions in an area. Anywhere within a reasonable distance of the site that provides access, recreational or tourist opportunities similar to those that are, or could be, provided on the site should be covered. This can be very important information; there is little purpose in developing sophisticated tourist facilities if the local area is already saturated with similar provisions. It is also important that nature conservation organisations avoid competing with one another.

3.4.3 Interpretation Provisions

For sites where interpretation is an important consideration this subsection should be included. The current interpretation use and facilities should be described; this includes information on the current beneficiaries or recipients, and the general purpose, or focus, of present interpretation. For example, are facilities intended for the interpretation of the site alone or for nature conservation in general? In some cases, it may also be appropriate to include accounts of earlier attempts at interpretation, whether successful or otherwise. Often, site managers will have experimented with various approaches to interpretation, and it may be possible to learn from their successes and failures.

3.4.4 Educational Use

For sites where educational use is an important consideration this subheading should be included. The current educational use of the site is described, including who uses it and for what purpose. When available, information on the number of individuals or organisations that use the site should be included. All current facilities are described. These will include, for example, the provision of guided educational visits, leaflets, education packs, education centres and the employment of dedicated staff.

3.5 Research Use and Facilities

An outline of any significant research that has been, or is being, carried out on the site should be provided. This should include any approved research projects that will be carried out in the future. Describe any research facilities that may be available, for example, some sites are equipped with a field laboratory. Include a note on the suitability of the site for research, for example, a site which is open to public use may not be suitable for certain types of research projects.

3.6 Landscape

Other sections in the plan will describe most of the components that make up the landscape. Later, and if appropriate, landscape will be considered in some detail during the evaluation process. The purpose of this section is to provide an objective description of features that form the landscape. In practice, this will often be a summary of visible features discussed under the previous headings. Include topography or landform, land cover and man-made elements. This section is descriptive and there should be no attempt to evaluate the landscape: that comes later.

Examples

(a) The area in general is low-lying, rolling hills leading to a coastal plateau. For the greater part, the land is arable with very large extensive modern fields. The site is the only significant area of woodland within the locality. It covers the main part of the highest land and dominates the view from most aspects. There are large numbers of dead elm trees. These are visible from a great distance. There are no buildings or other man-made objects on the site.

(b) The Kidepo Valley National Park comprises two broad, shallow valley systems, bounded on all sides, except the north-west boundary, by steep, rugged mountains. To the north-west, the valley system continues, beyond the confluence of the two principal water-courses, far into Sudan. Only low ridges or hills and the occasional more conical isolated volcanic peak break the relatively flat topography of the valley floors. Exposed volcanic plugs, which now take the form of rocky kopjes, are also a feature of the valleys.

The vegetation is predominantly savannah grassland with a sparse canopy of associated shrubs and low trees. The canopy is reduced as the valley sides rise gently towards the foot of the fringing mountains and become more arid. In the Kidepo Valley, close to the principal stream courses, Borassus palms become an important feature of the landscape.

Within the valley systems there are almost no man-made structures intruding into the landscape. In the southern central area of the Narus Valley, the buildings and structures that comprise the domestic, administrative and maintenance area at Apoka occupy a considerable area. Close to the southern boundary of the park, the ruins at Katurum mark the location of an intended luxury hotel. The venture failed before it reached completion and was consequently never opened. The only other constructed features are the out-stations that are mainly derived from natural materials, and which become visible only at close range. In the Narus Valley, and to a much reduced extent in the Kidepo Valley, roads are a necessary feature and a limited intrusion into the landscape. They are unpaved and their surfaces are constructed only from locally-derived, natural materials.

The mountains which fringe the perimeter of the park are abrupt, steep and rocky. It is a volcanic landscape, made up of peaks, ridges and deep valleys. The vegetation of the lower slopes is a continuation of the valley grassland and scrub, becoming more dominated by trees with increasing altitude.

Close to the higher summit ridges and peaks arid montane forest dominates. This is a rare and declining habitat, and is under considerable threat of modification from fire. In the deeper valleys, where the limited moisture will be retained for longer periods and where wildfire penetrates less frequently, a more substantial forest canopy persists. These valleys are important as refuges for the flora and fauna of the forest, which is seriously under threat from the continued and too frequent burning. (Alexander et al. 2000)

Fig. 12.3 Kidepo National Park, Uganda

3.7 Bibliography

The bibliography should be the most important section in the description. It will contain a reference to all papers, reports, journals, books, etc. used during the preparation of the plan. For some sites it may be useful to separate the bibliography into two sections:

3.7.1 Publications with specific relevance to the site
3.7.2 General reference works

It should also contain references to, and provide the location for, all relevant or useful published and unpublished information about the site.

References

Alexander, M., Hellawell, T. C., Tillotson, I. and Wheeler, D. (2000). *Kidepo Valley National Park Management Plan*. Uganda Wildlife Authority, Kampala, Uganda.
Brasnett, N. V. (1953). *Planned Management of Forests*. George Allen and Unwin, London.
JNCC (2004b). *Handbook for Phase 1 Habitat Survey – A Technique for Environmental Audit*. Joint Nature Conservation Committee, Peterborough, UK.
NCC (1989). *Guidelines for the Selection of Biological SSSIs*. Nature Conservancy Council, Peterborough, UK.

Rodwell, J. (1991a). *British Plant Communities, Woodlands and Scrub* Vol 1. Cambridge University Press, Cambridge.

Rodwell, J. (1991b). *British Plant Communities, Mires and Heaths* Vol 2. Cambridge University Press, Cambridge.

Rodwell, J. (1992). *British Plant Communities, Grasslands and Montane Communities* Vol 3. Cambridge University Press, Cambridge.

Rodwell, J. (1995). *British Plant Communities, Aquatic Communities, Swamps and Tall-herb Fens* Vol 4. Cambridge University Press, Cambridge.

Rodwell, J. (2000). *British Plant Communities, Maritime Communities and Vegetation of Open Habitat* Vol 5. Cambridge University Press, Cambridge.

Chapter 13
Features and Evaluation

Abstract Nature conservation features can be a habitat, a community or a population. Other features of interest can include geological, archaeological and historical features. For most sites, the presence of conservation features will have been the basis of site acquisition, selection or designation. Feature assessment or evaluation is simply the means of identifying, or confirming, which of the features should become the focus for the remainder of the planning process. There are two different approaches to identifying or selecting the important features on a site. The traditional approach was to use the Nature Conservation Review criteria for identifying important features. This chapter recommends an alternative: selection based on the use of the previously recognised status (local, national and international) of a feature. In some ways, this may be regarded as a consensus approach because it takes account of as wide a range of opinion as possible. In an ideal world, where resources are plentiful, all the features would be given some attention in the plan. Unfortunately, in reality, there are rarely sufficient resources even to manage the most important features. Consequently, the planner may have to be selective and, for example, in an extreme case, restrict management to features of national and international status. There will always be a need to draw a line somewhere. Sometimes, there are conflicts between features. These can often be resolved by understanding the relationship between the different site features. In most cases, one feature will be regarded as more important than another. Nature conservation management can be about creating opportunities for wildlife on seriously damaged or degraded areas, where little of the original flora or fauna has survived. The type of vegetation (plant communities) that can occupy and thrive in an area will be dictated, initially, by a range of natural factors. Once we understand the outcomes that nature, with and without human influence, will allow on a site, we need to decide what we want.

M. Alexander, *Management Planning for Nature Conservation: A Theoretical Basis & Practical Guide*, DOI 10.1007/978-94-007-5116-3_13,

13.1 Features

Nature conservation features can be a habitat, a community or a population. Other features of interest can include geological, geomorphological, archaeological and historical features.

The various approaches to nature conservation management were discussed in Chapter 9. There are two important divisions:

- Management planning by enabling process. This is most often applied to natural or wilderness areas.
- Management planning by defining outcomes for specified features.

The second approach is the most appropriate for managing cultural landscapes, IUCN category IV areas, Ramsar sites, European Natura 2000 sites, and any nature reserves or areas managed mainly, or in part, to maintain defined features in a specified condition. The focus of any plan for these sites will be the nature conservation features. These can be a habitat, a community or a population. For example, a National Nature Reserve in North Wales contains the following features:

- Oak-birch bilberry woodland, *Quercus petraea-Betula pubescens-Dicranum majus* woodland
- Alder-nettle flood-plain woodland, *Alnus glutinosa-Urtica dioica* woodland
- Deer grass wet heath, *Scirpus cespitosus-Erica tetralix* wet heath
- Silver studded blue butterflies *Plebejus argus*
- Lesser horseshoe bats *Rhinolophus hipposideros*
- Small red damselflies *Ceriagrion tenellum*

The Identification of Wildlife Features

For most sites, the presence of conservation features will have been the basis of site acquisition, selection or designation. This means that at some time in the past the site will have been evaluated and the most important features identified. When preparing, or substantially revising, a management plan, the list of features should be reconsidered to ensure that they are still relevant. There may also be additions.

The selection of features is not difficult. Once any site manager understands what is meant by features they will immediately be able to provide a list for their site. It is not possible to manage a site, with or without a formal plan, without having a clear idea of what is important. There are a few exceptions to this. These are most likely to involve restoration areas with no current wildlife interest, but where there is an intention is to create something. Planning for this type of site will be discussed later in this chapter.

Fig. 13.1 Bluebell wood

The first step is to prepare a provisional list of the features that are considered, for whatever reason, to be important. Ideally, all the features on a site should be considered. However, in practice, it may be necessary to concentrate on a shortlist of the most important features. These should be obvious from the site description and any previous evaluations. It is probably wiser to include rather than omit features, though this will obviously incur the penalty of having to assess them. Incidentally, it is not uncommon on smaller sites for a single habitat feature to occupy the entire site and, if there are no particularly important species, for this to be the only feature.

A controversial, and sometimes troublesome, issue arises when dealing with vegetation as a feature. This involves the level of definition that we apply to the feature. There may be choices as to whether the feature is defined at sub-community, community or habitat level. For statutory sites, legislation can define communities at each of these levels, so there is sometimes no choice but to follow the dictates of law. On sites where the features have no statutory status, it is generally best to define the feature at the level most appropriate for management. For example, a woodland habitat can contain a patchwork of different woodland communities, each being the consequence of different natural factors. There are no fences or other barriers separating the communities, and each individual community could be recognised as a feature. However, because management cannot be specifically directed at any given community and the woodland can only be managed as a whole, the most sensible approach is to define the feature at habitat level.

When preparing the shortlist for further evaluation, begin by including all statutory features. Beyond this, there are no hard and fast guidelines. Ideally, perhaps all habitats or plant communities should be included, but some sites can hold very small and insignificant areas of vegetation that simply do not justify any attention.

Making decisions about species that have no formal protection and when there is no requirement to report on population changes can sometimes be a problem. Which species should be included as features and which omitted? The best way of dealing with this is to ask a simple question: will all the management requirements of this species be met through managing the habitats on the site? Species management is, in most cases, achieved through maintaining habitats at favourable status. If there are no specific management requirements other than those that are already included in the management of the habitats then the species need not be included as a feature.

Fig. 13.2 Robin

When species are recognised as features on a site the habitats that support them should also be treated as features, regardless of the independent status of the habitat. Once again, this is because, in most instances, species protection is about ensuring that there is sufficient habitat in a condition that meets the needs of the species. In some instances, a habitat that is not considered important can support a number of very important species. For example, Case Study 3 considers a European protected site, an Alpine SAC, which contains large areas of commercially managed forest. The forest is not recognised as having any particular independent conservation value. However, it contains several bird species (black woodpecker *Dryocopus martius*, capercaillie *Tetrao urogallus*, hazel grouse *Bonasa bonasia,* Tengmalm's owl *Aegolius funereus*, black grouse *Tetrao tetrix*, three-toed woodpecker *Picoides tridactylus*) each of which is a SAC feature for the site. Each of the species is dependent on the forest and each has specific and different requirements. Taken together, these requirements identify a range of conditions that should be maintained in the forest. Attempting to deal with forest management for each of several different bird features would lead to unnecessary and confusing repetition. This is avoided by recognising the forest as a feature and taking account of the species requirements when preparing the objective for the forest.

13.2 Evaluation

Feature assessment or evaluation is simply the means of identifying, or confirming, which of the short-listed features should become the focus for the remainder of the planning process. It is about asking a question of each provisional feature in turn: is this feature, in its own right or in association with other features, sufficiently important to be regarded as one of the prime reasons for maintaining the protected area? Given that the process is about asking a question, the conclusion must be an answer to that question. Far too often this section can evolve into a rambling, inconclusive description of the feature.

Each feature identified in the provisional list is considered in turn. This means that the evaluation process is repeated several times. This may sound complicated, but it is far easier than trying to deal with everything at once.

There are two different approaches to identifying or selecting the important features on a site:

i. Selection based on the use of the Nature Conservation Review criteria for identifying important features. This is a derivative of an approach developed in Britain to identify the most important nature conservation sites (Ratcliffe 1977).
ii. Selection based on the use of the previously recognised status (local, national and international) of a feature. In some ways, this may be regarded as a consensus approach because it takes account of as wide a range of opinion as possible.

There is no suggestion that either approach can be regarded as scientific or objective. At best, they are an amalgamation of scientific value judgements (which interpret the significance of the available scientific information) and social value judgements (which take account of society's preferences and aspirations).

13.2.1 NCR Criteria

The UK Nature Conservation Review (NCR) criteria (Ratcliffe 1977) have been recognised as the standard or conventional approach to identifying the important site features for almost three decades. Although I no longer advocate the use of *all* the NCR criteria, I include all of them in this section because each clearly has some merit, and some people will be familiar with their use and will wish to continue using this approach. It is also important to be aware of all approaches in order to make informed decisions as to which is most relevant in any given circumstances.

The first, and perhaps most important, point to bear in mind is that the NCR criteria were developed for 'the selection of biological *sites* of national importance to nature conservation in Britain'. They were *not* developed for the identification of individual features within a site.

Some individuals and organisations support the use of the NCR criteria because they believe that the use of the criteria represents a scientific or objective method for identifying the important features. For example, the Royal Society of Edinburgh published a document which contained the following: 'The Nature Conservation Review (NCR) provides scientific criteria for selection of the majority of SSSIs and continues to be a sound and objective basis on which to designate SSSIs' (RSE 1998). A different view accepts that the evaluation process is a subjective expression of human preferences. This view is clearly supported in the text of The Nature Conservation Review. It states: 'A number of different criteria have, by general agreement and established practice, become accepted as a means of judging the nature conservation value of a defined area of land' (Ratcliffe 1977). Interestingly, it is difficult to locate any references to 'established practice' or any use of these criteria prior to the publication of the NCR. The criteria were designed to assess comparative site quality. That is, each site was compared with others of a similar type. When it comes to making decisions and choosing the key sites for inclusion in the national series the author makes the following statement: 'This is the most difficult part of the whole process to rationalise satisfactorily, since it is essentially subjective, even when based on a consensus view'.

The NCR criteria were first identified as a means of evaluating features in a management planning guide prepared for internal use by the Nature Conservancy Council (NCC 1981). There is no discussion or explanation to support their use in this context. Later, in *A Handbook for the Preparation of Management Plans* (NCC 1983), the NCR criteria are given considerable attention. Unfortunately, the text is confusing. The handbook recommends that the criteria should be applied to the site and its components (features). However, the text is a précis of the relevant sections taken from the NCR where the criteria are clearly intended for site selection and *not* the selection of individual features. A revised version of the guide was published in 1988: *Site Management Plans for Nature Conservation a working guide* (NCC 1988). This is even more confusing. It lists the criteria as a means of evaluation but follows this with a table that implies a different method of evaluation. (This is not dissimilar to the tables headed 'Identification/confirmation of the important features' which are introduced later in this chapter.) Despite their origins and the complete lack of any rationale for their use for identifying features, the NCR criteria became the method for selecting the important site features.

The use of the criteria spread very rapidly. By 1991 they appeared in the French planning guide, *Plans de gestion des reserves naturelles* (Ministere charge de l'Environmement 1991). In 1993, the RSPB in the UK adopted the criteria and also made it clear that the criteria should be applied to individual features (RSPB 1993), but in 2003 they abandoned the use of the criteria and adopted an approach which is very similar to that recommended in following section of this book (RSPB 2003). Management planning handbooks prepared by the Countryside Council for Wales (Alexander 1994 and 1996) also used the criteria. 1998 saw the publication in Estonia of *Guidelines for Development of Management Plans for Protected Areas* (Kaljuste 1998). Once again, the criteria are listed, without explanation, and with the interesting addition of cultural, religious, education, social and scientific values to the original NCR list. There was further confusion in 1999 when Eurosite published their *Toolkit for Management Planning* (Eurosite 1999). In this guide they claimed that 'experience throughout Europe[1] had shown that it is easier to define objectives for each site through a systematic consideration of the criteria'. However, the specific guidance given for each feature mixes feature and site evaluation.

It seems that the NCR criteria found their way into management planning guidance without any discussion or justification. At no time since they first appeared has anyone published anything that defends, questions or challenges their suitability in this context. They may simply have become self-perpetuating: something that people do because that is the way it has been done in the past.

Thus, the NCR criteria should, perhaps, only be used in a comparative or relative sense, and the process recognised as largely, if not entirely, subjective. The criteria were designed for one purpose: site selection; and used for another: feature evaluation. The most significant point is that, in nearly all circumstances, the important features on a site will have been previously recognised, often as a product of several different national or international evaluation or selection processes. Even where there are no formally recognised features on a site there may be sufficient general information (Red Data Books, for example) to guide the selection of features. There is only a purpose in applying these criteria when other information is not available, or perhaps to structure a discussion in support of other information.

For full details on how the criteria are applied to *site* selection read the NCR (Ratcliffe 1977). The following describes how the criteria might be used for feature selection.

The list contained in the NCR is:

Size
Diversity
Naturalness
Rarity
Fragility

[1] The Eurosite Toolkit did not provide any references to support the claim of experience throughout Europe.

Typicalness
Recorded history
Position in an ecological/geographical unit
Potential value
Intrinsic appeal

The list of criteria should not be regarded as fully comprehensive, and there is no suggestion that they will all be appropriate for all features on all sites. Only the criteria considered relevant and useful should be included, and additional criteria may be added if necessary.

It is important that the planner is not blinkered or constrained by the criteria. They are intended to stimulate, and even liberate, the thought process. The criteria are best regarded as a series of prompts that guide the individual through a structured discussion towards a conclusion about the status of a feature.

The criteria often overlap or are interdependent. For example, it is difficult to discuss fragility without considering rarity. Fragile sites are, by their very nature, rare sites.

The criteria should always be regarded as having positive, as well as negative, aspects. For example, high levels of biological diversity are usually valued on nature conservation sites, but, occasionally, high diversity can be the result of human intervention in a habitat that is naturally species-poor.

In their original context, the criteria were used in a comparative sense, that is, by comparing the quality of one site with others. Even when used in this context, it was recognised that because the judgements are comparative they are also relative. These are not absolute values. Given that many of the criteria merge or overlap, and that some of the values are contradictory, it is not possible to score the feature against each category, and it is absurd to assume that somehow totalling the scores will indicate anything. The criteria provide a focus for discussion, and no more.

The following notes are provided to aid the application of the various criteria. They are not intended as a comprehensive statement on their use. Once again, do not forget that the evaluation process is about asking questions and providing answers.

Size: In most cases, the importance of a feature will increase with size. However, size as a criterion must always be linked to other qualities. Small areas of high-quality habitat will often be more highly valued than large areas of low-quality habitat. Size is of particular importance where habitats are fragmented and populations isolated. The viability of small, and certainly small, isolated, features and sites is usually questionable. Very small populations are often extremely vulnerable and can become extinct simply through chance, even when appropriate management is applied. Some sites will contain a high proportion of, or even the entire, local, national or global population of a species. In these cases, regardless of how small the population, it may outweigh all other considerations.

Diversity: This criterion can be applied to physical, habitat, community and species diversity. There are clear relationships between each of these. Habitat diversity is dependent on the diversity of the physical environment. Different habitats contain different communities, and the number and variety of species varies from habitat to habitat.

The maintenance of biodiversity is usually regarded as one of the more important aims of nature conservation. This is largely because one of the most obvious and serious effects of human intervention on the environment has been the wholesale destruction of habitats and the extinction of species. Consequently, management is frequently carried out in order to maintain, or improve, site diversity. However, it must be recognised that there are occasions when high diversity is undesirable. For example, cut, over-drained, or otherwise modified, peat bogs will contain a greater diversity of communities and species than an intact, natural bog.

In general, naturally diverse habitats are highly valued. There are obvious and good reasons for this. However, there is some danger in ranking one natural habitat above another simply on the basis of the number of species that it contains. The obligation to maintain diversity is global in context, but there is no implication of responsibility to maximise diversity on any individual site, though there should be an obligation to obtain optimum diversity on all sites.

Naturalness: Natural is possibly the most important criterion, but it is a difficult concept since there is no widely accepted definition. Many will argue that natural is a state devoid of anthropogenic influence. But, when did people cease to be a natural component of their environment? Natural can be used in a relative sense. For example, it would not be unreasonable to claim that a natural sand dune is of greater value than a highly modified dune system covered with a commercial forestry plantation. Generally, but not always, the more natural a feature is the greater its nature conservation value will be. However, conservationists must also recognise that even highly modified habits can be extremely important for wildlife. Cultural or historic semi-natural habitats are also highly regarded. (See Chapter 9 for a more detailed discussion of naturalness.)

Rarity: This is the one aspect of nature conservation that has generally received most attention, and, as a consequence, we are usually aware of the most rare and endangered habitats and species on our sites. These will feature prominently in any management plan. Most often, it is the presence of rare habitats or species that leads us to selecting sites for nature conservation management. Rarity should not be a difficult criterion to apply: rely as far as possible on published, authoritative sources of information. Red Data Books (RDB) and national and international legislation and agreements are the best sources. It is essential that the difference between local, national and global rarity is appreciated. We should question the emphasis occasionally placed on conserving species in localities where they are rare as a consequence of natural factors when the species is abundant and secure within its natural range.

Fragility: To a greater or lesser extent all features demonstrate a degree of fragility. 'This criterion reflects the degree of sensitivity of habitats, communities and species to environmental change, and so involves a combination of intrinsic and extrinsic factors.' (Ratcliffe 1977)

Fragility should always be considered within a time scale. The degree to which the damage is permanent is a crucial consideration. Fragility is almost invariably linked to rarity: fragile features are, or soon become, rare. Thus, fragile features will often provide a focus for management. In other words, features considered fragile and rare will score highly in the evaluation process.

Do not always dismiss fragility as a negative factor. Many natural communities rely on disturbance for their survival. These, usually ephemeral, communities often occur during the early successional stages of dynamic habitats. The open communities in mobile sand dunes are a good example: if the community is stabilised they will be lost.

Species may also be fragile as a result of habitat change or destruction. Some have such specialised and complex requirements that a seemingly minor change can have devastating effects. Populations of some species may be naturally sturdy but have been, are, or may become, a specific target of human over-exploitation.

Typicalness: Sites are usually selected and valued because they contain the best example of a particular feature. The qualities that render a feature exceptional are most often the unusual or rare. It is also important that the typical and commonplace are not undervalued. This criterion is particularly useful for providing the justification for safeguarding the typical features in an area.

Recorded history: This criterion, although useful for selecting sites, has little if any value when applied to features. In its original context, the extent to which a site has been used for scientific study and research was considered important because the existence of a long-standing scientific record adds to the value of a site.

Position in an ecological/geographical unit: In its original context, this criterion was related to those of size and diversity. In the simplest sense, the value of a site increased if there was contiguity with others: in other words, sites which contribute to a larger ecological unit. This criterion is not particularly relevant when considering features within a site.

Potential value: Most features are, to a greater or lesser extent, imperfect. This criterion is used to assess the potential for improvement or restoration. Severely degraded features may have varying degrees of potential for improvement: some will have none at all, while others will have potential for total recovery given the appropriate management. The need to identify potential is crucial. There can be no justification for wasting resources in attempting to manage a degraded feature when the underlying reasons for the damage cannot be reversed.

Intrinsic appeal: 'There is finally the awkward philosophical point that different kinds of organisms do not rate equally in value because of bias in human interest, as regards numbers of people concerned' (Ratcliffe 1977). The NCR sought to address this problem and to ensure that less popular habitats and species were given adequate attention. The inclusion of this criterion in the original list is, therefore, confusing. If it is given equal weight to the other criteria it could undermine attempts to minimise the impact of human interest. This was, and remains, a difficult issue. Some people believe that intrinsic appeal is, in common with all the other criteria, an expression of human preferences and deserves equal weight.

13.2.2 *The Selection of Features Based on Previously Recognised Assessments*

Reminder: Feature assessment or evaluation is the process of identifying, or confirming, which of the previously short-listed features should become the focus for the remainder of the planning process. It is about asking a question of each provisional feature in turn: Is this feature, in its own right or in association with other features, sufficiently important to be regarded as one of the prime reasons for maintaining the protected area?

The individual features on a provisional list are each considered against a variety of different systems that have been previously used to define the status or importance of the feature. These evaluation systems can be international, national, local or organisational. For example, a Natura 2000 site will always contain some habitats or species of high European status. Given that a range of different evaluation systems may have been used to define the status of the feature, it is essential that each criterion is fully understood or defined by the planner. There are many different evaluation systems that can be included. The following examples are provided to give an indication of the different approaches that can be adopted.

Ideally, individual organisations should produce a list of criteria that meets their specific requirements and apply this to features on all their sites. This approach is described in an English Nature publication, *NNR Management Plans: a guide* (English Nature 2005). A similar approach was introduced by RSPB in UK (RSPB 2003). They used three categories for identifying important biological features:

- Features which are the prime reason for RSPB maintaining the reserve (RSPB Priority Features).
- Features for which RSPB have legal responsibilities and which will influence the management of the site.
- Features for which RSPB have legal responsibilities and which will *not* influence the management of the site.

Do not forget that the evaluation process is about *asking questions* and providing answers.

13.2.2.1 Criteria for Assessing Conservation Features:

Red Data Books: Is the feature included in any of the various Red Data Books for species or habitats? Many countries publish Red Data Books for specific groups of plants or animals. The species listed in the books are classified according to the perceived risk.

The International Union for Conservation of Nature and Natural Resources (IUCN) publishes and maintains a set of criteria with guidelines for classifying species at high risk of global extinction. They do not claim that the system is perfect since there is potential for over- or under-estimating risk (IUCN 2011).

The following are some of the red list categories as described in the guidelines. They are internationally recognised:

- Extinct (EX)
- Extinct in the Wild (EW)
- Endangered (EN)
- Vulnerable (VU)
- Near Threatened (NT)
- Least Concern (LC)

A full definition of each of these categories can be obtained from the IUCN website, and most individual RDBs describe the criteria used in their specific publication. Clearly, the higher the risk category of a species is the greater will be the concern and the justification for recognising it as a feature on a site.

International status: Does the feature have international status arising from, for example, CITES or Ramsar?

European legal status: Is the site a SAC or SPA? If yes, is the feature listed for the site? Or, if it is not a designated site, is this a feature listed in the Natura annexes? If the feature is listed for the site, it must be given full attention in the management plan.

National legal status: Does the feature have national status arising from local legislation? For example, UK domestic wildlife legislation provides for the designation of an important area as a Site of Special Scientific Interest (SSSI). On these sites, the features cited in the designation will be given full attention in the management plan.

Organisational values: Does the policy of the organisation responsible for the site identify any features as an internal priority? For example, an NGO responsible for bird protection may have prioritised lists of birds for protection.

Relationship with other features: A feature can be dependent on the presence of other features for its survival. The most important factor which influences the survival of all populations of species is the habitat, or habitats, which support it. Therefore, even when a habitat is generally not considered important enough to be recognised as a feature, it makes sense to treat it as a feature when it supports species that are a feature. For example, reed beds may be commonplace in an area and may not be considered important. However, if a site is notified because it contains a population of bitterns *Botaurus stellaris* which is entirely dependent on the reed bed, then the reed bed should also be included as a feature.

The feature from a wider perspective: Unfortunately, some site managers believe that their responsibility is to maximise biodiversity on their sites. Conservation management is about preventing, or at least minimising, loss of biodiversity. It is not about wanting everything everywhere: it is about ensuring that there is a place somewhere for everything. This means that we should have some means of ensuring that we focus on the features in our locality that are most important from a global or national perspective. We should not be too concerned about features that are better represented and protected elsewhere. This is not an easy section to deal with, and it is difficult to judge the significance of a feature unless we have some grasp of the wider perspective.

Ideally, the conservation of species and habitats would be based on strategies that take account of their conservation requirements within their entire range or extensive geographical region. Unfortunately, this is rarely happens, although there have been some good examples. There is little purpose in individual site managers trying to take account of the wider strategy if one does not exist. Nor is there any purpose in attempting to second-guess the outcome of a strategy. If there are no formal or published strategies, one possible alternative is to contact known experts (if there are any).

Aesthetic values (intrinsic appeal): (Note: intrinsic appeal and intrinsic value are different concepts.) Does the feature have aesthetic value? This was regarded as a difficult criterion when it appeared in the NCR. The issues remain unresolved and are a continuing area of debate. We could argue that nature conservation should be concerned with delivering something that the majority of us find appealing. If we allow ourselves to be over-influenced in this way, giving this criterion disproportionate attention, only those habitats or species that we find appealing would be given sufficient priority to ensure their protection. To some extent, this may already be the case. For example, in the UK The Royal Society for the Protection of Birds (RSPB) has 1,036,869 members, employs over 1,300 people and manages 182 nature reserves (for birds), covering 126,846 ha. By contrast, the British Arachnological Society (BAS) has 300 members, no employees and no reserves.

It is, perhaps, inevitable that species with an intrinsic appeal will gain an advantage. This may be acceptable from some points of view, but we must also understand that nature conservation is not simply about protecting the tiny minority of species that we happen to like.

Fig. 13.3 Yr Wyddfa and Glaslyn

Cultural values: Are there any cultural values associated with the feature? There are very many examples of plants or animals that have local, even regional, cultural significance. Many habitats or plant communities, such as hay meadows or heather moor, are important cultural artefacts. Coppice woodland is another good example. This habitat provides many obvious and well documented benefits for wildlife, but, in addition, through maintaining coppice woodland we also pay homage to our cultural heritage.

Landscape: Does the feature contribute to the wider landscape? This is particularly important in an area where the landscape is legally protected.

Other values: Different people will value features for a wide range of different, and sometimes apparently contradictory, reasons. The evaluation process can be extended to ensure that attention is given to a more comprehensive range of human values, and some of the NCR criteria, particularly naturalness, diversity, rarity and size, may be included. English Nature includes the following criteria as a means of providing an assessment of the current importance of a site: economic use, community interest, education and research (English Nature 2005). However, in addition to evaluating a site, these values could also be applied to individual features. There are so many different human values that could be included that it is not possible to provide a comprehensive list. My intention here is to highlight the fact that there is a diversity of values and that individuals should take some time to identify anything that may be relevant to the plan.

Fig. 13.4 Mawddach dawn

This section is best dealt with by preparing a table that lists the features and the range of criteria against which each will be considered. The following tables could be used:

The first example is the simplest or most basic approach. It is important that each criterion is clearly defined in the supporting text. There should be a definition of 'international', 'national' and 'local'(Table 13.1).

Table 13.1 A basic approach

Feature	International status	National status	Local status
Upland oak wood	/	/	/
Red squirrel		/	/
Song thrush			/

The following table is a more appropriate approach. It uses all the criteria that have been previously used to assess the site features. The following example is used by a UK conservation organisation (Table 13.2):

Table 13.2 A recommended approach

Feature	**RDB**	**International**	**European**	**National**	**UK BAP priority habitat/species**
Active raised bogs		Ramsar	SAC	SSSI	Yes
Saltmarsh communities			Marine SAC	SSSI	Yes
Sand dune communities				SSSI	Yes
Bryophyte assemblage of dunes				SSSI	Yes
Greenland white-fronted goose *Anser albifrons flavirostris*, wintering population	VU		SPA	SSSI	Yes
Otter *Lutra lutra*	NT		Marine SAC	SSSI	Yes
Red squirrel	NT			SSSI	Yes
Song thrush	LC				Yes

VU = Vulnerable, NT = Near threatened, LC = Least concern

In an ideal world, where resources were plentiful, all the features would be given some attention in the plan. Unfortunately, in reality, there are rarely sufficient resources even to manage the most important features. Consequently, the planner may have to be selective and, for example, in an extreme case, restrict management to features of national and international status. There will always be a need to draw a line somewhere.

13.2.3 Resolving Conflicts Between Features

Sometimes, there are conflicts between features. These can often be resolved by understanding the relationship between the different site features. A feature can have a considerable direct impact on another feature. For example, it is not impossible for both a predatory species and its prey to be features of equal standing. In extremely rare circumstances, one or other of the features must be sacrificed. An example of this is a site where a raised bog, an important habitat feature, is being restored. Unfortunately, the degraded bog provides a perfect habitat for nightjar *Caprimulgus europaeus*, which seek: '… dry open well drained habitats such as well spaced conifer woods, birch and poplar spinneys, scrub oak, heathery glades …' (Snow and Perrins 1998). As the bog recovers, the nightjar population will decline.

A feature can also have consequences for the management and the actual condition of another feature. This happens on sites where species have specific habitat requirements and both the species and the habitat are features. For example, in a northern forest there are two features: the forest itself and a population of grouse. The grouse require open areas for displaying males, high forest for nesting, and areas of dwarf willow for feeding hens prior to egg laying. These specific conditions will have to be reflected in the forest objective and, of course, the way in which the forest is managed. Thus, the grouse population is a factor that influences the way in which we manage the forest.

These conflicts are fortunately rare and can usually be accommodated in the planning process. In most cases, one feature will be regarded as more important than another.

13.2.4 Combining Features

Occasionally, there may be an advantage in combining several features and preparing a common objective. This will occur when features are not easily separated for monitoring or management purposes. Complex habitat mosaics, where each component qualifies as a feature, are good examples. Whenever it is expedient to combine features, include a detailed, well-considered justification in this section.

This also highlights the need to think ahead when confirming the features and the level at which they are defined. For example, vegetation can be defined at sub-community, community or habitat level. The level used to define features will usually determine the level at which their condition can be monitored.

13.2.5 Ranking or Prioritising Features

Ranking or prioritising features can be extremely difficult. Obviously, there will be no problem in ranking two features where one is of international importance and the other of limited local importance. Reasons for ranking could include situations where the safeguard of one feature threatens another, or when resources are so scarce that it is not possible to protect all the features. Under most circumstances, it is probably wise to regard all features that have survived a rigorous selection process as being equal.

13.2.6 Potential Features on Wildlife Creation Sites

Nature conservation management is sometimes about creating opportunities for wildlife on seriously damaged or degraded areas, where little of the original flora or fauna has survived. In very extreme situations, usually on brown-field or post-industrial sites and particularly following opencast or strip-mining, the land is scraped clean: any trace of buildings, waste tips, contaminated soils, etc. is removed. This usually also includes all references to our industrial heritage. The area is then re-profiled to produce a bland, featureless landscape, ready for something new.

Irrespective of how severely damaged a site may be, managers will invariably talk about habitat recreation. Recreation should mean that we are aware of something that once existed and that we intend to replicate whatever it was. A site will have been occupied by a succession of different habitats, including some where people had little, if any, influence, and there may also have been times when the site was occupied by highly modified farmland. Which of these past states should we choose for the future? In most cases, we do not have any reliable evidence to reveal what the past may have been. And, even when we think that we know what once existed, can we replicate the natural, social and economic climate that gave rise to that particular condition?

Many nature conservationists are setting aside the view that conservation should always be concerned with recreating or maintaining something that once occurred in the past and, instead, are beginning to manage places to optimise their future potential for wildlife. If we can break free from the past, these derelict sites could provide opportunities for new and creative conservation.

Traditionalist or otherwise, we will need to decide what we want for these sites. We can adopt one of two broad approaches. We could, if we had the courage, take the opportunity to experiment and allow nature to deliver whatever she can in these circumstances. Regrettably, most often, we will be obliged to adopt a more conventional approach, where the selection of biological features will follow some sort of landscape design which has been prepared to reflect the intended future use of the site. Future use could include anything from a nature reserve to an area for recreational activities. Providing there is some intention in the overall scheme to do something for wildlife then the preparation of a management plan that at least identifies the habitat features is justified.

So, how do we decide which habitats we want to occupy the site? First, we must understand that, if our intention is to create conditions that will improve or optimise opportunities for wildlife, there are limited possibilities. (Not everything can happen everywhere.)

> A plant community is an assemblage of plants with a distinct and unique composition and structure. The community contains species which coincide in space and time and have a shared and overlapping dependence on determining environmental factors, for example, climate, soil, biotic impacts and management. (Rodwell 2006)

The type of vegetation (plant communities) that can occupy and thrive in an area will be dictated, initially, by a range of natural factors. We have limited ability to change the natural factors and, even where we could, we need to consider the implications and cost. Perhaps this is one of the more significant differences between gardening and nature conservation. Gardeners will, to varying degrees, modify their land by controlling natural factors to provide artificial situations that support assemblages of exotic species which bear no relationship to the native flora. Nature conservationists, on the other hand, in general but perhaps not always, recognise and celebrate the limitations imposed by the natural factors that create the diversity of habitats. There are, of course, exceptions when intervention is necessary and desirable, for example, the application of lime to maintain particular kinds of hay meadow.

The following are some examples of natural factors that will influence the vegetation: altitude, slope, aspect, soil, geology, drainage, climate, grazing by wild animals and catastrophic events. We never start with a bare canvas: it has been primed and the background wash applied. The factors, in combination, will dictate the type or range of plant communities that *can* occupy an area. The specific communities that *will* occupy a place are a consequence of the combined influence of natural and controlled anthropogenic factors.

Once we understand the outcomes that nature, with and without human influence, will allow on a site, we need to decide what we want. There are no rules that can be applied here: within the realm of possibilities the choice is dictated mainly by human preference. However, decisions should always take account of the resource implications: we must strive to do the most for wildlife while using the least possible resource.

Each potential feature can, of course, be evaluated using one or other of the methods previously described, and the features likely to make the greatest contribution to wildlife can be selected as the future occupants of a site. A cost-benefit analysis could also be included.[2]

The following example, an extract from *British Plant Communities, Volume 3* (Rodwell 1992), illustrates the combined influences of natural factors and management. (Conservation management is invariably about controlling factors; control can mean removal, introduction, reduction or increase.) The feature is a particular kind of hay meadow found in England, the northern hay-meadow, *Anthoxanthum odoratum-Geranium sylvaticum* grassland. In addition to the following extract, the full original text provides considerably more information which describes and quantifies the conditions, including climate (temperature, wind, rainfall), altitude, slope, geology, geomorphology and soils, in the localities where this community can occur. It then describes the precise management that created and maintains this specific community.

> The Anthoxanthum-Geranium community is an upland grassland confined to areas where traditional treatment has been applied in a **harsh sub-montane climate.** It is most characteristic of **brown soils** on **level to moderate sloping sites** and is now almost entirely restricted to a few valley heads, between **200 and 400 m**, in northern England.
>
> The Anthoxanthum-Geranium community is essentially a hay-meadow and comprises part of the 'in-by' land of Pennine and Lakes hill farms. These valley fields are grazed in winter, mainly by sheep, except in very unfavourable weather when stock are kept indoors. In late April to early May, the meadows are shut up for hay and the stock, apart from animals in poor condition, transferred to the 'out-by' summer grazing on the open moorland. Mowing takes place generally in late July to early August though, in unfavourable seasons, it may be delayed as late as September. The aftermath is then grazed once more until the weather deteriorates. Traditionally, the meadows have been given a light dressing of farmyard manure after being shut up and it is this, together with liming, which has helped maintain the richness and diversity of the sub-community. (Rodwell 1992)

Note: The natural factors are shown in **bold** and the anthropogenic factors (human influences) are underlined.

---

[2]Cost-benefit analysis is a relatively simple technique for deciding whether to do something. As its name suggests, it involves simply adding up the value of the benefits of a course of action and subtracting the costs associated with it.

13.2.7 Summary Description of the Feature

Once the evaluation is complete and all the features have been identified, there is merit in preparing a succinct description of each feature (generally no more than one or two sentences). As with all sections of the plan, the description should be written in plain language. The purpose of the description is to provide the reader with a clear understanding of what the feature is. For common species this is obvious and easy, as most people will recognise a species from its name. However, some rare or obscure species that do not have common names will require a supporting explanation. Photographs can, of course, be included. Habitats and communities may be more demanding and require longer descriptions.

References

Alexander, M. (1994). *Management Planning Handbook.* Countryside Council for Wales, Bangor, Wales.

Alexander, M. (1996). *A Guide to the Production of Management Plans for Nature Conservation and Protected Areas*. Countryside Council for Wales, Bangor, Wales.

English Nature (2005). *NNR Management Plans: A Guide*. English Nature, Peterborough, UK.

Eurosite (1999). *Toolkit for Management Planning.* Eurosite, Tilburg, The Netherlands.

IUCN (2011). *The IUCN Red List of Threatened Species. Version 2011.2.* http://www.iucnredlist.org.

Kaljuste, T. (1998). *Guidelines for Development of Management Plans for Protected Areas.* Ministry of Environment of Estonia, Tallinn, Estonia.

Ministere Charge de l'Environmement (1991). *Plans de gestion des reserves Naturelles.* Direction de la Protection de la Nature, Paris.

NCC (1981). *A Handbook for the Preparation of Management Plans for National Nature Reserves in Wales.* Nature Conservancy Council, Wales, Bangor, UK.

NCC (1983). *A Handbook for the Preparation of Management Plans.* Nature Conservancy Council, Peterborough, UK.

NCC (1988). *Site Management Plans for Nature Conservation, A Working Guide.* Nature Conservancy Council, Peterborough, UK.

Ratcliffe, D. A. (ed.) (1977). *A Nature Conservation Review.* Cambridge University Press, Cambridge.

Rodwell, J. (1992). *British Plant Communities, Grasslands and Montane Communities*, Vol 3. Cambridge University Press, Cambridge.

Rodwell, J. (2006). Personal communication.

Royal Society of Edinburgh (1998). *People and Nature: A new approach to SSSI designations in Scotland.* Response to Scottish Office consultation on SSSIs, Edinburgh.

RSPB (1993). *RSPB Management Plan Guidance Notes,* version 3. Internal publication, The Royal Society for the Protection of Birds, Sandy, UK.

RSPB (2003). *RSPB management plan guidance notes* version 6. Internal publication, The Royal Society for the Protection of Birds, Sandy, UK.

Snow, D. W. and Perrins, C. M. (1998). *The Birds of the Western Palearctic, concise edition.* Oxford University Press, Oxford, UK.

Chapter 14
Factors

Abstract A factor is anything that has the potential to influence or change a feature, or to affect the way in which a feature is managed. These influences may exist, or have existed, at any time in the past, present or future. Factors can be natural or anthropogenic in origin, and they can be internal (on-site) or external (off-site). This chapter introduces the concept and application of factors in management planning. Factors will be revisited at several key stages in the planning process for each feature: the selection of attributes for features, the selection of performance indicators for features and the management rationale. To avoid unnecessary repetition, a master list of all the factors is prepared at an early stage in the plan. The list should contain all the factors that have affected, are affecting, or may in the future affect, any of the features on a site. Once a master list has been prepared, it can be used to ensure that all the relevant factors are considered for each feature. The management of habitats and species is nearly always about controlling factors, or taking remedial action following the impact of a factor. Control means the removal, maintenance, adjustment or application of factors, either directly or indirectly. Factors can have a positive or negative influence. Some factors, for example, invasive alien species, will always be negative. Others, such as grazing, can be positive or negative. Our ability to achieve conservation objectives will always be constrained by our ability to control factors. We can never be certain that we have identified all the factors, and we should not assume that we fully understand the implications of each factor. However, management planning is a process, and we can only react to what is known and understood at any given time. Time will reveal our errors and failures, and then we can take different actions.

M. Alexander, *Management Planning for Nature Conservation: A Theoretical Basis & Practical Guide*, DOI 10.1007/978-94-007-5116-3_14,

14.1 Background

N. V. Brasnett's book, *Planned Management of Forests* (Brasnett 1953), was clearly an important influence (if not always acknowledged) in early British planning guides. Although Brasnett does not specifically use the word 'factor', he wrote very clearly about the influence of factors on habitats:

> Any association of plants is a living entity, the result of the interaction of soil, climate, surrounding vegetation, and of any interference by man or other animals, as by cutting, burning, grazing, and so on. Vegetation is not static but changes gradually, as for instance when one association of species produces conditions favourable for the germination and development of the young of other species.

Some of the earlier references to factors in nature conservation management are contained in the University College London *Handbook for the Preparation of Management Plans for Nature Reserves* (Wood and Warren 1976, 1978). Apart from recognising the need to list the factors, the handbooks provide no guidance on how they should be used in the planning process.

The Nature Conservancy Council produced a *Handbook for the Preparation of Management Plans* (NCC 1983) for internal use. This was followed by a published version, *Site Management Plans for Nature Conservation* (NCC 1988). Both refer to 'factors influencing management'. The 1988 guide contains the following statement:

> The management of all sites is constrained by obligations, trends and outside influences which have to be identified, and their effects upon future site management recognised, before the operational objectives can be formulated. These factors may also be considered when preparing work programmes in Stage 3 of a plan.

(Stage 3 was the Prescription or Action Plan.)

It is interesting that, although the authors recognise that factors have two quite separate functions in a plan, they fail to provide any explanation of the statement: 'These factors may also be considered when preparing work programmes'. There is no reference to factors in the section on preparing work programmes. Some attention is given to the use of factors as influences, i.e. they are used to 'consider how the ideal objectives of management can be achieved, or if necessary modified'. Although the material is sketchy and ambiguous, it gives the impression that the ideal objective can be compromised or diminished by accepting the constraints imposed by factors. This is certainly the meaning assumed by the authors of countless plans which have been based on this guide.

The same approach was adopted by Eurosite (1996, 1999). There is no ambiguity in their definition of factors:

> In reality the manager will not be able to achieve all the ideal objectives because of a number of influencing factors. These are called constraints and modifiers.

In this approach the 'constraints' and 'modifiers' are applied to the ideal objectives which are *moderated* and then called 'operational objectives'. Factors are not used in any other sections of the plan.

The Countryside Council for Wales also published a planning guide in 1996 (Alexander 1996). This adopts a very different approach in which ideal and operational objectives are abandoned along with the idea that conservation objectives should be modified by factors. Factors are listed under seven headings, and in the rationale section of the plan they are applied to each feature in turn. The most important factors, i.e. those that have the potential to influence or change a feature, are identified. These factors would require surveillance or, where appropriate, operational limits are applied so that the factors can be monitored. Most significant of all, factors are used as the means of identifying the management required for the features.

The more recent CCW guide to management planning (Alexander 2003) and the CMS planning guide (Alexander 2005) both place significant emphasis on the relevance of factors to management planning. These guides:

- Recognise that factors are agents of change.
- Recognise the relationship between factors and the selection of the attributes[1] which demonstrate that change is taking place.
- Recognise that, if the limits within which a factor is considered acceptable can be identified, the factor with limits can be used as a performance indicator.
- Recognise that monitoring or surveillance projects must be established for all the important factors.
- Emphasise that conservation management is mainly about controlling factors and that, consequently, factors should be used as the focus for the identification of management activities. (A feature cannot be considered to be at Favourable Conservation Status unless the factors are under control.)

The USA approaches to management planning place considerable emphasis on controlling factors. The Limits of Acceptable Change (LAC) system (Stankey et al. 1984), used most widely in the USA, is a good example. The LAC process is based on the recognition that all recreational use of wilderness causes some impact (recreational use is a factor). The LAC process, in its original form, does not define objectives (goals or targets) and gives no indication of what the desired outcomes might be.

A more recent and highly developed approach to planning, which has the control of factors at its core, is described in *Measures of Success, Designing, Managing, and Monitoring Conservation and Development Projects* (Margoluis and Salafsky 1998). Although not specifically mentioned, their planning methodology appears to be rather more relevant to natural habitats or ecosystems, though many of the elements would be equally applicable to cultural landscapes. In their approach, factors are divided into three categories:

- **Direct threats**. Factors that immediately affect biodiversity (the target condition) or physically cause its destruction.
- **Indirect threats**. Factors that underlie or lead to the direct threats.

[1] An attribute is a characteristic of a feature that can be monitored to provide evidence about the condition of the feature.

– **Contributing factors**. Factors that are not classified as indirect or direct threats but that somehow affect the target condition. Opportunities are included in this category. (Opportunities are factors that potentially have a positive effect on your target condition.)

The planning process begins with the development of a 'conceptual model' or diagram which identifies the relationships between the conservation target, the direct threats and the indirect threats. This approach is also described in the *Wildlife Conservation Society Technical Manual 2* (WCS 2006a). Their model has two additional components, a 'goal' and 'intervention'. A goal is 'a visionary, relatively general, but brief statement of intent (e.g. "Conserve wildlife and their habitats over the long term")'.

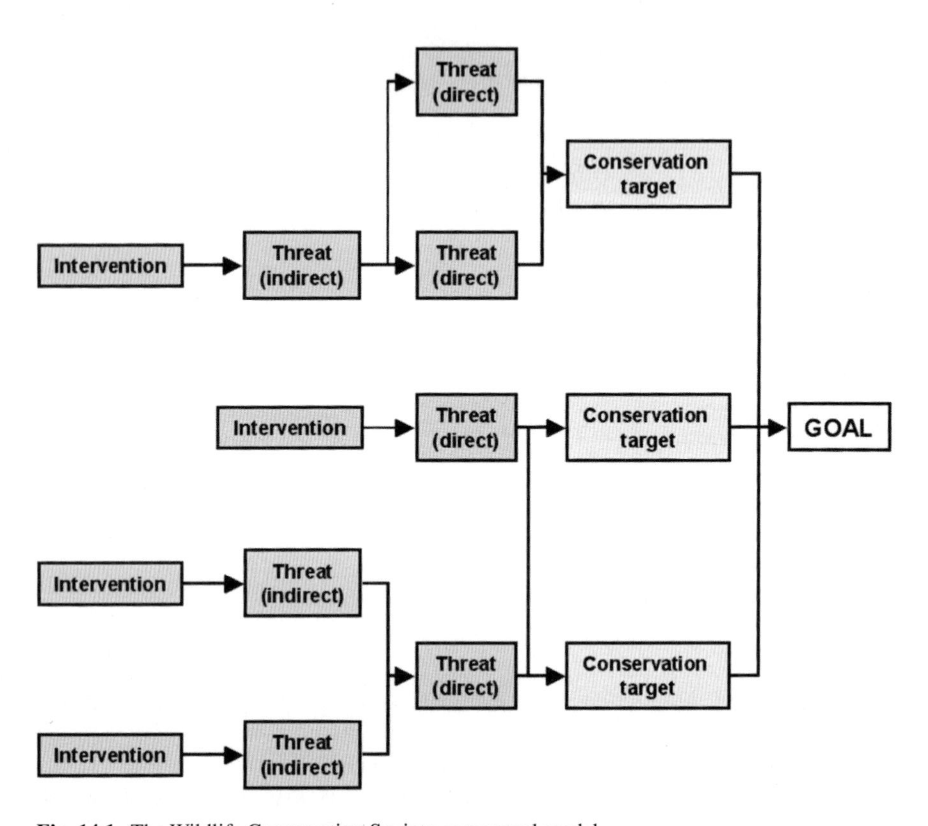

Fig. 14.1 The Wildlife Conservation Society conceptual model

This can be a useful planning tool. It helps to visualise or establish the sometimes complex relationships between factors and features.

Note: Any comparison of different approaches to planning is complicated because there is no standard language or universally agreed definitions. In this example, when the American words are translated into the language used in this book, 'goals' are 'policies', 'intervention' is 'management' and 'threats' are 'factors'.

14.2 Factors and Planning Defined Outcomes (Features Approach)

Factors are considered at several key stages in the planning process for each feature. These are:

The Selection of Attributes for Features

Quantified attributes are used as performance indicators to provide evidence about the condition of a feature. The selection of the attributes should, to some extent, be guided by the presence of factors. While factors are the influences that can change or maintain a feature, attributes reflect the changes that take place, or the conditions that prevail as a consequence of these influences. For example, in Britain many traditional permanent pastures and meadows are managed by taking a hay crop in the summer followed by light aftermath winter grazing. Typical of this type of vegetation are the knapweed meadows, *Cynosurus cristatus-Centaurea nigra* grassland. In this context, both **haymaking** and **winter grazing** are factors, and it is these, along with a number of natural factors, that are essential for the maintenance of this community. If other factors, the application of **chemical fertilisers** and a change from winter to **year-round grazing,** are introduced the grassland will change. The <u>diversity of the sward</u> will decline and <u>grasses</u> will increase; <u>ryegrass</u> Lolium perenne and <u>white clover</u> Trifolium repens will become abundant. If this inappropriate management continues, the original community will be replaced by ryegrass pasture, *Lolium perenne-Cynosurus cristatus* grassland. (Rodwell 1992)

(Note: **Factors** are shown in bold and <u>attributes</u> are underlined)

The Selection of Performance Indicators for Features

For a feature, habitat or species, to be at Favourable Conservation Status (FCS) the factors must be kept under control. This means that, for all features where the objective is FCS, all the factors that could change the feature must be monitored, directly or indirectly, to ensure that they are under control. Indirect monitoring is achieved by using attributes (see above). There are occasions, though not many, when the levels or limits of tolerance to a factor are known and where the impact of the factor is difficult to measure. In these cases, the factor can be monitored directly. These factors are always anthropogenic influences, and the most frequently encountered examples are invasive alien species. An upper limit, which represents our tolerance to the factor, is specified, and this provides the performance indicator.

The Management Rationale

The management rationale is the stage in the management process (repeated for each feature) where all the management requirements are identified. All the actions or projects required to ensure that the factors are kept under control are identified. The influences that all the other factors will have on the management of the feature are also considered. This can be complicated by the fact that, while an individual factor may have only a limited impact on a feature, several factors in combination can become a significant issue. This means that factors should be considered both individually and collectively.

14.3 Factors Can Be Positive or Negative

Factors are influences which can be negative or positive or both. Factors can change, becoming positive or negative depending on the intensity of their influence. For example: grazing is a factor; both over-grazing and under-grazing are negative influences; grazing at an appropriate level has a positive influence. A factor will also influence different features in different ways. For example, riverbank erosion may destroy grassland, but the same process will create and maintain shingle banks.

When factors are considered for individual features it is important that the influences, both positive and negative, are recognised. However, in the first instance, when preparing a master list of factors, there is no need to consider how a factor is likely to affect features on a site.

14.4 Types of Factors

There are many good reasons for specifying a standard range of headings and subheadings that can be used as an aide-memoire to help identify the wide range of factors that have potential to influence the management of features. There are many different ways in which the list of headings could be structured. An approach which works well in a wide variety of different situations begins with a small number of broadly defined main headings, each of which can, if necessary, be divided into any number of subheadings. The number of subdivisions will increase in proportion to the complexity of the site and, in particular, to the number of different factors affecting the features on a site. The main headings might be adequate without subdivision on small, simple sites, but on large, complex sites there may be a justification for several tiers of subheadings.

There are significant advantages in arranging the factors under at least four main headings. Some factors will appear in more than one location. For example, invasive species can be a factor whether they are internal or external.

- Internal anthropogenic factors
- Internal natural factors
- External anthropogenic factors
- External natural factors

14.4.1 Internal and External Factors

There is a need to distinguish between internal and external factors. This is mainly because internal factors are usually, though not always, controlled by direct on-site intervention, while, in contrast, external factors are rarely controlled through direct action. Occasional exceptions include the control of alien invasive species, for example, rhododendron *Rhododendron ponticum*. This species is controlled on land adjacent to National Nature Reserves in North Wales to prevent it from spreading and infesting the reserves. The indirect control of external factors is usually through influencing others, informally or formally, for example, by providing evidence when developments are planned.

Regardless of our ability to control external factors, we cannot ignore them. Where there is evidence to demonstrate that external factors are damaging a feature, and particularly when this happens on statutory sites, the evidence may be used to help justify political or legislative changes.

Fig. 14.2 External factor

External global factors, for example, climatic change, are extremely difficult to deal with. These all-pervading influences will probably have a greater effect on our ability to conserve wildlife than a combination of all the other factors. In circumstances where change is taking place and defensive measures are possible, there may be justification in taking action if the impact of global change can be delayed. There is, of course, a counter argument: if these changes are taking place, why not accept the inevitable? However, nature conservation should be about doing our best in any situation. At the very least, we must attempt to slow down the rate of environmental degradation and the consequential losses of habitat and species. By keeping options open for as long as possible we may provide some choices for the future.

14.4.2 Anthropogenic and Natural Factors

The division between natural and anthropogenic factors (human influences or the consequence of human activities) is also significant. This can be an extremely difficult division, both in practical terms and from a philosophical perspective. It is often impossible to differentiate between changes to a feature which are the consequence of natural processes and those which are a consequence of anthropogenic processes or a combination of the two. The more significant issue, which was discussed in Chapter 9, is the definition of 'natural' or the concept of 'natural' which excludes humanity as a component. Unfortunately, there is no consistency in the way in which 'natural' is defined or applied. All decisions concerning the objectives, i.e. what we want to achieve, are made at a different stage in the planning process, and that is when the consequences of conservation ethics should be considered. This is not the place to revisit that debate. However, the division between natural and anthropogenic is important because it will help to differentiate between factors which are regarded as having a positive influence and those which are considered negative.

The management of wilderness, and other situations where habitats are allowed to develop in response to natural factors, is usually concerned with controlling or removing anthropogenic influences. In contrast, the cultural landscapes of the Old World are most often managed by controlling natural processes or factors. For example, hay meadows are highly regarded semi-natural grassland communities, and yet the maintenance of these features is entirely dependent on our ability to suppress natural processes. A combination of mowing, grazing and fertilising prevents the regeneration of scrub and maintains soil fertility. If we can accept that 'natural' can be used in relative terms (that is, some features will be more natural or subject to less human influence than others), then at the natural end of the spectrum human influence will be mainly negative and natural influences mainly positive. The converse is also true. When managing cultural habitats some human influence will be positive while natural processes can be negative.

14.4.3 Features as Factors

Some of the most important natural factors encountered on a site will be the features. This is because each individual feature has the potential to influence the management of the other features. The relationship between features is only very rarely a significant problem. Species and habitats coexist for good reasons and they are often interdependent. Whenever a species is recognised as an important feature on a site, the habitat that supports it will always be one of the most important factors. It is not unusual for both the species and the supporting habitat to be recognised as important features of the same site.

Complications will occasionally arise when a species feature and the habitat feature require conflicting management. This can happen when a habitat that has been damaged is recovering. For example, the restoration of a raised bog (habitat feature) will have a negative impact on a nightjar *Caprimulgus europaeus* population (species feature). The nightjars thrive in the previously degraded bog, but the restored bog will not provide suitable habitat. In situations where the requirements of a species are in conflict with the habitat that supports it, the first step is to decide which is the most important. If it is the species, the condition of the habitat may have to be compromised, and, of course, the opposite action will be taken when the habitat is more important.

This issue will occasionally emerge when dealing with species that are dependent on successional or ephemeral habitats. When planning the management of these, it is essential that the relationship between the species and the successional communities is understood. For example, on a large dune system the dune slack communities provide an ideal habitat for many species of orchid *Orchidaceae*. The orchids are a feature and so is the dune habitat (Fig. 14.3). Mobile dunes continually inundate the orchid slacks and, in order to protect the orchids, managers may attempt to stabilise the dunes. However, dune slacks are created by mobile dunes and blowouts, and they must be regarded as ephemeral communities. In time, even in the absence of mobile dunes, a slack will change and eventually no longer support orchids. Management must recognise that, over time, these communities will move within the site or, in some cases, another site may become more suitable for them. It is important that they exist but not where they exist.

Fig 14.3 Morfa Harlech a mobile coastal dune system

There will, of course, also be some occasions, particularly when managing very rare and threatened species, where the habitat will have to be modified or maintained in an early successional state to meet the requirements of the species.

14.4.4 Anthropogenic Factors

14.4.4.1 Legislation and Policy

Legislation and policy are such important factors that they should be included as subheadings in all management plans. Legislation and policies are described and discussed at an early stage in the planning process. Both will have a very significant influence on the selection of the features and the development of objectives for the features. They also influence the management of the features and consequently should be regarded as factors.

Wildlife legislation, although intended to protect wildlife, can occasionally limit our ability to carry out management. Employment and health and safety legislation can also severely restrict our ability to manage sites.

14.4.4.2 Health and Safety

This is a subsection of legislation, but it is so important that it must not be overlooked in any plan. Almost every management activity will require a risk assessment and many will require expensive safety equipment and procedures. Health and safety considerations are not a direct or primary factor, but there will always be implications for the way in which management is carried out. In some extreme cases, it will not be possible to provide adequate safety measures and work will not be possible. This has led to sites being abandoned.

14.4.4.3 Obligations Non-legal

Obligations which have no legal basis can arise for a variety of reasons. Some are obvious, for example, the need to maintain good relationships with neighbours and the public in general. Most obligations of this type will arise from traditional uses and activities which, although there may be no legal basis for them, carry a moral obligation.

14.4.4.4 Owners and Occupiers

Many protected areas are owned or occupied by other people. It is essential that their interests are taken into account and, as far as possible, safeguarded. It is equally important that liaison and all other management activities relating to owners and occupiers are identified and included in the plan. Any attempt to complete this section without some level of communication with owners and occupiers will probably fail. Ideally, this section should contain a statement about their aspirations for the site: for example, they may wish to continue their present use or to increase utilisation. In some cases, it may only be possible to gain an indication of future intent based on their current and past practices. All that is really necessary at this stage is a decision to include or exclude owners/occupiers as factors. When dealing specifically with the individual features at the rationale stage the discussion should focus on the extent to which owner/occupier activities are compatible with managing the features or how they will influence our ability to manage the features.

14.4.4.5 Stakeholder and Public Interest

A stakeholder is any individual, group or community living within the influence of the site or likely to be affected by a management decision or action, and any individual, group or community likely to influence the management of the site. Stakeholder interests will usually have implications for site management. They cover a broad spectrum, ranging from the interests of the local individual or community to organised national, or even international, interest.

In ideal circumstances, a management plan should have a section and objective for stakeholder management. Stakeholder interests and involvement will vary enormously from site to site, and, obviously, the attention given to the subject should be appropriate to individual circumstances. Stakeholders should at least be considered as factors, even if they are not included anywhere else in the plan. They can have both a negative and a positive impact on site management. There are many sites where management would not be possible without the direct involvement of stakeholders, and there are some sites where stakeholder activities are a serious threat to features.

Strictly, visitors, tourists and people who use a site for leisure activities are stakeholders. However, given that providing for visitors is usually a specific and separate management activity, it is recommended that factors arising from this form of public use are dealt with separately.

14.4.4.6 Public Use – Access or Tourism

For many protected sites, public access or tourism and the provision of opportunities for leisure activities can be very important. Occasionally, it is the most important purpose of management. With few, if any, exceptions, people will have some impact on the site features. In other words, they, and more particularly their activities, are factors. All aspects of public use which are likely to impact, either directly or indirectly, on the features should be identified as factors.

Fig. 14.4 Public use

Fig. 14.5 Newport Wetlands Nature Reserve

14.4.4.7 Past Intervention/Land Use

The past human utilisation of a site will sometimes be the most important factor that influences management and the selection of attributes. It is not always necessary to have a precise understanding of past management, but it is important that the consequences are recognised. Obviously, it is not possible to change past management, but conservation is often remedial, i.e. management to make good damage which is the consequence of past activities. For example, peat was cut on many raised bogs in the past. Although the activity may have ceased, the impact, usually a lowering of the water table, will continue to threaten the bog. Remedial management is required to block all drainage channels and reinstate the water table. In addition, trees which may have become established on the drier surface will have to be cleared.

There are many examples where an appreciation of past human utilisation can help us understand why some features exist in an area. For example in parts of North Wales, there was a very important tanning industry (Fig. 14.6). Cattle hides were tanned in a solution obtained from the bark of oak trees. This helps us understand why, in this area, we have so many oak woods, mostly monocultures with little age variation in the canopy.

Sometimes past intervention is the most important positive factor on a site. This is nearly always the case when the features are plagioclimatic habitats or communities, for example, hay meadows and pastures. In these situations, the key to managing the future often lies in an understanding of the past.

Fig. 14.6 An ancient tannery

14.4.4.8 Physical Considerations/Constraints

These can be quite significant, for example, a site may be so remote and inaccessible that management is impossible. Sites on mountain slopes can be inaccessible to machinery. Sometimes, when managing a bog, for example, it is difficult or prohibitively expensive to carry out management works.

14.4.4.9 Resources

The availability of resources will obviously influence our ability to manage sites and features. Ideally, a management plan will be used as a bidding document. It sets out the objectives, along with a costed action plan. Senior managers or donor organisations would then decide on the level of resource that they would make available for management. If this were always the case, 'resources' would not be a significant factor.

Unfortunately, in most circumstances, this does not happen; conservation management is generally under-resourced. This does not diminish the use of a management plan as a bidding tool. In fact, quite the opposite is true. In a resource-deprived environment, extraordinary levels of care are required to ensure that everything is justified before resources are allocated. It is important that organisations or individuals responsible for managing sites are aware of the actual cost of management. They can then make decisions about limiting resources with a full understanding of the consequences of their decisions. In the first instance, resources should not be included as a factor. Management is identified according to need and is costed. If the required resources are not made available, resources are later applied as a factor and the proposed management activity is abandoned or modified. This will often mean that it takes longer to meet the objectives or that management is less efficient. Inadequate management resources usually lead to very expensive solutions.

14.4.4.10 Size & Connectivity

Connectivity is the re-establishment of linkages between isolated fragments of habitat. Although connectivity is an essential consideration for many plans, it is not easy to decide where it should be included in a plan. I have decided to include connectivity as a factor because, if it is an issue on a site, it will have a significant impact on the management of the features. It will also have some relevance to the selection of features.

George Peterken (1996) does not suggest that the application of habitat variety (Williams 1964) and equilibrium hypotheses (MacArthur and Wilson 1967) to British woodlands has been proven. He does, however, suggest that they will help us to understand the management of woodlands. It is tempting to suggest that if these hypotheses might apply to British woodland they might also apply to other fragmented and isolated habitats. The simplest interpretation is:

- Larger areas will have greater diversity of species than smaller areas (everything else being equal).
- The number of species should eventually reach a steady state due to the balance between immigration and extinction.
- The greater the isolation the lower the likelihood that an area will acquire new species.

David Quammen (1996) employs a very powerful metaphor to illuminate the consequences of fragmentation and isolation.

> Let's start by imagining a fine Persian carpet and a hunting knife. The carpet is twelve feet by eighteen. That gives us 216 square feet of continuous woven material. When we're finished cutting, we measure the individual pieces, total them up – and find that there's still nearly 216 square feet of recognizably carpet-like stuff. But what does it amount to? Have we got thirty-six nice Persian throw rugs? No. All we're left with is three dozen ragged fragments, each one worthless and commencing to come apart. … An ecosystem is a tapestry of species and relationships. Chop away a section, isolate that section, and there arises a problem of unravelling.

The size of an area of habitat and its degree of isolation will obviously have significant implications for sustainability. These factors will influence the management of a habitat. For example, they may lead to the development of an acquisition strategy if the area of the habitat within a site is considered too small to be viable or to the establishment of corridors or linkages with other areas of similar habitat when the site is isolated.

14.5 The Preparation of a Master List of Factors

Factors are considered for each feature at several key stages in the planning process. However, an individual factor can have implications for many different features on a site; for some it will be a positive influence, for others negative. To avoid unnecessary repetition, a master list of all the factors is prepared at an early stage in the plan. The list should contain all the factors that have affected, are affecting or may in the future affect any of the features on a site. Once a master list has been prepared, it can be used to ensure that all the relevant factors are considered for each feature.

The creation of the master list of factors is similar to any other form of brainstorming. It is important that our thinking is free and not constrained. We can organise our thinking by dividing the list into various categories: this is to facilitate, but not constrain, the process.

The following table contains examples of the different factors that may be used at various times in the management plan. The main reason for including a list is that it provides planners with an aide-memoire: a list of prompts which will help to ensure that factors are not unintentionally omitted. In reality, it is virtually impossible to produce a list that covers everything and, given that there are so many different factors, the list would be very long and unwieldy. Some of the subheadings will be important on all sites, while some will rarely be encountered. The lists are not definitive and the factors could be categorised under a variety of different, but equally valid, headings (Table 14.1).

Fig. 14.7 Himalayan balsam – an invasive species in the UK

Table 14.1 Examples of factors

Examples of factors		
Main headings	Subheadings (examples)	Examples of factors for a coastal sand dune site
Internal anthropogenic factors	Owners'/occupiers' objectives Stakeholders Traditional legal rights, e.g. grazing, fishing, hunting Tenure Past land use/management (not conservation management) Cultural values, e.g. archaeological or historic monuments Tourism/access Recreational activities Illegal activities, e.g. hunting, off-road vehicles, fires, collecting Alien invasive species Pollution – airborne and waterborne Safety, e.g. old mine or quarry workings Lack of management expertise Grazing by uncontrolled domestic stock	The occupier grazes the site with cattle and sheep Rabbits Local wildfowlers legally use the site Military training area during second world war Low key tourist amenity beach Marine litter and beach cleaning activities Occasional off-road vehicles Alien invasive species: sea buckthorn, Japanese knotweed
Internal natural factors	Physical considerations/constraints, e.g. isolation, steeply sloping ground, micro-climate Geomorphological processes, e.g. sand deposition, riverbank erosion Water levels Safety, e.g. cliffs, bogs, animals Grazing by wild animals	Internal mobility and redistribution of sand in the dune system Natural succession of dune communities Dangerous currents in the estuary Dangerous sand cliffs
External anthropogenic factors	Stakeholders Tourism Recreational activities Alien invasive species Landscape considerations Sea level changes Global climate change (specific and known consequences) Agricultural practices Forestry	Dependence of some local stakeholders, particularly those reliant on tourism Large and popular golf course adjacent to the site Large, uncontrolled population of sea buckthorn in adjacent estuary Some hard coast engineering which may interrupt sand supply Commercial conifer plantation with potential to lower the water table Increasing sea levels as a consequence of global warming

(continued)

Table 14.1 (continued)

Examples of factors		
Main headings	Subheadings (examples)	Examples of factors for a coastal sand dune site
External natural factors	Geomorphological processes, e.g. longshore drift on coastal sites. Water supply/levels, e.g. river catchments outside site boundary.	The sand supply (glacial in origin and not an infinite resource)
Legislation	Health and safety legislation Access legislation Public liability Wildlife legislation	All management operations, including the use of vehicles, must be undertaken by trained and certificated personnel The Occupiers' Liability Act requires that all management infrastructure is safe and does not place any visitor at risk The site is a SAC, SPA, SSSI and National Nature Reserve
Policy	Access policy Stakeholder policy Wildlife policy	The management of the reserve is consistent with organisational policy

14.6 Primary and Secondary Factors

The division of factors into 'primary and 'secondary' is another useful, but not essential, process that can help to ensure that all the important factors have been identified. It can be used as the first stage in linking factors to attributes and factors to management. It must be stressed that the divisions are not absolute and, occasionally, a factor will span more than one division or can be placed in more than one division.

The master list of factors is the source of information used to populate the divisions, but there is a need to be selective. The master list is created by brainstorming. This will inevitably identify many factors that are not relevant to the planning process. These are mainly factors that have not had, and will not have, any impact on the feature. Much more significantly, brainstorming will usually identify a number of natural factors which are unchanging, for example, soils, aspect, slope and altitude: these underlying factors are the reasons why a particular plant community or habitat can exist in any given location. Clearly, they will only change in extreme circumstances, and so, for planning purposes, they can be set aside. The factors that are relevant to planning are those that meet the definition given at the beginning of this chapter: the factors that have the potential to influence or change a feature, or to affect the way in which a feature is managed. These influences may exist, or have existed, at any time in the past, present or future.

14.6.1 Primary Factors

Primary factors will always have a direct influence on a feature and, in common with many other factors, the influence can be positive or negative. Some will be changing, or have the potential to change, a feature. They will always require direct or indirect control, and they should be monitored, either directly or indirectly (the latter through monitoring attributes). Where they cannot be directly monitored, surveillance projects should be established. Examples of primary factors include grazing, invasive alien species, uncontrolled hunting, pollution, burning and offshore dredging.

There is another group of primary factors, those which, although no longer active, were at some time in the past responsible for changing a feature. It may not be possible to gain a complete understanding of what happened in the past, and this is not always important. It is the consequences of past factors - how they influence future management - that require attention. Nearly all the examples of this type of factor are past human intervention or management. For example, many woodlands were felled during the First or Second World Wars (felling is the factor). The consequences of this are: insufficient dead wood; a young and even-aged canopy structure; diminished species diversity in the canopy; and (in an oak wood) suppressed natural regeneration of oak. In other words, the status of the woodland is unfavourable. It is obviously not possible to monitor the factor, but the attributes (all underlined), which represent the changes that were a consequence of the factor, can be monitored.

14.6.2 Secondary Factors

Secondary factors have an indirect influence on a feature. They will have implications for our ability to manage features. For example, one of the primary factors which has a negative influence on an oak wood is the presence of beech *Fagus sylvatica* in the canopy. A nature reserve has, over many years, become so infested with beech that it dominates the canopy. The most efficient and effective way of removing the beech would be to fell all the trees in as short a time as possible. However, there is a secondary factor: the woodland reserve sits within an area where the landscape is legally protected. The landscape designation is concerned with maintaining woodland and does not differentiate between desirable and undesirable canopy species. The consequence is that the beech trees will have to be removed gradually over a very long period of time. The eventual outcome will be the same; it will just take longer to get there.

14.6.3 The Relationship Between Primary and Secondary Factors and Features

The Wildlife Society's conceptual model was introduced at the beginning of this chapter. The diagrams on the following pages are a modified version, which is more suited to demonstrating a features approach to planning.

The early sections of the management planning process described by Margoluis and Salafsky (1998) are based on the 'development of a conceptual model'. Modelling may be a useful planning tool, but there are serious limitations, mainly because the relationships between primary and secondary factors can be very complicated and difficult to illustrate. In addition, the relationship between the different features is difficult to display on a simple model. It is, as Margoluis & Salafsky claim, hard work, and it is sometimes quite difficult to construct the model. They provide the following guidance:

> The best way to arrange the factors and target condition in a diagram is to cut out small pieces of paper (self-sticking memo notes work very well for this) and write the factors and target conditions (Features) on them. Lay them out on a table or on the ground. It is best if you do this over a very big piece of paper so that you can draw the arrows that connect them together.

A model is perhaps a rather grand word for a rather simple diagram. A diagram can help to visualise or describe the relationships between factors, attributes and a feature. Attempts to include more than one feature will usually generate a hopelessly overcomplicated diagram. It is important that the model is regarded as provisional throughout the planning process. It can, and should, be revised whenever the need arises (Figs. 14.8 and 14.9).

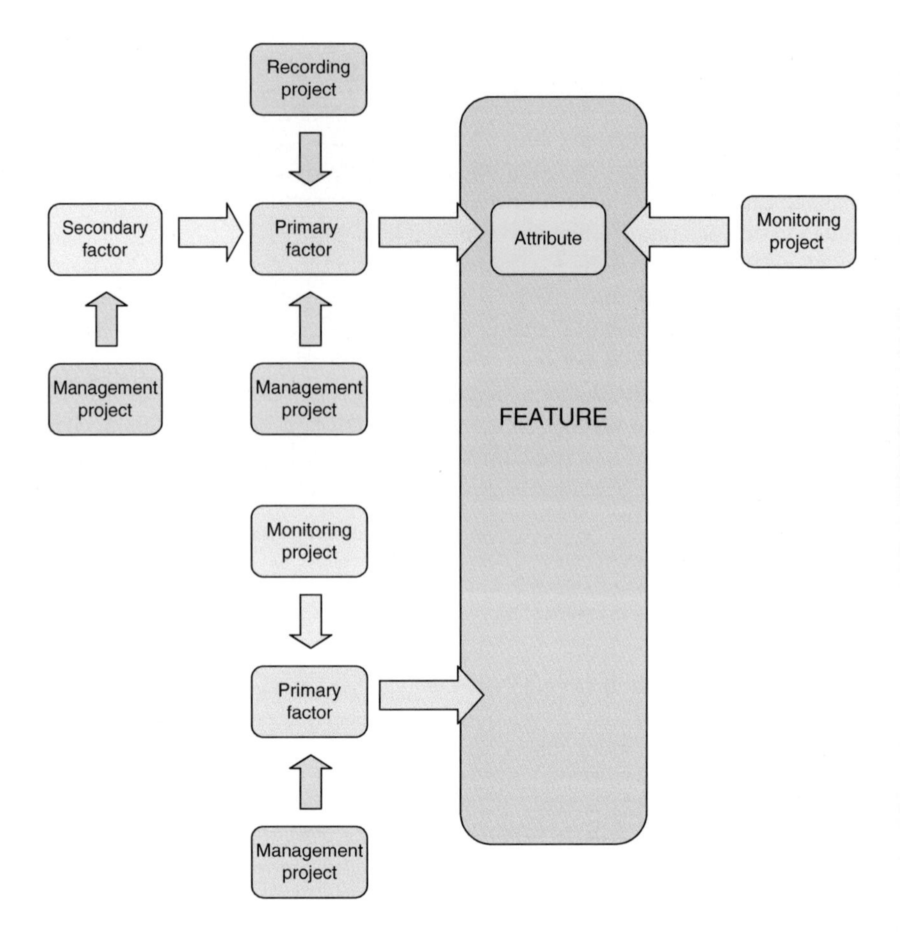

- **Primary factors** will have a direct influence on the feature: as the factor changes so will the condition of the feature.
- **Secondary factors** will not have a direct influence on the feature but will influence primary factors, or our ability to manage the primary factors.
- **An attribute** is a characteristic of a feature that can be monitored to provide evidence about the condition of the feature. Attributes change in response to the influence of a factor or factors.
- **Management projects** are the activities or interventions that are employed to control the factors. All management projects are recorded.
- **Monitoring** a feature can be *direct,* when an attribute is monitored, or *indirect*, when a factor is or monitored.
- **Recording projects** are generally applied to factors where the range of acceptable levels is unknown, i.e. they cannot be monitored.

Fig. 14.8 The relationship between factors, attributes, management, monitoring, surveillance & recording

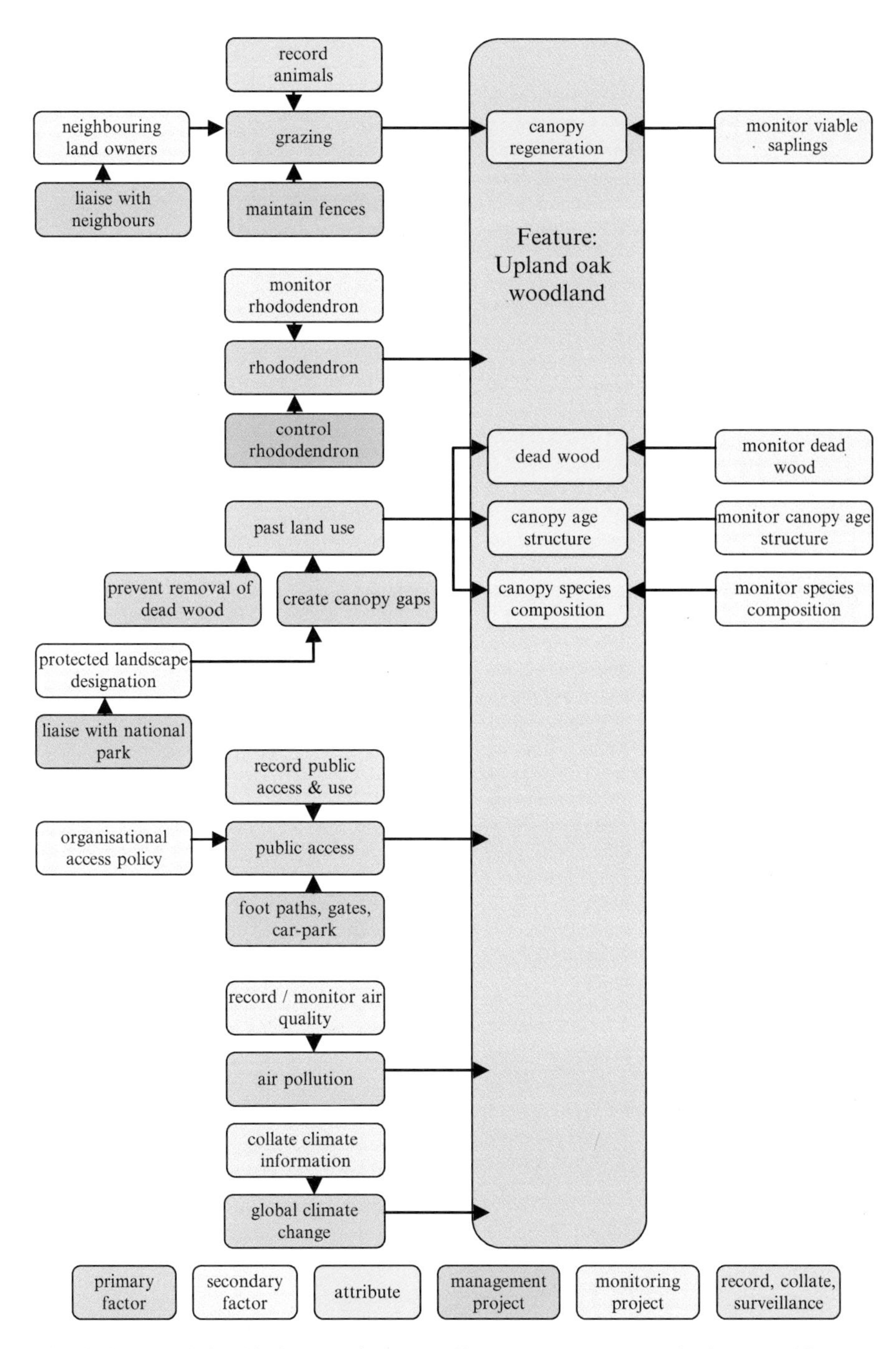

Fig. 14.9 The relationship between factors, attributes, management, monitoring, surveillance & recording in an upland oak woodland

References

Alexander, M. (1996). *A Guide to the Production of Management Plans for Nature Conservation and Protected Areas*. Countryside Council for Wales, Bangor, Wales.

Alexander, M. (2003). *CCW Guide to Management Planning for SSSIs, Natura 2000 Sites and Ramsar Sites*. Countryside Council for Wales, Bangor, Wales.

Alexander, M. (2005). *The CMS Guide to Management Planning*. The CMS Consortium, Talgarth, Wales.

Brasnett, N. V. (1953). *Planned Management of Forests*. George Allen and Unwin, London

Eurosite (1996). *Management Plans for Protected and Managed Natural and Semi-natural Areas*. Eurosite, Tilburg, The Netherlands.

Eurosite (1999). *Toolkit for Management Planning*. Eurosite, Tilburg, The Netherlands

MacArthur, R. H. and Wilson, E. O. (1967). *The Theory of Island Biogeography*. Princeton University Press Princeton, NJ, USA.

Margoluis, R. and Salafsky, N. (1998). *Measures of Success: Designing, Managing, and Monitoring Conservation and Development Projects*. Island Press, Washington DC.

NCC (1983). *A Handbook for the Preparation of Management Plans*. Nature Conservancy Council, Peterborough, UK.

NCC (1988). *Site Management Plans for Nature Conservation, A Working Guide*. Nature Conservancy Council, Peterborough, UK.

Peterken, G. F. (1996). *Natural Woodland, Ecology and Conservation in Northern Temperate Regions*. Cambridge University Press, Cambridge, UK.

Quammen, D. (1996). *The Song of the Dodo*. Pimlico, London, UK.

Rodwell, J. (1992). *British Plant Communities, Volume 3 Grasslands and Montane Communities*. Cambridge University Press, Cambridge.

Stankey, G. H., McCool, S. F. and Stokes, G.L. (1984). *Limits of acceptable change: A new framework for managing the Bob Marshall Wilderness*. Western Wildlands, 10(3), 33–37

Wildlife Conservation Society (2006a). *Living Landscapes Technical Manual 2 Creating Conceptual Models – A Tool for Thinking Strategically*, New York.

Williams, C. B. (1964). *Patterns in the Balance of Nature*. Academic Press, London.

Wood, J. B. and Warren, A. (ed.) (1976). *A Handbook for the Preparation of Management Plans – Conservation Course Format, Revision 1*. University College London, Discussion Papers in Conservation, No. 18. London.

Wood, J. B. and Warren, A. (ed.) (1978). *A Handbook for the Preparation of Management Plans – Conservation Course Format, Revision 2*. University College London, Discussion Papers in Conservation, No. 18. London.

Chapter 15
Objectives and Performance Indicators for Biological Features

Abstract Objectives should lie at the very heart of a management plan; they are the outcomes of management and the single most important component of any plan. An objective is the description of something that we want to achieve. Wildlife outcomes are habitats, communities or populations at a favourable status. Objectives must be quantified so that they can be monitored. This is quite a tall order: an objective is a multi-purpose statement that describes the required outcome of a feature (something that we want to achieve) using both plain and quantified scientific language. The solution is to prepare composite statements that combine a vision for the feature with quantified and measurable performance indicators. A number of performance indicators can be used to quantify the objective and provide the evidence that a feature is in a favourable condition or otherwise. Two different kinds of performance indicators are used to monitor an objective. Specified limits define the degree to which the value of a performance indicator is allowed to fluctuate without creating any cause for concern. In many ways, specified limits can be regarded as limits of confidence. When the values of all the performance indicators fall within the specified limits, we can be confident that the feature is at Favourable Conservation Status.

15.1 Background

Many management plans fail to provide any indication of what the outcomes of management might be, and for the few that actually describe outcomes most of the definitions are so vague that they are almost meaningless. A variety of different words are used in management plans to describe the management outcomes, for example, goals, aims, targets, policies and objectives. Although the choice of words that are used is perhaps not very important, their meaning should be clear, but, regrettably, it rarely is. Often, the expression of objectives in a plan is stratified and hierarchical, with, for example, goals or aims leading to strategic objectives,

M. Alexander, *Management Planning for Nature Conservation: A Theoretical Basis & Practical Guide*, DOI 10.1007/978-94-007-5116-3_15,

followed by operational objectives. These structures can be very complicated and difficult to understand. They may disguise the fact that the plan is an inconclusive arrangement of facts and figures, with lists of actions but no clearly defined outcomes.

In 1976, the University College London (UCL), in collaboration with Nature Conservancy Council (NCC) staff, produced a *Handbook for the Preparation of Management Plans for Nature Reserves – Revision 1* (Wood and Warren 1976). This was followed in 1978 by a more detailed *Revision 2* (Wood and Warren 1978). These handbooks were the progenitors of most UK, and many European, planning guides (NCC 1983; Ministere charge de l'Environmement 1991; RSPB 1993; Kaljuste 1998; Eurosite 1999). The UCL handbook introduced the concept of a two-stage objective:

> Long-term objectives relating to the maintenance of the overall biological and physiographic importance of the reserve and of its principal features. Such objectives may not be achievable in the period of operation of the current plan. … They should be major considerations that will stand for at least five years, and could in some cases be made without consideration of constraints: i.e. an ideal state.
>
> Short-term objectives should be those that are attainable over a time span of 3–5 years. They may relate to specific parts of the reserve.

Although not entirely clear from the text, the implication (mainly because of an example of an objective given in the handbook) is that a long-term objective describes the desired state of a site. The short-term objectives are, in fact, outline management prescriptions. The emphasis in the guide is on the preparation of an action plan and the grouping of projects by aim. An 'aim' could be for obligatory management, public access, survey, habitats in which the features occur, or a single important species. For some reason, habitats are not considered as features in their own right.

The UCL approach was further developed in *A Handbook for the Preparation of Management Plans* (NCC 1983). In this version, the two-stage process begins with an ideal objective and then moves on to take account of the impact of constraints, trends and influences (factors). It formulates operational objectives, described as: 'a statement of the intended management of a site'. The only definition of an ideal objective is: 'This section provides a statement of the ideal objectives of management considered necessary to protect the special nature conservation interest of the site.' This is followed by two examples: 'to remove sheep grazing from the core of the mountain massif, to allow seral succession to take place'; 'to completely eradicate rhododendron from a large woodland site and the surrounding countryside'. Therefore, in this context, an ideal objective is no longer an indication of the 'desired state' (Wood and Warren 1976, 1978) but is, in fact, a management activity. The NCC definition of an ideal objective might imply that there is an intention to 'protect the special nature conservation interest', and we can assume from the examples of objectives that the actions are intended to achieve an outcome of some kind. However, there is nothing in the guide that recognises, or even implies, a need to describe the desired outcomes for the site or features. The use of the word 'objective' to mean 'activity' is confirmed in the definition of operational objectives: 'The operational objectives are a statement of the intended management of the site and can be seen as the strategic framework for the management of the resource…'

The NCC handbook (1983) was followed by a published version *Site Management Plans for Nature Conservation* (NCC 1988). This retains the two-stage approach to objectives. The definitions of 'ideal objective' and 'operational objective' are much the same as the original handbook, but the examples provided to illustrate the definition suggest something quite different. One example of an ideal objective is: 'Management of the heath to ensure a better spread of age classes within the site, both to increase diversity and increase suitable habitat for marsh gentian *G. pneumonanthe,* silver studded blue *Pargus,* bog bush cricket *M. brachyptera* and chough *P. pyrrhocorax.*' This is quite confusing. The definition of an objective clearly implies that it is something that describes a management activity, but the example conflicts with the definition since it describes an incomplete, but nevertheless recognisable, outcome.

Between 1992 and 1994 the Countryside Council for Wales[1] revised their approach to management planning. In 1994 they published a management planning handbook (Alexander 1994). Although this was originally intended for internal use, the handbook became widely used in Britain and elsewhere. This handbook contains the following: 'Objectives should not be prescriptive. They should be a statement of purpose and not method.' This guide maintains the division between ideal and operational objectives, and gives examples where ideal objectives are modified as a consequence of anthropogenic factors to become diminished operational objectives.

By 1996, a single objective had replaced the two-tier approach (Alexander 1996). This was in part a reaction to the excessive compromise implied in the two tier approach, but was mainly influenced by a report commissioned in Britain by JNCC: *Common standards for monitoring SSSIs* (Rowell 1993). This change was possibly one of the most important advances in the development of UK conservation objectives. The new definition stated that, 'An objective is a composite statement which describes, by defining the target condition of each of the selected attributes, the overall favourable condition of a feature' (Alexander 1996). In retrospect, this was a step too far. It meant that an objective for a feature was reduced to a quantified list of the few attributes that could be measured. This may have been understood by those with a specific expertise in the feature but generally failed to address a wider audience.

In 2000, the Conservation Management System (CMS) Consortium published the *CMS Management Planning Guide for Nature Reserves and Protected Areas* (Alexander 2000). This document combines a revised version of the preceding 1996 definition of an objective with a vision for the feature written in plain language. This makes the objective accessible to a wide audience and, more importantly, aligns the objective with the European definition of Favourable Conservation Status (FCS).

Elsewhere, the picture was, and remains, confusing. Many plans contain excellent site descriptions and comprehensive lists and descriptions of the management actions,

[1] The Countryside Council for Wales was the successor in Wales of the Nature Conservancy Council.

but the critical information, i.e. what they want to achieve, is missing. Where an indication of outcomes is given, it is usually vague or ill-defined and is certainly not recognisable or measurable.

The Limits of Acceptable Change planning system, widely used in the USA, is built on 11 principles. The first is, 'appropriate management depends upon objectives' (McCool 1989). However, 'objective' appears to have a very different meaning:

> Management objectives provide an answer to the question of how much change is acceptable by deciding what types of recreation experience a particular recreation area should provide, the feel of naturalness of environmental conditions, the kind of experience offered, and the intensity of management practices. (Manning 1986)

This system was designed to manage recreation in wilderness areas and so, in this context, 'objectives' define the recreation outcomes and not wildlife outcomes.

Remaining in the USA, the publication *Measures of Success* (Margoluis and Salafsky 1998) is a guide to planning conservation projects. The plan structure applies a tiered approach to defining outcomes. It begins with a goal, with associated objectives and activities. The book is based on four different planning scenarios, including a savannah ecosystem. The following example demonstrates their approach:

> **GOAL**: To conserve the grassland and savannah ecosystems of Kalimara National Park.
>
> **OBJECTIVE**: To reduce by 90% incidents of illegal hunting inside that park and the wildlife management areas by the end of the project.
>
> ***Activity 1*** Hold meetings with local trophy game hunting operators to clarify park boundaries, hunting restrictions and penalties for unlawful hunting.
>
> ***Activity 2*** Provide National Park Service guards to accompany and monitor all trophy game hunts.
>
> ***Activity 3*** Hold community meetings to discuss hunting restrictions in the park and penalties for unlawful hunting.
>
> ***Activity 4*** Work with local community leaders to develop a community based self policing system for monitoring illegal hunting in the park.

This goal, with its associated objective and activities, does not stand alone. Monitoring strategies linked to the goal are developed later in the plan:

> **Monitoring strategy 1** Compare the number of elephants and rhinos killed before and after project interventions.
>
> **Monitoring strategy 2** Measure the change in the number of encounters with hunters inside the park over time.
>
> **Monitoring strategy 3** Compare the levels of illegal hunting by trophy operators in Kalimara National park to a neighbouring park.

The approach in the above example does not give an adequate indication of what the outcomes might be. The goal is: 'To conserve the grassland and savannah ecosystems'. What does conserve mean, and what will the savannah look like when the ecosystem has been conserved? The associated objective, which we can only assume is one of many for this goal, is, in fact, a management action: the reduction of illegal hunting.

The activities are things that they plan to do to reduce hunting. Two of the three monitoring strategies deal with measuring the effectiveness of the management activity in reducing hunting. The remaining monitoring strategy measures changes in the number of elephants and rhinos killed before and after management controls, but it does not give any indication of how many elephants or rhinos they want in the park.

It would be unreasonable to be overly critical of the approach to planning presented by Margoluis and Salafsky since examples given in their guide are hypothetical and, I assume, simplified to illustrate the issues.

In sharp contrast to most USA publications, *Monitoring Plant and Animal Populations* (Elzinga et al. 2001) recommends an approach to setting objectives which is very similar to the approach introduced by Alexander (1996). Their objectives describe the desired condition of the resource (resource being a habitat or population, in other words, a feature). Attributes are used to quantify and describe the qualities required of a resource. These objectives differ from the approach advocated in this book in two significant ways. They include: 'Action; the verb of your objective (e.g., increase, decrease, maintain)', and 'Time frame; the time needed for the management strategy to prove effective'. These issues are discussed later in this chapter.

The following examples of objectives are taken from published management plans that have some form of official approval:

Tasmania Parks and Wildlife Service

The objectives of conservation areas are:

- To conserve natural biological diversity
- To conserve geological diversity
- To preserve the quality of water and protect catchments
  (Tasmania Parks and Wildlife Service 2003)

There is nothing in the plan that describes what 'conserve' might mean in this context.

Eagletail Mountains Wilderness Management Plan

Objective 1: Maintain and enhance the wilderness values of naturalness, outstanding opportunities for solitude and primitive recreation, and protect special features of the Eagletail Mountain Wilderness by:

- Rehabilitating the impacts of three closed vehicle tracks….
- Notifying the State of Arizona of federal Water Rights for any available water…
- Eliminating unauthorised motor vehicle access….
- Improving opportunities for recreation while preserving naturalness…
  (Arizona State Office 1995)

This is similar to the example from Margoluis and Salafsky, but this plan does not contain an explanation of what they mean by 'wilderness values of naturalness'.

Management Plan for the Wicklow Mountains National Park 2005–2009

> **Objective 1**: To maintain and where possible enhance the ecological value of all natural and semi-natural habitats and geological features within WMNP – blanket bog, heath, lakes and rivers, woodlands, exposed rock, grasslands and scrub.
>
> (National Parks & Wildlife Service 2005)

What is 'ecological value', and when will they know that it has been enhanced?

The 'maintain and enhance' approach is found everywhere and, unfortunately, represents the norm in management planning. So often, planners avoid deciding and describing what they want to achieve. A further example, which illustrates this weakness, is taken from a management plan which is provided as an exemplar in the *Eurosite Toolkit for Management Planning* (Eurosite 1999): 'To maintain and enhance the diversity of all natural and semi natural habitats.' What does this mean? If the objective was simply to maintain the diversity, then we might conclude that whatever condition prevailed at the time when the objective was written should be maintained. But it also indicates that diversity should be enhanced, which implies that the present state is not acceptable. There is nothing in the objective, or in the complete plan from which it comes, that provides any indication of what the current or enhanced state may be. Since the objective is open to endless interpretation, how will it be possible to know when the objective has been achieved? Even where an objective is to maintain something in its present condition, that condition must be described.

A Eurosite publication, *Management Planning for Protected Areas, a guide for practitioners and their bosses* (Idle and Bines 2005), unequivocally advocates a return to planning by specifying activities. They dismiss the approach to setting objectives contained in the earlier Eurosite Toolkit and suggest that the traditional method is, 'often lengthy and time-consuming but also may not be easily understood by stakeholders…' The guide provides two examples of objectives; these are described as 'solutions':

1. To ensure that there is sufficient grazing to prevent invasion by scrub.
2. To ensure that water tables do not fall.

This planning system then moves directly to identifying the work programme. With the exception of the site description, which specifically mentions the features that make the natural area important, there are no indications in this guide that a management plan should contain any reference to the management outcomes.

15.2 Definition of an Objective

An objective is, or should be, the description of something that we want to achieve. These are the outcomes of management. Wildlife outcomes are habitats, communities or populations at a favourable status.

Objectives should always define management outcomes. It would appear that the majority of management plans, and even some recent planning guides, do not recognise a need for objectives (or anything else) that define the management outcome. There are, however, a few notable exceptions, including:

The IUCN Guide (Thomas and Middleton 2003), where the role of an objective is described as: 'statements of outcomes rather than how to achieve them'.

The Ramsar Guidelines for Planning (Ramsar 2002): 'An objective is a description of something that should be achieved…'

Monitoring Plant and Animal Populations (Elzinga et al. 2001): 'Value, amount of change, or trend that you are striving to achieve for particular population or indicator. Objectives may also set a limit on the extent of an undesirable change.'

My firm belief is that management plans must contain objectives, and that an objective must be a description of something that we want to achieve. Clearly, what we want will change with time and so will the objectives. The concept and consequence of using an adaptive approach is described in the section on adaptive planning in Chapter 6. Adaptive management is only possible if we know what we are trying to achieve. This is a fundamental component of any planning process applied to any area of human endeavour. Could an architect produce a plan for constructing a building if it described only the actions, and did not provide a detailed description of what the completed building would look like? It is only when we know what we are trying to achieve that we can determine whether or not our actions are appropriate.

15.2.1 Objectives Are Composite Statements

Objectives contain two basic components: a vision which describes in plain language the outcome or condition that we require for a feature, and performance indicators which are monitored to provide the evidence that will be used to determine whether the condition that we require is being met or otherwise.

15.2.2 SMART Objectives

There is a view that management plans should contain SMART objectives (Clarke and Mount 1998; Eurosite 1999). There are many variations and different definitions of SMART, but generally the acronym stands for:

Specific
Measurable

Achievable
Relevant
Time-based

The use of SMART objectives is widespread; they are commonly encountered in the business world and just about everywhere else. The definition of a SMART objective as applied to business, can, with modifications, be applied to wildlife objectives. The SMART definition is particularly relevant to objectives that define outcomes for habitats and species. I apologise to the author of the original definition of SMART for the many liberties that I have taken in modifying the definition for use in management planning for wildlife.

Specific

Objectives for features must specifically address the feature. An objective must be written to include each of the features identified as being important during the preceding evaluation.[2] Specific also implies that objectives should be clearly defined and should not be open to different interpretations. This is particularly important when preparing objectives for statutory sites. We live in a litigious society and objectives for these sites must be sufficiently robust and specific to stand up to legal challenge. Business objectives are often concerned with defining an activity; they can be prescriptive (about doing things). Wildlife objectives must not be prescriptive. This is one of the most important rules or tests that should be applied to an objective. Wildlife objectives define the condition required of a feature, and not the actions taken to obtain or maintain that condition.

Adaptive management is only possible when there is a clear distinction between what we want and what we do. Management actions are adapted or changed according to the condition of a feature. The management required to return a damaged feature to the required condition can be very different to the management necessary to maintain that condition. For example, when managing grassland that has become invaded by scrub, and where the objective is to maintain grassland, the scrub is initially cleared by cutting or burning, and this is followed by the application of heavy grazing pressure. Later, when the scrub has been successfully removed, lighter grazing can be introduced to maintain the grassland.

Measurable

If objectives are not measurable, how will we ever know that they are being achieved? Clearly, objectives for conservation features must be quantified and measurable. Adaptive management is only possible when a judgement can be made about management effectiveness. Management can be considered effective when an

[2] For Natura 2000 sites an objective must be written for each Natura feature.

outcome or objective is measured (monitored) and found to be in a favourable condition. Unfortunately, the objectives used in the majority of management plans are not measurable.

Achievable or Aspirational

Objectives should be achievable. This is business talk and in that context is perhaps an obvious statement. Can there be any purpose in pursuing unobtainable objectives? Many commercial or business definitions of SMART suggest that sufficient resources must be available for an objective to be considered achievable. However, this is not appropriate and should not apply to objectives for nature conservation. Provided that an outcome could be achieved if resources were available, then the objective should be considered achievable. In the world of nature conservation we must recognise that it may take decades, even centuries, to obtain our objectives, and that long before we reach our goal the objective may have changed. Whenever we achieve a condition that we consider favourable there should be a long-term commitment to maintaining that condition. ('Long-term' will be discussed under the 'time-based' heading.) There is also an argument that objectives need not be achievable, that they should, in fact, be aspirational. If we reach for the treetops we may only collide with the trunk, but if we aim for the stars we'll soar above the trees.

Relevant

Objectives must be relevant and must comply with the strategies, policies and legal obligations that govern the organisation responsible for managing the site or feature. This should also be taken to mean that objectives should, in the context of the governance, be desirable.

Time-Based

Business objectives are usually time-based. The objective will contain a date for the task to be started and, if it is a short-term project, when it will be completed. Start times can be relevant in wildlife management and would usually be the time when an objective is approved and adopted for a feature. Management plans can occasionally be written for a specified period, often 5 years; they are then rewritten or revised. Such plans sometimes contain short-term objectives that describe a condition that should be achieved within the planning period. An example of this approach is described in *A Handbook for the Preparation of Management Plans* (NCC 1983). The handbook recommends that long-term objectives should be for at least 25 years or longer, but gives no explanation of why this period was chosen.

Occasionally, and particularly when dealing with restoration sites, there are good reasons for considering short-term or intermediate objectives. For example, on brown-field sites where there is an intention to establish woodland, an objective could be to ensure that the entire area is planted with trees within 5 years. These short-term, time-based objectives tend to be rather prescriptive, but they are extremely useful when there is a need to demonstrate achievement, particularly when politics are involved. Short-term objectives are only valid if they define a progression of conditions that is entirely consistent with obtaining a long-term objective.

However, in most instances, if we recognise that our commitment to nature should be endless and not time-based, and that management planning is an adaptive process intended to optimise opportunities for wildlife, the concept of a time-based objective takes on a different meaning. First of all, the objective for a feature should be a description of the condition that we want to achieve, and thereafter maintain, in the long term. However, there is no widely accepted definition[3] of long term, there is certainly no forever, and we cannot predict how long any particular condition can or should persist. Long term is in the mind of the beholder; it is as far ahead as anyone can envisage.

We recognise that all natural features will change (of this we can be certain) and that the degree and direction of change is not always predictable. An objective can do no more than reflect our values, knowledge and aspirations at the time that it is written. The cyclic adaptive planning process was developed as a response to these issues, and an essential element of the process is the mandatory requirement to review the objectives at intervals. The length of the interval will be determined by our confidence in the objective, and this will be influenced by a range of different issues. The most important are:

- Our knowledge and understanding of a feature; we often have to manage features, species and habitats when there is very little available information.
- The natural dynamics of a feature. Some habitat features, for example sand dunes, can be very dynamic, even ephemeral; other features can be very stable.
- The quality of the scientific evidence that is available.[4]
- Our direct experience of, and competence in, managing the feature.
- Changing environmental factors, for example, global climate change.
- Changing human values and perceptions.

Each time an objective is reviewed a date, which is a reflection of the confidence of the review team, should be set for the next review. It is likely that the period between reviews will vary. Review does not necessarily mean that an objective will be discarded or even modified. In many cases, the review may confirm that an objective is appropriate and should remain unchanged.

[3] Although there is no widely accepted definition of long term, some authors have suggested what it might mean. For example, Sinclair et al. (2006), writing about ecosystems, conclude, *'In most cases planning should be for 30–50 year periods or longer'*.

[4] A paper, *'Evidence of Effectiveness'* (Pullin 2002), provides a compelling case for evidence-based conservation.

Note: Whenever objectives are time-based they should be introduced with an indication of the period for which they will remain valid without review.

15.2.3 An Objective Must Be Communicable

An objective must be easily understood by the intended audience. Management plans, and particularly objectives, are about communicating our intentions, sometimes to a very wide audience, many of whom will not be scientists or conservationists. In addition to informing others, the objective must also provide a clear and unequivocal guide for reserve managers. Objectives must also be quantified so that they can be monitored. This is quite a tall order; an objective is a multi-purpose statement that describes the required outcome of a feature (something that we want to achieve) using both plain and quantified scientific language. The solution is to prepare composite statements that combine a *vision* for the feature with quantified and measurable *performance indicators*. The first part of the objective, the vision, is a portrait in words that should create a picture in the reader's mind of what we want to achieve. Visions should be written using plain language. They must never be patronising, but they should not contain difficult or obscure scientific language. For example, the use of scientific species names should be avoided whenever possible. It is important, however, that the quality of the information conveyed by the vision is not diminished as a consequence of using plain language. The quantified and measurable performance indicators, which accompany the vision, provide the evidence that is used to assess the status of the feature. The performance indicators, in contrast to the vision, should be written using precise scientific language, which will include scientific names.

15.2.4 Objectives Are Best Written in the Present Tense

An objective for a feature is a description of a conservation outcome. Objectives for many features, and particularly for habitats, will specify the conditions required for a variety of different attributes. For example, an objective for woodland can describe the canopy species, the ground flora, the age structure of the trees, the amount of dead wood, regeneration, etc. At any given time, some of these conditions may be acceptable and others not. For this reason, an objective is best written in the present, and not the future, tense. Otherwise, an objective will contain a mixture of tenses: future for attributes that are currently unfavourable and present for attributes that are favourable, and these will change and potentially alternate. Elzinga et al. (2001) suggest that an objective should contain a 'verb', for example, increase, decrease or maintain. If this advice is taken, there will be a need to change the objective as the condition of the feature changes. Clearly, it is not possible to increase or decrease for ever. This information is essential whenever the objective is reviewed, and particularly when assessing the management requirements, but this does not mean that it must be part of the objective.

15.3 Favourable Conservation Status (FCS)

The main aim of the European Habitats Directive is to promote the maintenance of biodiversity by requiring Member States to take measures to maintain or restore natural habitats and wild species listed on the Annexes to the Directive at a Favourable Conservation Status, introducing robust protection for those habitats and species of European importance. In applying these measures Member States are required to take account of economic, social and cultural requirements, as well as regional and local characteristics.

The definition of FCS for habitats and species, as used in this book, is based on, and is consistent with, the statutory definition of FCS for habitats and species given in Article 1 of the Habitats Directive (Council Directive 92/43/EEC of the 21st May 1992 on the conservation of natural habitats and of wild fauna and flora. [*Official Journal of the European Communities* OJ no. L206, 22.7.92, p.7.]) (Table 15.1).

IMPORTANT: It must be stressed that a decision to use FCS at site level is made entirely for practical purposes. There is *no* legal requirement to do so. The practical application of the concept can provide an extremely useful, and entirely appropriate, basis for defining the desired status of habitats and species at any geographical scale, from the entire geographical range to a defined area within a site.

Table 15.1 Definition of Favourable Conservation Status used in this book

Definition of Favourable Conservation Status

Habitat features

For a habitat feature to be considered to be at FCS, ALL of the following must be true:

- The area of the habitat must be stable in the long term, or increasing.
- Its quality (including ecological structure and function) must be maintained.
- Any typical species must also be at FCS, as defined below.
- The factors that affect the habitat, including its typical species, must be under control.

Species features

For a species feature to be considered to be at FCS, ALL of the following must be true:

- The size of the population must be maintained or increasing.
- The population must be sustainable in the long term.
- The range of the population must not be contracting.
- Sufficient habitat must exist to support the population in the long term.
- The factors that affect the species, or its habitat, must be under control.

15.4 Visions for Features

Writing an objective for a feature will always be challenging, but it is much easier when the vision is based on the definition of Favourable Conservation Status. FCS is an uncomplicated and common sense expression of what we should attempt to achieve for all important features. It is a generic statement that could be applied anywhere but should not, in its original raw form, be used as an objective. Some organisations are, regrettably, adopting a rather bureaucratic approach to writing objectives. By this I mean that organisations can become too concerned with pursuing administrative output targets, for example, setting targets for the completion of management plans, regardless of quality. These management plans frequently use generic objectives, for example, 'to maintain feature X at Favourable Conservation Status'. It is very important that objectives are site-specific. Our commitment to maintaining biodiversity must include an obligation to ensure that local distinctiveness is maintained. Generic objectives that can be applied everywhere have very limited value anywhere. This does not mean that we have to start afresh each time we write an objective. We should use examples of objectives prepared earlier or elsewhere to help formulate objectives for new management plans, but this must be done intelligently. An objective must be tailored to meet the particular conservation values of a feature in any specific location.

An objective can be built around the FCS definition by dealing with each section of the definition in turn. Before beginning to create a structured objective, it helps to jot down, in any order, the qualities or attributes of the feature that are clearly desirable. Consider the current condition of the feature on the site. If any part, or parts, of the feature appear to be in the required condition, this provides an excellent starting point for deciding what favourable might mean. In situations where features are not in a favourable condition, the question should be: why is the feature unfavourable, and what is the difference between what we see and what we want to see? Experience from other similar places where the feature is considered to be favourable may help, but do not forget the importance of local distinctiveness.

15.4.1 Visions for Habitats

For a habitat to be at FCS its size must be stable or increasing. This is a very obvious requirement. In addition to the area occupied by a feature, its distribution can also be extremely important. So, an objective should begin with some indication of the size and distribution of the feature. In short, how much do we want and where do we want it? Often, the easiest and best way of indicating where something should be is to use a map; there is no reason to restrict the expression of an objective to the written word. Maps and illustrations should be used whenever they will help to clarify the objective.

Specifying the desired location of features is occasionally complicated. There are situations where more than one important habitat feature occupies the same area of a site, and where there is a requirement to obtain FCS for each of these features. For example, consider a site which contains two important habitat features; scrub and grassland, could each occupy the same space. At any one time the site will contain areas of grassland which are free of scrub, areas newly colonised by scrub, and areas of mature scrub. The total area occupied by scrub will not change, but the distribution of scrub over the site must change. This means that the objective for scrub should specify an upper limit and lower limit for the extent of the scrub, but there would be no purpose in expressing limits for the grassland as it will occupy the remainder of the site. The precise location of the communities at any time may not be an issue. However, there can be occasions, often as a consequence of associated species, where the actual distribution of the communities is important. For example, there may be reason to ensure that the scrub is distributed around the edges of a site.

It is essential that the natural diversity within individual habitats is maintained. This can be a problem where small, isolated sites contain fragments of habitats. Diversity within a habitat is most often the product of size, since small areas can only provide a limited variety of conditions. There should be no assumption that there is a responsibility to achieve all potential variations everywhere. Places should not be managed in isolation but within the context of the dynamic bio-geographical distribution of species and habitats. Ideally, a management strategy that takes a much wider perspective should be developed. The aim of the strategy would be to obtain diversity over a series of sites and not within a single site.

Once a habitat has been *quantified* and located, the required *quality* must be specified. The temptation may be to provide exhaustive lists of species that are considered important, but species lists are more likely to confuse than inform. It is better to focus on the most important species, or groups of species, both the desirable and undesirable. There will be some species that are indicators of the required conditions and other species which indicate that a change is taking place.

Nature conservation is often about maintaining highly-valued, semi-natural communities, such as grassland. These communities are the product of intervention or management and, in most cases, they can be precisely defined. Conversely, when dealing with more natural habitats there should be an acceptance that natural processes can deliver a variety of, sometimes unpredictable, conditions, so that precise descriptions of favourable condition are meaningless. However, even in these cases, it is necessary to provide some indication, at least in very broad terms, of what might be considered acceptable.

There are no compelling reasons to quantify the feature at the vision stage; that will come later when the performance indicators are identified. However, given that this statement is meant to help readers gain a picture of what the site will look like when the objective has been achieved, the inclusion of some quantified description can help.

Fig. 15.1 An upland oak woodland

The process of developing a vision based on FCS is best explained by working through a few examples:

Vision for a Small, Wet, Upland, Acid, Oak Woodland

The woodland habitat is the only feature on this nature reserve (Fig. 15.1). The site was previously intensively managed: it was heavily grazed by sheep and some timber was extracted. The intention for the future is that minimum-intervention management will enable the woodland to develop into a high forest. The definition of FCS has been used as a framework to help develop this objective. Each part of the definition of FCS is shown in the left-hand column, followed by the relevant text from the objective. Where the text requires explanation a note has been included.

Table 15.2 A vision for a woodland

Definition of FCS	Vision for an upland acidic oak woodland	Notes
The area of the habitat must be stable in the long term, or increasing.	The entire site is covered by a high forest, broadleaf woodland.	Objectives can include a map showing the distribution of the woodland within the site.
Its quality (including ecological structure and function) must be maintained	The woodland is naturally regenerating, with plenty of seedlings and saplings particularly in the canopy gaps. There is a changing or dynamic pattern of canopy gaps created naturally by wind throw or as trees die.	The woodland processes, death, decay and regeneration, are easily observable surrogates which demonstrate that a system is functional. A diverse and dynamic woodland structure, i.e. trees of all ages with replacements in the field layer containing the typical species (see below), demonstrates that ecological structure is being maintained.
Any typical species must also be at FCS.	The woodland has a canopy and shrub layer that includes locally native trees of all ages, with an abundance of standing and fallen dead wood to provide habitat for invertebrates, fungi and other woodland species. The field and ground layers will be a patchwork of the characteristic vegetation communities developed in response to local soil conditions. These will include areas dominated by heather or bilberry, or a mixture of the two, areas dominated by tussocks of wavy hair-grass or purple moor-grass, and others dominated by brown bent grass and sweet vernal grass with abundant bluebells.	There is no widely accepted definition of typical. Typical species could be those which define the habitat or community. So, for a woodland of this kind the canopy and shrub layer species are obviously important. Occasionally, there are good reasons for naming specific species, as demonstrated in the description of the field layer. However, if it is not necessary, avoid naming individual species. This example deliberately talks imprecisely of 'locally native species'. This is because the woodland habitat is dynamic and will change over time. The canopy was until recently dominated by oak *Quercus spp.*

(continued)

Table 15.2 (continued)

Definition of FCS	Vision for an upland acidic oak woodland	Notes
Any typical species must also be at FCS -continued	There will also be quite heavily grazed areas of more grassy vegetation. Steep rock faces and boulder sides will be adorned with mosses and liverworts and filmy ferns. The lichen flora will vary naturally depending on the chemical properties of the rock and tree trunks within the woodland. Trees with lungwort and associated species will be fairly common, especially on the well-lit woodland margins.	But, following a severe gale when most of canopy species were blown down, the canopy that is re-establishing may be dominated by birch *Betula* spp. This is a perfectly natural and desirable situation. In circumstances where management is mainly about enabling natural processes the objectives must not be too precisely defined.
The factors that affect the habitat, including its typical species, must be under control	The woodland does not contain any rhododendron or other invasive alien species with the exception of occasional beech and sycamore. There will be periodic light grazing by sheep and very occasionally by cattle. This will help maintain the ground and field layer vegetation, but will not prevent tree regeneration.	The definition of Favourable Conservation Status is concerned with the future: habitats and populations should be sustainable in the long term. The most reliable evidence that can be used to demonstrate that there is probably a future for a feature is that the factors which are most likely to change the feature are under control. It is important not to overlook the factors when preparing the vision, and it is appropriate that some are mentioned. However, it is probably wiser to deal with the factors when identifying the performance indicators and not to over-complicate the objective at this stage.

15.4.2 Visions for Extremely Dynamic Features

Some habitat features will be very dynamic and unpredictable, and we recognise that a very wide range of future conditions will be regarded as acceptable. For example, on a coastal sand dune system the specific composition and structure of the vegetation may not be an issue, providing that the following conditions are met:

- The system consists of a dynamic, shifting mosaic of sand dune communities (the individual communities can be described), where the actual composition and structure is governed by natural processes.

- Regardless of how the feature evolves, a sufficient area of sand dune habitat exists to support the full complement of dependent plant communities and typical dune species. This should include any such species that are features of the site in their own right. (The presence of species which are important features in their own right can limit the scope for accepting natural change.)
- The distribution of plant communities and populations of typical species are also governed by natural processes (again, provided this is compatible with the obligation to maintain species populations that are themselves features of the site).
- The factors that influence, or may influence, the sand dune system are under control.

Fig. 15.2 Morfa Harlech, a dynamic sand dune system

Fig. 15.3 A guillemot ledge

15.4.3 Visions for Species

It is invariably easier to write a vision for the population of a species than it is to write a vision for a habitat. This is because, in most cases, there is significantly less that can be said about a population. We can describe the size and distribution of a population, the site-specific factors and, in exceptional circumstances, the age structure, survival and productivity rates, and that is all.

The following example is for a small Welsh island which has extremely important colonies of seabirds. One of the Natura 2000 SPA features is a population of guillemots *Uria aalge*. Guillemots are members of the auk *Alcidae* family. They are quite large, cliff-nesting, diving seabirds with a northern distribution in Europe. This example is chosen because of the exceptionally good information that is available for this particular site. In this sense it is atypical. For most species features there is very little information available on population dynamics, and for many there are no reliable means of assessing population size or trends. This example tackles one of the more difficult, though widely encountered, problems in species management: the management of a protected area where the most important species are mobile or migratory, and where they depend on an area for only part of their annual life cycle.

Table 15.3 Vision for guillemots

Definition of Favourable Conservation Status	Vision for guillemots
The size of the population must be maintained or increasing.	Skomer Island is a very important breeding site for a large, robust and resilient population of guillemots. The size of the population is stable or increasing (in 2011 the population was 21,688).
The range of the population must not be contracting.	The distribution of the colonies (shown on the attached map) is maintained or increasing.
The population must be sustainable in the long term.	At least 80% of the breeding adults survive from 1 year to the next, and at least 70% of the breeding pairs raise a chick each year. This will help to ensure the long-term survival of the population.
Sufficient habitat must exist to support the population in the long term, and the factors that affect the species, or its habitat, must be under control	The safe nesting sites and secure breeding environment are protected. There are no ground predators and the impact of predatory birds is insignificant. The size and range of the population are not restricted or threatened, directly or indirectly, by any human activity on the island. The nesting colonies are not disturbed from the sea by boats or other human activities during the breeding season.

Discussion of the example given in Table 15.3: The most significant factor as far as any species is concerned is the habitat that supports it. However, because habitats are so important for species, the definition of FCS gives them specific attention and deals with them separately from the other factors. This is confusing because, at a later stage, the definition of FCS states that the factors affecting the habitat must also be under control. At an earlier stage (Chapter 13), the need to treat the habitat that supports an important species as an independent feature, even when it does not qualify as a feature, was explained (see also Case Study 3). This example for guillemots is an exception to that general guidance. The nesting habitats are rocky cliffs where the vegetation is irrelevant; all that matters is that the area is free from excessive human disturbance and ground predators. Consequently, in this case there is no need to write an objective for the habitat. The guillemots use the island as a place to breed and nothing more. They spend the greater part of the year offshore. Adults are present at sea, but reasonably close to the breeding colonies, throughout the year. Younger birds disperse widely over a larger area in the Atlantic. Therefore, the important habitat is the surrounding sea and wider ocean. Clearly, we cannot write an objective for the southern Irish Sea or the Atlantic Ocean. In this particular circumstance, the sea immediately surrounding the island is a marine nature reserve. Together, the terrestrial and marine reserve can do no more than make a contribution towards protecting the species from local human disturbance. Two of the most important anthropogenic factors are marine pollution, particularly oil spills, and commercial fishing, but, apart from identifying the factors and recognising their potential impact, there is not much else that can be done in the local management plan. These are global problems, and they must be dealt with at that level. This does not in any way negate the value of an objective when the control of factors lies outside the remit of site management. Information from the individual sites, and the failure or otherwise to meet local objectives, will inform politicians and others responsible for policy and legislation.

15.4.4 The Level of Definition for an Objective

The level at which an objective is defined will vary from place to place and feature to feature. At one extreme, consider a small area of semi-natural grassland which is the by-product of past agricultural intervention and is now regarded as being important for both wildlife and cultural reasons. The objective for this feature will be very detailed and will contain performance indicators that provide precise specifications for sward height, species composition, etc. In contrast, an objective for a savannah ecosystem in an African park will have a significantly less defined objective. For example:

> The vegetation consists of a dynamic shifting mosaic of savannah woodland, scrub and grassland, for which the composition and structure is governed mainly by natural processes. The large and extensive tracts of habitat will provide a secure, sustainable future for robust populations of elephants, large and small ungulates, including buffalo, common zebra, Rothschild's giraffe, eland, hartebeest, klipspringer, Guenther's dik-dik, mountain reedbuck, lesser kudu and greater kudu. Secure populations of predators, including lion, leopard, cheetah, striped hyena, spotted hyena and African hunting dog will also thrive in the park. (Alexander et al. 2000)

The level of definition for an objective will be influenced by:

- Naturalness: highly-modified, plagioclimatic habitats that are dependent on continual management will require much tighter definition than natural habitats, where the outcome is not necessarily predictable.
- The size, complexity and dynamics of a feature.
- The resource available to manage and monitor a feature. When resources are very restricted, it is often not possible to do anything other than measure a limited number of performance indicators that will provide a general indication of the condition of a feature. In the savannah ecosystem given as an example above, resource limitations restrict monitoring to key mammals and management

to controlling poachers and other illegal activities. Detailed objectives for vegetation would be completely irrelevant in this situation.
- The availability of reliable scientific information and experience of the feature.

It is so important that management planning is recognised as a process that will develop and adapt. Even where there may be initial constraints because of limited resource or science, these obstacles may be overcome with time. An objective can begin life as a simple, general statement with very few performance indicators, but, as our knowledge, understanding and resources improve, so the objective will evolve.

15.5 Performance Indicators

If we try to apply the SMART test of an objective to any of the examples of visions given above they will fail. With a few minor exceptions, the visions do not quantify the features in a way that makes it possible for the objective to be measured. It is neither possible nor necessary to quantify every aspect of a feature, and quantification is only part of the issue; there is little purpose in quantifying something if it cannot be measured.

A number of performance indicators can be used to quantify the objective and provide the evidence that a feature is in a favourable condition or otherwise. The evidence will not be sufficient to allow a conclusion to be proven beyond any reasonable doubt, but we are dealing with wildlife and not criminal law. A feature can only be considered to be favourable when the values of all the performance indicators fall within the specified range. A balance must be struck between having sufficient performance indicators to minimise the risk of errors and the cost implications of having too many. All performance indicators must be monitored – that is their entire purpose – but monitoring can be very expensive, and there are inadequate resources for nature conservation. This is further complicated because there is potential for two types of error. The first, generally the consequence of too few indicators, is when we assume that a feature is favourable when in fact it is not. The second, generally the consequence of too many indicators, is where we assume something is unfavourable when in fact it is favourable. The more indicators that are tested the greater the scope for a false assumption that all is not well. The first type of error is potentially very serious; a failure to recognise that a feature is unfavourable could lead to its destruction. The second type of error is usually less damaging; the consequences are most likely to be the implementation of unnecessary management.

15.5.1 Favourable Condition and Favourable Conservation Status

So far, the case for performance indicators which provide evidence that a feature is at *favourable condition*, or otherwise, has been discussed. This is the condition of a feature when the desired outcome has been achieved. The example of a vision for a woodland describes the favourable condition of the wood. This is rather like a snapshot taken at some point in time, but it gives no indication of the factors that must be under control for the condition of the woodland to be considered sustainable. Two of the factors which affect the woodland are grazing and invasive alien species. For a feature to be at Favourable Conservation *Status*, the *condition* of the feature must be favourable, and this condition must be sustainable in the long term. An objective based on FCS must, therefore, deal with both aspects of the definition and, consequently, two different kinds of performance indicators are used to monitor an objective. These are:

- Quantified attributes with limits which, when monitored, provide evidence about the condition of a feature.
- Factors with limits which, when monitored, provide the evidence that the factors are under control or otherwise.

There is a slight complication. Factors are the agents of change, and attributes are the characteristics of a feature which change as a consequence of the factors. Consequently, the selection of attributes as performance indicators should, to some extent, be guided by the presence of factors. This also means that the evidence that can be used to demonstrate that a factor is under control can be obtained directly by measuring the factor, or indirectly by measuring the attribute which changes as a consequence of the factor. The only difficult issue is that there is a need to introduce both factors and attributes at the same time. Clearly, this is impossible so, following some sort of logic, attributes will be introduced first. However, you might want, occasionally, to refer to the section on factors while reading about attributes.

15.5.2 Attributes

There are two commonly used definitions of an attribute:

> **Attributes** are the characteristics, qualities or properties of a feature which are inherent to, and inseparable from, the feature.
>
> - Attributes should be indicators of the general condition of a feature.
> - Attributes must be measurable.
> - Attributes should be informative about something other than themselves.
> - Attributes should, whenever possible, be indicators of the future rather than the past.
> - Attributes can be economical surrogates. (Alexander 1996)
>
> **Attribute:** A characteristic of a habitat, biotope, community, or population of a species which most economically provides an indication of the condition of the interest feature to which it applies. (JNCC 1998)

Both carry much the same meaning, but the first is both tautological and long-winded while the second is rather convoluted. It is time for a new, improved definition:

An attribute is a characteristic of a feature that can be monitored to provide evidence about the condition of the feature.[5]

[5] In this context, a feature can be a habitat, biotope, plant community or population of a species. However, the concept can also be applied to many other non-biological features, for example, geology, geomorphology, archaeology.

Examples of attributes:

For species

Quantity

- The size of a population, for example:
  - The total number of individuals present
  - The total number of breeding adults
  - The population at a specified point in an annual cycle
- The distribution of a population

Quality

- Adult survival rates
- Productivity
- Age structure
- Sex ratio

For habitats

Quantity

- The size of the area occupied by the habitat, or by one or more constituent communities
- The distribution of the habitat, or of one or more constituent communities

Quality

- Physical structure (a wide range of attributes is possible here, and they are very feature-specific)
- Presence, abundance, relative proportions, distribution of individual species, or groups of species, indicative of condition
- Presence, abundance, relative proportions, distribution of individual species, or groups of species, indicative of change

Additional attributes for woodland

Quality

- Tree and shrub layer canopy cover
- Tree and shrub canopy composition
- Canopy gap creation rate
- Tree regeneration
- Age structure of trees
- Volume of dead wood
- Field and ground layer composition

15.5.2.1 Selecting Attributes

The best guide for the selection of attributes is the definition of FCS; this has already been used to construct the vision. It is important that there is consistency between the vision and the choice of attributes. By this I mean that the attributes that are selected as performance indicators should also have been mentioned in the vision. This is best explained by returning to the examples previously used for guillemots and woodland (Tables 15.4 and 15.5).

Table 15.4 A vision for Guillemots with attributes and factors

Definition of Favourable Conservation Status	**Vision for guillemots** (attributes are shown in **bold** and the factors are underlined)
The size of the population must be maintained or increasing.	Skomer Island is a very important breeding site for a large, robust and resilient population of guillemots. **The size of the population** is stable or increasing (in 2011 the population was 21,688).
The range of the population must not be contracting.	The **distribution of the colonies** (shown on the attached map) is maintained or increasing.
The population must be sustainable in the long term.	At least **80% of the breeding adults survive** from 1 year to the next, and at least **70% of the breeding pairs raise a chick each year**. This will help to ensure the long-term survival of the population.
Sufficient habitat must exist to support the population in the long term, and the factors that affect the species, or its habitat, must be under control	The safe nesting sites and secure breeding environment are protected. There are no ground predators and the impact of predatory birds is insignificant. The size and range of the population are not restricted or threatened, directly or indirectly, by any human activity on the island. The nesting colonies are not disturbed from the sea by boats or other human activities during the breeding season.

Table 15.5 Vision for an oak woodland with attributes and factors

Definition of Favourable Conservation Status for a habitat	Vision for an upland acidic oak woodland (**attributes** are shown in bold and the factors are underlined)
The area of the habitat must be stable in the long term, or increasing.	The **entire site** is covered by a high forest, broadleaf woodland.
Its quality (including ecological structure and function) must be maintained.	The woodland is **naturally regenerating,** with plenty of **seedlings and saplings** particularly in the canopy gaps. There is a **changing or dynamic pattern of canopy gaps** created naturally by wind throw or as trees die.
Any typical species must also be at FCS.	The woodland has a **canopy and shrub layer** that includes **locally native trees** of **all ages,** with an abundance of standing and fallen **dead wood** to provide habitat for **invertebrates, fungi** and other woodland species. The **field and ground layers** will be a patchwork of the characteristic vegetation communities, developed in response to local soil conditions. These will include areas dominated by **heather, or bilberry**, or a mixture of the two, areas dominated by **tussocks of wavy hair-grass or purple moor-grass,** and others dominated by **brown bent grass** and **sweet vernal grass** with abundant bluebells. There will also be quite heavily grazed areas of more grassy vegetation. Steep rock faces and boulder sides will be adorned with **mosses and liverworts and filmy ferns**. The **lichen flora** will vary naturally depending on the chemical properties of the rock and tree trunks within the woodland. Trees with **lungwort and associated species** will be fairly common, especially on the well-lit woodland margins.
The factors that affect the habitat, including its typical species, must be under control.	The woodland does not contain any rhododendron or any other invasive alien species with the exception of occasional beech and sycamore. There will be periodic light grazing by sheep and very occasionally by cattle. This will help maintain the ground and field layer vegetation but will not prevent tree regeneration.

15.5.2.2 Attributes Should Be Indicators of Future Change

It is essential that attributes tell us that a change is taking place before a feature is seriously damaged, and not that a change has taken place. This was clearly illustrated in the example for the guillemot population. The only really important issue is that the guillemot population is maintained or increasing. If this is the case, why be concerned about anything else? Guillemots are long-lived birds. They have a low productivity rate, producing only a single egg clutch. These birds are characteristic of species living in relatively stable environments and having relatively stable populations (Birkhead and Perrins 1997). Two of the early indicators of potential decline in a population are a reduction in the adult survival rate or in the productivity rate. The key issue is that there is a significant delay of 6–7 years between the young guillemots fledging and their recruitment to the colony as breeding adults. The consequence is that if one attribute, the size of the population, is used in isolation to make a judgement, it would be possible to come to the wrong conclusion and assume that the population is favourable when in fact it is not. This is because, although the total population of breeding birds could be stable for up to 7 years, they might not be producing any young which would later be recruited into the population.

15.5.2.3 The Selection of Attributes Can Be Guided by Factors

All the important factors, and particularly the primary factors, have been identified by this stage in the plan (see Chapter 14). The selection of attributes should, to some extent, be guided by the presence of factors. While factors are the influences that can change or maintain a feature, attributes reflect the changes that take place, or the conditions that prevail as a consequence of these influences. This means that the evidence to demonstrate that a factor is under control can be obtained directly, by measuring the factor, or indirectly, by measuring the attribute that changes as a consequence of the factor.

Factors that no longer have any influence on a feature will also influence the choice of attributes. The past management of a site will sometimes be the most important factor that influences the selection of attributes. It is not always necessary to have a precise understanding of past management, but it is important that the consequences are recognised. An oak-birch bilberry woodland, *Quercus petraea-Betula pubescens-Dicranum majus* woodland, with a long history of human intervention and management, provides a good example. At an early stage, it was utilised for the production of a wide range of woodland products. Later, it became an important source of bark for the leather tanning industry. It was almost completely clear-felled during the First World War and then, most recently, became a nature reserve. There is little detail of earlier management but some of the consequences are obvious. The canopy trees are young and even-aged with no veteran specimens. There are no canopy gaps and, consequently, no oak *Quercus* regeneration. There is insufficient dead wood. It is obvious that, as a consequence of past management, the woodland is currently in an unfavourable state. Some changes towards favourable condition will take place naturally, and others may require intervention. A number of attributes can be selected to demonstrate that the required change is taking place. These include age structure of the canopy, the gap creation rate, the regeneration of canopy species, the presence of veteran trees and the volume of dead wood.

An understanding of the relationship between attributes and factors is fundamental to selecting relevant attributes. This relationship is illustrated in the guillemot example. Two of the important factors are *'human activity on the island'* and *'disturbance from the sea by boats or other human activities'*. These could threaten the size and distribution of the population on the island directly or indirectly. Both factors can be measured, but, because the levels of disturbance that the guillemots will tolerate are not known, it will not be possible to set meaningful limits to human activities. Consequently, it will not be possible to monitor human activities, but a surveillance project can be established to measure any trends or changes in human activity.

The important question is: how could the guillemot population change as a consequence of excessive disturbance? The most likely changes will be to the distribution and size of the individual colonies. Consequently, there is a need monitor both these attributes. So, for example, if a particular guillemot colony shows any sign of decline that is not reflected elsewhere on the island and there is an associated change of human activities in the vicinity it would be reasonable to conclude that there is a relationship between the two that would justify management action.

The relationship between factors and the selection of attributes for habitats and plant communities is also very important. It is clearly impossible to measure everything in a plant community, let alone a habitat. Even if we could, it would be prohibitively expensive and quite unnecessary. There is a need to focus on a limited range of attributes which together can provide sufficient evidence to reveal the condition of the feature.

Where we are aware of factors, and understand their impact on a habitat, it is often possible to predict the nature of the changes that are likely to take place, and to select attributes and set targets for them on that basis. For example, the application of artificial fertilizer to a traditional hay meadow would lead to an increase in some undesirable species, such as rye-grass *Lolium perenne* and white clover *Trifolium repens,* and a corresponding loss of desirable species.[6] Both groups of species are attributes that can be monitored and will provide useful performance indicators. In short, species as attributes can be divided into two main groups: those that are indicators of change, and those that are indicators of the condition required of a feature.

15.5.2.4 The Use of Species as Indicators of Habitat Condition

Species which are indicators of the condition required of a community, or are indicative of change, offer opportunities for the economical monitoring of communities. Quite simply, the presence of a small range of species, or even an individual species, may

[6] A detailed account of attributes for UK grassland and all other habitats is available from JNCC (JNCC 2004).

indicate that a community is likely to be in a favourable condition. Conversely, an increase in, or appearance of, other species could indicate that the community is becoming unfavourable.

But we must be cautious; there are many pitfalls. Indicator species have been used for many different purposes:

- Species whose presence indicates the presence of other species
- Keystone species
- Species that indicate habitat damage
- Species that indicate particular environmental conditions
- Species indicative of environmental change
- Species indicative of management intervention

(Lindenmayer et al. 2000)

Some indicator species are used in a very specific context. For example, ancient woodland indicators are species that are usually common in ancient woodlands and that spread relatively slowly by vegetative means (Peterken 1996). These indicators have to be used with caution. They can only contribute to the evidence that a woodland is ancient: their presence alone, in the absence of other supporting evidence, tells us very little. Most of the ancient woodland indicators, bluebell *Hyacinthoides non-scripta* is a common example, can thrive outside woodland.

Many planners have tried to use 'keystone species' as indicators or as a focus for monitoring. I include this section to illustrate the danger of using concepts or approaches that are poorly understood. The definition of keystone species: an organism that has a significant influence on the ecosystem it occupies that is disproportionately large compared to its abundance or biomass.

Hatton-Ellis (2011) completed a literature review of keystone species. His main conclusions were:

- Despite 43 years of research, the concept remains unproven.
- The term is often used with a lack of rigour in the literature, and surprisingly few studies have adequately tested whether given species genuinely perform a keystone role in their ecosystem.
- Numerous studies have shown that individual species may have important ecosystem effects, but remarkably few demonstrate the keystone status of a single species at an ecosystem level.

Hatton-Ellis also suggests that in the highly modified cultural landscapes of the Old World, where many habitats are entirely dependent on management, the keystone species is, in fact, man. Most significant of all, he is concerned that the keystone species concept may itself be an illusion. Illusion or not, it is clearly a useful concept that can help us understand ecosystems, but can the concept be applied to nature conservation management and planning?

Once again, we encounter a topic where there are different, and conflicting, views. Clearly, there are conservation scientists and managers who believe that they can use keystone species as indicators of the general condition of a habitat. I agree entirely with Hatton-Ellis when he suggests that the concept is used with a lack of rigour. The following is typical of the many examples of this. Readers may draw their own conclusions.

The Conservation Measures Partnership use, for demonstration purposes, the management plan for the Appalachian Ohio Forest Conservation Area. They use a number of 'monitoring indicators' to assess the quality of the forest. One of the indicators is the breeding populations of cerulean warblers *Dendroica cerulea*, a small American warbler. This is described in the plan as a keystone species. The first questions must be: is this 'a species that has a significant influence on the ecosystem'? How will the ecosystem change, or be discernibly different, if small populations of this tiny, insectivorous warbler are absent? In other words, what can changes in the population of this species tell us about changes in the condition of the forest ecosystem?

The population is changing:

> The cerulean warbler's population is dropping faster than any other warbler species in the United States. Between 1966 and 1999, it declined an average of 4% per year throughout its eastern US breeding range for a total population loss of 70%. (Hamel 2000)

The next question must be, why is this change taking place? There are certainly adverse factors in the breeding habitat, for example, loss and fragmentation of the forest habitat, nest predation by the brown-headed cowbird *Molothrus ater,* and a brood parasite which lays its egg in the nests of other small passerines.

But these are not the only factors. The cerulean warbler is a migratory species. In winter, they are generally restricted to mid-elevation forests on the eastern slopes of the Andes Mountains. These forests are among the most threatened in South America and are being rapidly degraded by coffee, tea and cacao production. Unfortunately, we know very little about the ecology of cerulean warblers during the non-breeding season. Evidence suggests that events on breeding, stopover and wintering grounds may be causing the decline (Hamel 2000).

Even if the warbler is a keystone species, can it be used as an indicator of the condition of the forest, when changes in the population cannot be attributed to any specific and understood factor or factors in the Appalachian Forest? It is essential that when we select an indicator species we understand what it can tell us about the condition of habitat and that we can differentiate between local on-site factors, global factors and offsite factors. The population of cerulean warblers is changing, probably because of the combined impact of these three different types of factors. The impact of each is unknown.

Remember, this is not about monitoring a population of a seriously endangered species. It is about using that species as an indicator of the condition of the forest.

15.5.2.5 Attributes Must Be Quantifiable and Measurable

Attributes must be quantifiable and measurable so that they can be monitored; that is their entire purpose. When making the initial selection of attributes, it is important to consider, and describe in outline, how the attribute will be monitored. The details of the monitoring methodology can be left until later.

15.5.2.6 Insufficient Attributes

Ideally, a manager should have sufficient evidence about the condition of a feature to make management decisions with some confidence, but, quite often, it is not possible to identify enough attributes that can be monitored. When the difficulty arises because of a shortfall of information this should be noted, and relevant surveys or research should be planned. In the meantime, the plan should carry a warning that the list of attributes is insufficient. This is not as unsatisfactory as it may appear; planning is a process and shortfalls can be addressed at a later stage. While it may be highly desirable to deal in certainties, this is not a luxury that conservation managers can afford. Management decisions are made, in fact can only be made, on the basis of the best available information.

It is essential that the reasons for selecting each of the attributes that will be used as performance indicators are clearly explained in the management plan. This should include an explanation of why an attribute has been selected, what information it is intended to convey, and what, if any, is the relationship between the attribute and the factors.

15.5.2.7 UK Guidance for Common Standards Monitoring

In the UK, the Joint Nature Conservancy Committee (JNCC) publishes on their web site (http://www.jncc.gov.uk/page-2201) guidance on monitoring habitats and species. The guide contains an introduction followed by advice on the identification of attributes for British habitats and species (JNCC 2004).

The guidance is intended to ensure a consistent approach to selecting attributes and to monitoring across the UK. Consistency of approach enables data to be aggregated for the production of regional and national reports, and also allows the comparison of monitoring results from different parts of the UK. Common Standards Monitoring was designed to provide the UK agencies with an ability to report on the condition of features on statutory sites. It was *not* designed to meet the broader monitoring requirement for management planning or site management.

This is an important source of information for use in the UK, but it must be used intelligently. When individuals are pressed for time there is an understandable, but unfortunate, tendency to take and apply guidance without considering the extent to which it is relevant in any given situation. For this reason, generic approaches must always be used with extreme caution. If something is intended for use everywhere, can it be used without modification anywhere? The need for site-specific objectives is recognised; the habitat sections refer to attributes that are indicators of local distinctiveness. These are generally site-specific quality indicators. (An alternative approach, which takes account of local conditions and factors to ensure that each individual attribute reflects local distinctiveness, is advocated in this book.)

The JNCC guidance was prepared for the UK statutory nature conservation agencies, but it can be used by other organisations. Anyone engaged in the preparation of objectives for habitats or species in the UK would be unwise to ignore the JNCC document, but it was not intended for use outside the UK. Excellent though most of this information is, it is only a guide, and not something that should be regarded as a definitive standard to be applied without modification in all situations. As with all guides, take from it the information that is relevant and ignore the remainder.

Specific guidance is available for the following habitats and species: coastal habitats, freshwater habitats and species, lowland grassland, lowland heathland, lowland wetland, marine, upland habitats, woodland, earth science, reptiles and amphibians, birds, invertebrates, marine mammals, terrestrial mammals, vascular plants, bryophytes and lichens.

15.5.2.8 An Objective Is More Than the Sum of Its Performance Indicators

The exclusive use of attributes that can be monitored is an extremely dubious basis for defining objectives. In the UK, some organisations have adopted an approach to setting management objectives which is entirely based on those attributes that can be monitored. For example, English Nature published an internal guide to management planning for their National Nature Reserves (2005). It includes a vision for an entire NNR without providing any detailed description of the condition or status that is required for the features. This guidance requires that the attributes identified through the UK Common Standards Monitoring Process are used to define favourable condition. The attributes recommended by UK Common Standards were designed to enable the standard *reporting* of the condition of features throughout UK. The consequence is that the expression of an objective is guided by, and limited to, something that can be monitored: if it can't be counted it doesn't count.

Monitoring can only ever provide some of the evidence that we use to determine the condition or status of a feature. Regardless of our intentions, there is usually little that we can monitor and even less that we can afford to monitor. The guillemot example used in this chapter illustrates an almost ideal approach to monitoring a species. The attributes are: overall population size, size of the population as sampled in study plots, distribution of colonies, the annual survival rate of breeding adults, and the annual breeding success. Unfortunately, because of the extremely high costs, this level of monitoring cannot be afforded everywhere. This approach is only possible on one out of six National Nature Reserves in Wales with guillemot populations. This is not to suggest that any less is expected of the guillemot colonies on the five remaining NNRs: the objectives are very similar for all sites. The only significant variation is the quantification.

15.5.3 Specified Limits

15.5.3.1 Background

There is considerable confusion concerning the use of limits in management plans, and most of the problems arise because the definition of limits varies depending on where the concept is being applied and by whom. During the 1980s similar ideas were developed in both the USA and UK. They share a common starting point in that they both recognise that there will be environmental change, but the most significant similarity is that they both used the word 'limits'.

The most widely known system, Limits of Acceptable Change (LAC), was originally developed to manage recreation in designated wilderness areas of the USA. It came from the basic premise that change is a natural, inevitable consequence of recreational use. The LAC system is mainly concerned with defining the carrying capacity of an area. Its development claimed a shift in focus from 'how much use' to 'how much change' (Stankey et al. 1984). However, the examples of limits provided by Stankey are actually concerned with 'how much use':

- Number of other parties met per day while travelling
- Number of other parties camped within sight or sound per day
- Percentage of trail systems with severe erosion miles with multiple trails…

In the UK, the use of specified limits was initially outlined in a *Handbook for the Preparation of Management Plans for Nature Reserves* (Wood and Warren 1978). The UK approach was different in that it was concerned with specifying the limits for the condition of features beyond which management intervention becomes necessary. Somehow, these two concepts, LACs and specified limits, became entwined. There is no problem in the USA because the numerous publications about LACs are consistent with the original Stankey definition. But elsewhere many authors unfortunately use 'LAC' with a very different meaning. Some publications, for example, *A Guide to the Production of Management Plans for Nature Reserves and Protected Areas* (Alexander 1996), quite inappropriately provide a UK definition for LACs which bears little resemblance to the Stankey version.

The IUCN *Guidelines for Management Planning of Protected Areas* recognises the original definition of LACs, and claims that, 'the benefits of this approach are now recognised in the wider planning process'. The guidelines contain the following:

> LACs are designed to identify the point at which changes in the resource brought about by *another management objective* have exceeded levels that can be tolerated. A LAC contains 'standards' that express minimally accepted conditions (but not desired conditions or unacceptable conditions). The implication is that the condition can be allowed to deteriorate until this minimally acceptable condition is in danger of being reached. At this point management should intervene to prevent a further deterioration. (Thomas and Middleton 2003)

This version is, in fact, very different to the more widely understood definition of LACs and to specified limits. It is quite difficult to understand the assertion that LACs can only be applied to changes in the resource brought about by another management objective. The guide does not provide a reason or discussion.

In 2000, the Conservation Management System partnership published *A Management Planning Guide for Nature Reserves and Protected Areas* (Alexander 2000). This guide recognised the confusion that had arisen as a consequence of the many different and conflicting definitions of LACs. LACs are a USA concept and their original definition should be regarded as definitive. When a different (even though vaguely similar) approach is adopted, this really should be given a different name. Consequently, the CMS guide resurrects 'specified limits' (Woods & Allen 1978; NCC 1981, 1983, 1988). The Ramsar Convention on Wetlands published *New Guidelines for Management Planning for Ramsar Sites and Other Wetlands* (Ramsar 2002). This uses the following definition:

> Specified limits define the degree to which the value of a performance indicator is allowed to fluctuate without creating any cause for concern.

This definition of 'specified limits' is used for the remainder of this book.

15.5.3.2 Specified Limits for Attributes[7]

In ideal circumstances, attributes require two values: an upper limit and a lower limit. Unfortunately, it is not always possible to define both limits. Specified limits were developed in recognition of the inherent dynamics and cyclical change in populations and communities, and in acknowledgement of the fact that such variation is often acceptable in conservation terms. In reality, there are very few features for which the inherent fluctuations are fully understood. For a population, the lower limit might be the threshold beyond which that population will cease to be viable. However, even if the viability threshold is known, it is at best incautious and at worst foolhardy to set a lower limit close to the point of possible extinction. The upper limit could be the point at which a population might begin to threaten another important feature, or where a population becomes so large that it risks compromising the habitat that supports it. Often, upper limits may be unnecessary. In many ways, specified limits can be regarded as limits of confidence. When the value of all attributes falls within the specified limits, we can be confident that the feature is in a *favourable condition*, and if all factors are also within their limits we can conclude that the feature is at *Favourable Conservation Status.*

It is important to remember that the identification of specified limits will always require a degree of judgement. Firstly, it is rare to have robust empirical datasets that show the inherent variability of features from which specified limits can be directly derived. The best that can be done in many cases is to set limits using expert judgement (expert in terms of the feature generally and in terms of knowledge of the site), backed up by some form of peer review and corporate ownership gained through the management planning approval process. Conservation objectives are about what we want on sites and this is not necessarily what we currently have. Specified limits are primarily value judgements rather than scientifically derived figures.

What happens when a limit is exceeded?

The key to understanding how limits work is to understand how we should respond to a limit when it is exceeded. Attributes with limits represent *part* of the evidence required in order to judge whether or not an objective is being met. Part, because, when taken alone, the values of the attributes describe the condition of a feature: they can tell us whether it is acceptable or otherwise. Objectives are concerned with defining the *status* of a feature, and so additional evidence is required to dem-

[7] This and the remaining sections on limits are adapted from Alexander (2005).

onstrate that the factors are under *control*. For the condition of a feature to be considered favourable, the values of *all* the attributes must fall within the specified limits. However, for a feature to be considered unfavourable, only one limit need be exceeded. When this happens, the following procedure should be adopted:

- The monitoring project and the data collected must be checked to ensure that there are no errors. If everything is in order proceed to the next step. If not, the monitoring project should be amended and any decision deferred until the monitoring project has been corrected.
- If a change has taken place and the limit has been exceeded, the reason for the change must be established. Changes happen because of the impact of a factor, or factors, or the lack of appropriate management. Where the reason for a change is known, remedial management can be carried out to deal with the factor, or to improve management.
- When a change has taken place and the reason is unknown a research project should be established to identify the cause.
- Do not forget the precautionary principle: we do not need conclusive, scientific proof in order to take an action to protect a feature.

15.5.3.3 Monitoring Attributes

Whenever attributes are identified they must be monitored: that is their entire purpose. Monitoring attributes provides some of the evidence that is used in the assessment of the conservation status of the features.

> Monitoring is surveillance undertaken to ensure that formulated standards are being maintained (JNCC 1998).

There can be no monitoring without planning and no planning without monitoring. It is a bold statement, but it can be easily defended. If there is no monitoring, it is not possible to know that the features are in the required condition, and there is no means of knowing that management is appropriate. It is the planning process that determines the condition required for the features, i.e. the formulated standard, and without a standard there can be no monitoring.

This approach to management planning is unique in that monitoring is an integral component of the planning process and particularly of the objectives. There is, of course, a penalty: a monitoring project must be developed for every performance indicator.

Quantified attributes with limits have now been added to the following examples of an objective for a guillemot population and a simplified version of the objective for a woodland. Perhaps the most obvious omission in the guillemot example is the absence of ‘upper limits’. This is not uncommon; it is much more difficult to decide when there is more than enough of something than to recognise when there is too little. The notes that follow the tables relate to the quantified attributes which require an explanation. Although these are site-specific and real examples, the notes are relevant more generally.

Table 15.6 Vision for guillemots with quantified attributes

Definition of Favourable Conservation Status	Vision for guillemots (**attributes** are marked in bold)	Quantified attributes with limits
The size of the population must be maintained or increasing	Skomer Island is a very important breeding site for a large, robust and resilient population of guillemots. The **size of the population** is stable or increasing (in 2011 the population was 21,688).	**1. The total island population;** *Lower limit:* 3 consecutive years of at least 18,000 individuals **2. The study plot population;** *Lower limit:* 3 consecutive years with at least 8% of the UK study population total
The range of the population must not be contracting.	The **distribution of the colonies**, shown on the attached map, is maintained or increasing.	**3. The distribution of the colonies;** *Lower limit:* 3 consecutive years with at least 90% of existing area occupied by the colonies as shown on the attached map
The population must be sustainable in the long term	At least **80% of the breeding adults survive** from 1 year to the next, and **at least 70% of the breeding pairs raise a chick** each year. This will help ensure the long-term survival of the population.	**4. The annual survival rate of breeding adults;** *lower limit:* 3 consecutive years with a survival rate of at least 80% **5. The annual breeding success;** *Lower limit:* 3 consecutive years of at least 0.7 chicks per breeding pair

Notes:

The total island population

As with all other attributes for guillemots, the total population will be measured over 3 years. This is because there is a need to avoid over-reacting to a short-term aberration. For example, a particularly stormy year can have a significant and immediate impact on the colonies, but there may be no long-term consequences. Also, when attempting to count an entire population, and in this case quite a large population, there is a high probability that there will be errors. The trend measured over a longer period is much more relevant and reliable.

The lower limit of 18,000 given in this example is much lower that the actual population, the population was 21,688 individuals in 2011. This is not to suggest that we are prepared to lose over 10%. The specified limit takes account of the fact that the population is counted only once a year, and in some areas counts are made from a boat at sea. Consequently, the method is inherently inaccurate.

The study plot population

This is a rather specific example, but it illustrates two important points. Almost 30 years ago, a number of guillemot colonies throughout the UK were established as study populations (Wilson 1992). They were intended to ensure that a representative sample of the UK population was accurately monitored using a common methodology which enabled the results from all locations to be combined and compared. This means that local variations which may be the consequence of site- or locality-specific factors can be detected. This information is far more reliable and significant than data collected in isolation from any specific location. This approach is particularly important when attempting to specify limits for migratory species, such as wintering populations of wildfowl. The number of birds present at a particular location can be a response to weather systems throughout northern Europe and beyond, so setting a limit for total attendance is almost meaningless. A much more meaningful approach is to express a limit that represents the number of birds on any particular site as a proportion of the birds from a wider geographical range.

The guillemot study plots are a small, but representative, sample of the total population. The plots can be counted accurately and consistently; data from samples are much more reliable than counts of entire populations. Problems obviously arise when attempts are made to use samples to calculate the total population, but, in most cases, this is not necessary. Conservation management can respond to changes or trends in populations without being too concerned with totals.

Fig. 15.4 Guillemot population trends

The distribution of the colonies
Earlier, the potential for human disturbance to have a local effect on individual colonies was discussed. With all attributes it is important that they give an indication of potential future trends before any significant damage occurs. If the only attributes monitored were the total island population and the study plot populations, it would be possible, particularly when the overall population is increasing, to miss the impact of disturbance to a local colony. However, if the impact of a particular human activity is not recognised at an early stage it may continue to be tolerated, with damaging consequences. The solution is to monitor the distribution of all colonies, or those considered most vulnerable. A photographic record would possibly be sufficient.

The annual survival rate of breeding adults & the annual breeding success
Previously, the need to identify attributes that provide an early warning of changes that may take place was discussed, and two of the early indicators of potential decline in a population were described: a reduction in the adult survival rate and a reduction in the productivity rate. Adult survival rate and productivity rate are the attributes: both can be measured, but it is prohibitively expensive for most guillemot populations and, in fact, in most situations where species are managed as special features. The approach adopted for guillemots is to concentrate efforts on a typical site and to assume that the results will be representative of population trends within a much larger area.

I have included this guillemot example because it is one of the most complete species objectives available. There is no suggestion that this approach should be applied to all species management, though, of course, in ideal circumstances, this is what should be done.

Fig. 15.5 Guillemot breeding success

Table 15.7 Vision for an oak woodland with quantified attributes

Definition of Favourable Conservation Status for a habitat	**Vision for an upland acidic oak woodland (attributes** are shown in bold)	**Quantified attributes with limits**
The area of the habitat must be stable in the long term, or increasing.	The **entire site** is covered by a high forest, broadleaf woodland which has a closed canopy.	**1. Extent of the woodland.** *Lower limit:* 135 ha (see map)
Its quality (including ecological structure and function) must be maintained	The woodland is naturally regenerating, with plenty of **seedlings and saplings** particularly in the canopy gaps. There is a changing or dynamic pattern of **canopy gaps** created naturally by wind-throw or as trees die.	**2. Canopy cover (within the woodland).** *Upper limit*: 90% *Lower limit:* 75% **3. Canopy gap creation rate.** *Upper limit:* 0.5% of the canopy per annum *Lower limit*: 0.25% of the canopy per annum **4. Natural regeneration of canopy trees (in gaps).** *Upper limit:* not required *Lower limit:* 2 viable saplings per 0.01 Ha of gap
Any typical species must also be at FCS.	The woodland has a **canopy and shrub layer** that includes locally native trees of all ages with an abundance of standing and fallen **dead wood** to provide habitat for invertebrates, fungi and other woodland species.	**5. Species composition of the canopy.** *Upper limit:* not required *Lower limit:* 90% locally native species **6. The volume of dead wood (fallen trees and branches, dead branches on living trees, and standing dead trees).** *Upper limit:* not required *Lower limit*: 30 cubic metres per hectare

Discussion of the example given in Table 15.7:

Extent of the woodland

This should always be quantified in an objective; the definition of FCS requires that a habitat is stable or increasing in area. The extent of a habitat is best shown on a map.

Canopy cover, canopy gap creation rate & natural regeneration of canopy trees (in gaps)

Natural regeneration in oak woodlands is only possible when there is enough light for the seedlings and saplings. Woodland that contains a dynamic, changing pattern of gaps will also, over time, deliver a structurally diverse canopy which will provide opportunities for a wide range of associated species. There is, therefore, a need to ensure that there are sufficient gaps in the canopy to provide opportunities for the regeneration and that the gaps fill with the required canopy species. Three different attributes are monitored to provide the evidence that this is actually happening: First, the overall canopy cover is defined; that is, how much of the canopy should be open at any given time. The rate at which gaps should be created is specified (this attribute will also contain a definition of 'gap'). Finally, the presence of viable saplings in the gaps is monitored. Monitoring saplings is not strictly necessary; if gaps are created and the overall canopy cover does not fall below the lower limit, this must mean that the gaps are being filled. In other words, the same thing is being measured twice. This often happens, and in this case the justification is the time scale. The presence of saplings will indicate that conditions are right for regeneration long before the fact that a gap is filling can be measured. This, once again, is an example of an early warning system.

Another common reason for using two or more attributes to monitor the same thing is in circumstances when there is limited confidence in the evidence provided by either attribute. For example, the monitoring of the abundance or trends in a population of marsh fritillary butterflies *Euphydryas aurini* is based on a methodology derived from the UK national butterfly monitoring scheme (Pollard and Yates 1993), where adult butterflies are recorded within a defined band along a fixed transect. Thus, the first attribute monitored is the number of adults recorded along a fixed transect. For many reasons, this method is not always reliable. For example, butterfly activity, which is severely restricted by weather conditions, is spasmodic and may not coincide with sampling times. A second attribute, the number of larval webs recorded along a fixed transect, is therefore monitored. This is also intended to measure population abundance and trends. Both methods have different potential for error, but taken together they increase the probability that the evidence is reliable. (See Case Study 4 for further details.)

Species composition of the canopy

This is a very obvious attribute; the most important species in any woodland are most often the canopy trees.

The volume of dead wood

Dead wood is included as an attribute because it is an extremely useful surrogate, i.e. the presence of dead wood will indicate potential for the presence of a wide range of typical woodland species, including beetles, fungi, epiphytic lichens and hole-nesting birds (Peterken 1993). The general view concerning the use of environmental or habitat-condition surrogates for species, and particularly for invertebrates, is that it is certainly better than nothing. Reliance on the presence of a broadly defined habitat, for example woodland, as a surrogate to indicate the presence of a particular species is too crude a measure. It must be supplemented by other surrogates, such as dead wood, which provide more specifically for the species in question (Samways 2005). Dead wood is also a good measure of the ecological structure of a woodland; the presence of too little or too much dead wood will be indicative of a dysfunctional or inappropriately managed woodland. So, although I associated this attribute with the section of FCS which relates to the need to ensure the presence of typical species in a habitat, it could also be taken as a measure of the ecological structure and function of a habitat.

Fig. 15.6 Dead wood in a living ancient tree

15.5.4 *The Selection and Use of Factors as Performance Indicators*

Definition

A factor is anything that has the potential to influence or change a feature, or to affect the way in which a feature is managed. These influences may exist, or have existed, at any time in the past, present or future. Factors can be natural or anthropogenic in origin, and they can be internal (on-site) or external (off-site).

The definition of FCS states that the factors must be under control for both habitat and species. The reason for this is obvious from the preceding definition of a factor. An objective based on FCS must, therefore, include some means of defining when factors are under control. The relationship between attributes and factors was discussed earlier in this chapter. For most factors, attributes are used as performance indicators to provide evidence of whether the factor is under control. However, the impact of some factors cannot be measured indirectly through the attributes. This happens when attributes either cannot be identified or cannot be monitored. There are also occasions when it is much more efficient to monitor the factor. These factors are almost exclusively anthropogenic in origin and include, for example, invasive alien species, pollution and water quality.

There are some factors which, although very important, cannot be monitored. These are the factors that influence our ability to manage a feature, for example, a shortfall of resources, difficult terrain, legislation or the lack of legislation. Past management or human intervention is often an extremely important factor, but, obviously, this type of factor cannot be monitored.

15.5.4.1 The Relationship Between Factors and Attributes

The relationship between factors and attributes has already been discussed, but there is a need to develop the discussion further at this stage. While factors are the influences that can change or maintain a feature, attributes reflect the changes that take place, or the conditions that prevail because of these influences. Some of the attributes will have been selected as a consequence of particular factors, and the measurement of these will act as a surrogate in place of directly measuring the factor. For example, on a raised bog one of the most important factors is the water table. Although expensive, it is possible to measure the water table. However, unless the precise relationship between the water table and the condition of the bog vegetation is understood, managers will not have sufficient confidence to rely on water table measurements alone when making decisions. One of the more obvious consequences of a low water table is that trees, usually birch *Betula,* will begin to survive in the drier conditions. Birch trees are an attribute of the bog. They occur naturally, but are suppressed by a high water table and do not survive beyond the seedling or early sapling stage. Ideally, in these and similar circumstances, the attribute (trees) is monitored and the factor (water table) is recorded. The water table can only be recorded because, when the required height range of the water table is unknown, there is no standard that can be measured.[8] In circumstances where the limits within which a factor is acceptable are known, the factor can be monitored and used as a performance indicator. So, for a bog where the relationship between water levels and the condition of the vegetation is understood (i.e. we know when we have too much or too little water, and the water level can be measured) the water level could be used as a performance indicator.

Reliance on using factors in place of attributes can place features at risk. For example, it is common practice to set very specific grazing levels for a range of different habitats. The number and type of animals, along with the timing of grazing, is specified, with an assumption that this will deliver, or maintain, a habitat in a favourable condition. These decisions are, or at least should be, based on experience or on evidence obtained from a similar situation where a particular management activity has delivered something desirable. How can this approach possibly fail? It will fail because there are many unpredictable and variable factors. For example, when managing grassland, the grazing levels that are appropriate during wet years, when the vegetation grows through the summer, are completely inappropriate during periods of drought. The only way to be sure that vegetation is in a favourable condition is to monitor the attributes that are a measure of the condition of the feature. In this grassland example, the most important attributes will be sward height and species composition. The response to any undesirable changes to the condition of the vegetation will be to adjust management; this is simply another way of saying 'control the factors'. However, it is essential that the factors, in this example grazing levels, are recorded.

15.5.4.2 Factors Can Be Monitored Directly, Indirectly or Both

There are occasions, though not many, when the levels or limits of tolerance to a factor are known. In these cases, the factor can be monitored directly. These factors are always anthropogenic influences, and the most frequently encountered examples are invasive alien species. The impact of most invasive species, although obvious, is often difficult to describe and measure. In most circumstances, we will have no, or very low, tolerance of the presence of alien invasive species. This is because once they have a foothold they spread rapidly and can become impossible to control. Clearly, when dealing with this type of factor there is little point in attempting to find attributes that measure the impact of the factor. Instead, we specify an upper limit which represents our tolerance to the factor.

When managing habitats that are sustained through natural processes, for example, a coastal dune system, our objective could be to enable the habitat to develop

[8] Reminder: monitoring is surveillance undertaken to ensure that formulated standards are being maintained.

in response to natural processes. The definition of a factor is anything that has the potential to influence or change a feature, or to affect the way in which a feature is managed. Clearly, this means that natural processes are factors. An example of a natural factor for a coastal dune ecosystem is the rate of sand deposition. The dunes as a whole, and particularly the early successional stages, are dependent on sand deposition. This natural factor can be variable for entirely natural reasons. Unfortunately, coastal processes are often modified by human influences, such as offshore dredging, coastal engineering, etc. This is not unusual; natural factors or processes are quite often influenced by anthropogenic factors. The primary factor is the natural rate of sand deposition and the secondary factors are the anthropogenic influences.

Let us assume, in this case, that management has decided to accept or tolerate natural processes. It is recognised that the dune systems can be ephemeral features; the natural sand supply may become depleted and the system may erode. However, in contrast, there will be no tolerance of changes to the sand supply which are the consequence of anthropogenic factors, particularly when these might be controllable. The problem is the perceived need to differentiate between natural and anthropogenic factors. How can management differentiate between changes in sand deposition which are the consequences of either natural or anthropogenic factors or a combination of both? This may, of course, be impossible, but there is a potential solution. The important issue is that the primary factor – which is the rate of sand deposition – can be easily measured and, once limits are specified, it can be monitored. If the limits are exceeded, the manager will know that there has been a change which is sufficient to give cause for concern. This is the point at which there is an actual need to differentiate between natural change and change brought about by anthropogenic factors. In addition to monitoring the sand, all human influences (dredging, soft and hard engineering, etc.) in the coastal cell are recorded. If the specified limits for the rate of sand deposition are exceeded, and there is a direct and demonstrable correlation with human influences or engineering works, this will provide the evidence required to justify a detailed investigation and, if necessary, the suspension of the works.

As a general rule, the factors that directly impact on a feature are monitored as performance indicators, and the secondary factors that indirectly impact on a feature (that is, by influencing other factors) are recorded or placed under surveillance. It can be the secondary factors that provide the focus for management activities.

It should now be clear that all the important factors must be monitored, indirectly, directly or both. Even when attributes provide an indirect means of monitoring the impact of a factor, there is also a need to measure the factor directly in some way. This is because attributes will change as a consequence of a single factor or the combined influence of several factors. When a change takes place it is essential that we understand which factors are responsible. So, even when a factor cannot be monitored because it is not possible to set limits, some way must be found to measure or record the factor. Over time, an understanding of the relationship between factors and a feature will be developed and management effectiveness will improve.

15.5.4.3 Public Use as a Factor

(This is an extremely important section as it provides the essential link between planning to protect features and planning to provide for access or tourism).

Fig. 15.7 Public use as a factor

For many protected sites, public access or tourism and the provision of opportunities for leisure activities can be very important. Occasionally, it is the most important purpose of management. With few, if any, exceptions, people will have some impact on the site features. In other words, they, and more particularly their activities, are *factors*. However, the key role of any protected area or nature reserve is to ensure that wildlife is safeguarded against the excesses of uncontrolled human behaviour. This is true even when the prime purpose of a site is to provide for people. If the wildlife, countryside or wilderness quality that attract people to a site are lost or damaged, people may stop visiting, or the quality of their experience may be diminished. As a consequence, human activity must be controlled, but there may also have to be some limited compromises, and areas of habitat may have to sacrificed to provide the infrastructure necessary to accommodate people (for example, paths, roads, parking facilities, accommodation, information centres, etc.).

Some aspects of public use can have serious and obvious consequences for wildlife features, for example, climbing on cliffs used by seabirds, dog walking (emptying) in sensitive botanical sites, wildfowling where the feature is a wintering population of wildfowl. Where the activity is changing, or has obvious potential to change, a feature, these activities should be recognised as factors which must be kept under control. This type of factor is often monitored directly, i.e. specified limits are used to define our tolerance and provide a performance indicator.

Fig. 15.8 Public use is an important factor

It is much more difficult to deal with human use when there is not such a direct or obvious impact on a feature or features. From an ethical, and sometimes legal, position, it is difficult to rationalise a situation where an area is declared a nature reserve and the consequence is that subsequent public use of the area damages the wildlife. An appropriate response in these circumstances would be to consider the precautionary approach, as established in Principle 15 of the 1992 Rio Declaration on Environment and Development adopted by the United Nations Conference on Environment and Development (UNCED), which affirms that:

> In order to protect the environment, the precautionary approach shall be widely applied by States according to their capabilities. Where there are threats of serious or irreversible damage, lack of full scientific certainty shall not be used as a reason for postponing cost-effective measures to prevent environmental degradation.

If the precautionary principle is invoked, there is no need for scientific proof in order to restrict human use or any specific activities if there is a reason to believe that they are a threat.

In essence, the precautionary principle is about not taking chances with our environment. So, logically, when applying the principle to the carrying capacity of a feature, there should be an obligation to prove with full scientific certainty that an activity will not cause any damage before an activity or level of activity is permitted.

Turning to Europe, the legal situation for Natura 2000 sites is clearly set out in an official European Commission document, *Managing Natura 2000 sites: The provisions of Article 6 of the 'Habitats' Directive 92/43/EEC* (2000). Although the legal implications of Article 6 only apply to Natura sites, the interpretation is widely relevant. The article states:

> Member States shall take appropriate steps to avoid, in the special areas of conservation, the deterioration of natural habitats and the habitats of species as well as disturbances of the species for which the areas have been designated, in so far as such disturbance could be significant in relation to the objectives of this directive.

In addition to the article itself, the document contains two particularly appropriate passages:

> This article should be interpreted as requiring Member States to take all the appropriate actions which it may reasonably be expected to take, to ensure that no significant deterioration or disturbance occurs.

> In addition, it is not necessary to prove that there will be a real significant effect, but the likelihood alone ('could be') is enough to justify corrective measures. This can be considered consistent with the prevention and precautionary principles.

Having made the case for recommending that the precautionary principle should be applied to the actual and potential public use of an area, the next step is to consider the need for performance indicators. Earlier in this section the point was made that factors can only be monitored when the limits within which they can be tolerated are known. For many, if not most, human activities, particularly leisure activities, the limits are not known. The precise impact that public use will have on features is also rarely understood and so the potential for using attributes as surrogates is very limited. The obvious way forward is to limit or manage access and human activities rather than expecting the wildlife to adapt. The management required to control human activities will be identified and described in the 'rationale' section, which follows later in the planning process. Compliance monitoring or formal recording will be used to ensure that appropriate management is in place.

15.5.4.4 Specified Limits for Factors

Specified limits are applied to factors in precisely the same way that they are used with attributes. If we express the limits within which a factor is considered acceptable, we have provided a performance indicator. Limits are an early warning system that should trigger action before it is too late. They are used to express the range of values within which a factor can be considered beneficial to, or does not threaten, a feature.

Limits require an upper or lower limit, or both. In general, upper limits are applied to undesirable factors – they define our maximum tolerance – and lower limits are applied to positive factors. In reality, there are few occasions where the impact of a factor is sufficiently well understood to enable us to set both upper and lower limits with any confidence. In most cases, the best that we can achieve is to set a lower limit for positive factors and an upper limit for negative factors. Limits should only be set at the current level of influence of a factor if that level is considered compatible with the achievement of our objective for the feature.

When the value of the factor falls outside the specified limits at least we have evidence to suggest that management is inappropriate and, more importantly, that the status of the feature may deteriorate and can no longer be considered favourable.

Examples of specified limits for factors are given in the text boxes. The following demonstrates the use of limits where the factor is the recreational use of a site and where the features are populations of cliff nesting birds. Extracts from an information leaflet for climbers:

> The cliffs of south Pembrokeshire are within the Pembrokeshire Coast National Park. They are of international importance for cliff-nesting birds and provide some of the best sea-cliff climbing in the country. This leaflet gives details of agreed seasonal climbing restrictions established to protect nesting and feeding sites. These are essential for the continued existence of the colonies of seabirds and for the chough, peregrines and ravens which nest here. The climbing areas are within Sites of Special Scientific Interest (SSSI) and European Special Protection Areas (SPA).
>
> Do not climb between red markers during restricted periods.
> North Pembrokeshire seasonal restrictions: 1st February to 1st August.
> Check signs on site.
> Restrictions may vary or be lifted early.
>
> These restrictions have been agreed between the British Mountaineering Council (BMC), Pembrokeshire Climbing Club, National Trust (NT), Ministry of Defence, Wildlife Trust South and West Wales, Countryside Council for Wales (CCW) and the Pembrokeshire Coast National Park Authority.

The 'limits', referred to as restrictions in the leaflet, are linked to management actions, including the production and distribution of the leaflet, liaison to gain the cooperation of the climbers, and employment of a ranger to patrol the cliffs. Compliance with the limits is monitored by the ranger and, providing there are no significant breaches, this factor is considered to be under control.

Fig. 15.9 Limits on climbing

15.5.4.5 Monitoring Factors

Monitoring factors requires exactly the same care and consideration required for attributes (see preceding section). Monitoring is only possible when the factor is quantifiable. Recording, surveillance, or indeed research, will be required when the relationship between a feature and a factor is unclear. Even if all the factors are within limits, this must not be taken as conclusive evidence that a feature is at FCS. The attributes must also be within specified limits.

In addition to monitoring the factors, there is also a need to demonstrate that the factor can be kept under control. In other words, we need to have some confidence that management is sustainable. Whenever factors are monitored and whenever management is reviewed this question should be addressed. There are many secondary factors that can influence our ability to sustain management, for example, short-term tenure agreements, failing resources, changes in agricultural policy and global climate change.

This is demonstrated by returning to the example of a woodland objective; with the addition of the factors with limits the objective is now complete.

Table 15.8 Vision for an oak woodland with quantified factors

Definition of Favourable Conservation Status for a habitat	Vision for an upland acidic oak woodland (the factors are underlined)	Quantified factors with limits
The factors that affect the habitat, including its typical species, must be under control.	The woodland does not contain any rhododendron or any other invasive alien species with the exception of occasional beech and sycamore. There will be periodic light grazing by sheep and very occasionally by cattle. This will help maintain the ground and field layer vegetation but will not prevent tree regeneration.	**7. Rhododendron** *Upper limit:* no flowering rhododendron *Lower limit:* not required **8. Sycamore** *Upper limit:* 5% of the canopy *Lower limit:* not required

Discussion of the example given in Table 15.8:

Rhododendron *Rhododendron ponticum*

The limit applied to rhododendron requires some explanation. The obvious question is: why tolerate any rhododendron when this is such an aggressive, alien, invasive species? Clearly, the aim should be to eradicate the species from the site. The problem is that it is very difficult, probably impossible, to prove that there is no rhododendron in an area. The seedlings are extremely small for the first few years and even when rhododendron reaches the sapling stage it is very difficult to locate, particularly when growing in dense woodland. Once the plants begin to flower they can be easily seen, because the pink flowers stand out against the green of the woodland. Monitoring is then very simple: unskilled individuals can wander through the wood searching for flowers and, once flowers are seen, all the rhododendron, flowering and non-flowering, can be controlled. If this process is maintained, sooner or later the species will be eradicated. This example clearly demonstrates the need to consider how a factor will be monitored before setting a limit.

Sycamore *Acer pseudoplatanus* (Fig. 15.10)

This is an alien species in the UK, though, recently, its presence has been tolerated, particularly in places where it is not very invasive and will not dominate the woodland canopy. The woodland site used in this example is located on very impoverished acid soils that are not at all suitable for sycamore. The limit is, therefore, probably not necessary, but why take chances? Setting limits for invasive species, particularly where there is some tolerance, is nearly always an arbitrary decision. How do we decide when there is too much? A simple and pragmatic approach is to set a level at a point beyond which management will become difficult or prohibitively expensive. In other words, invasive species are always kept at a controllable level.

This approach to setting objectives that include performance indicators facilitates targeted and specific monitoring projects. However, this can never replace the presence of experienced and vigilant reserve staff. It will never be possible to identify all the factors or predict all the consequential changes to a feature. It is therefore important to maintain a continuity of presence on a site that is sufficient to detect any unforeseen changes.

Fig. 15.10 Sycamore

15.5.4.6 Management as a Performance Indicator

Two different kinds of performance indicators that can be used to provide the evidence that a feature is at FCS have been described, but there is potentially a third area from which evidence may be drawn.

The conservation management of habitats and species is mainly, if not entirely, achieved through controlling factors, and, more specifically, the consequences of human intervention, past, present and future. Our ability to achieve conservation objectives will always be constrained by our ability to manage factors. This could suggest that evidence showing that appropriate management is in place can be used to support a claim that a feature is at FCS. The obvious question is: how do we know when management is appropriate? The answer: when the feature is maintained at Favourable Conservation Status. Some organisations have, in the past, indicated that if management is in place then they can conclude that the feature is, or will become, favourable. Unfortunately, they have not suggested the use of, or a definition for, 'appropriate'. This is not to say that the fact that management is in place is irrelevant. When assessing the status of a feature, it is extremely important to look for evidence to demonstrate that a factor is under control (factors with limits), and also for evidence that gives some confidence that the factors will be kept under control in the future. Therefore, we need to know that management is in place and that it can be maintained in the future. The simple fact that a feature is being managed can be taken to suggest that, at the very least, attempts are being made to control the factors. If the management process is adaptive, there is also an implication that, sooner or later, it will become appropriate.

At a later stage in the planning process the management requirements are identified. All management actions should be clearly described, and all these activities will be recorded. In some cases, organisations will use compliance monitoring or audit to ensure that the work is being carried out in compliance with the management plan.

15.6 Testing Objectives

The following is a series of questions that can be used to test objectives in a management plan. If the answer is yes to all the questions then the objective is probably fit for purpose.

- Do the objectives describe the outcomes, at least for the important conservation features?
- Are the objectives specific; do they directly address the features; are they clearly defined, and not open to interpretation?
- Do the objectives define the outcome required of a feature and not the actions taken to obtain or maintain that condition, nor a mix of the two?
- Has the objective been quantified (performance indicators) and can it be monitored?
- Is the objective relevant; does it comply with the strategies, policies and legal obligations that govern the organisation responsible for managing the site or feature?
- Is the objective communicable; can it be easily understood by the intended audience?

In addition, where the objective is to obtain Favourable Conservation Status the following should be considered:

For habitat features:

- Has the area of the habitat been specified?
- Are there sufficient performance indicators to define the quality (including ecological structure and function) of the feature?
- Are there sufficient performance indicators (surrogates are acceptable) to provide the evidence that typical species are at FCS?
- Are there sufficient performance indicators to demonstrate that the factors are under control? (The evidence can be direct or indirect from attributes.)

For species features:

- Has the size of the population been specified, or are performance indictors that can be used to monitor population trends included?
- Are there sufficient performance indicators to provide evidence that the population is sustainable in the long term?
- Has the range of the population been defined, or is there a performance indictor that can be used to monitor changes in the range?
- Has the habitat which supports the population been given adequate attention? (In most cases, an objective that meets this test should have been prepared for the habitats that support the species.)
- Are there sufficient performance indicators to demonstrate that the factors are under control? (The evidence can be direct or indirect from attributes.)

For sites where natural processes will determine the outcomes and where defining the outcomes is not appropriate or possible:

- Does the plan identify sufficient evidence (performance indicators) that can be used to demonstrate that the ecosystem, habitats and species are developing in an appropriate direction?

References

Alexander, M. (2003). *CCW Guide to Management Planning for SSSIs, Natura 2000 Sites and Ramsar Sites*. Countryside Council for Wales, Bangor, Wales.

Alexander, M. (1994). *Management Planning Handbook*. Countryside Council for Wales, Bangor, Wales.

Alexander, M. (2005). *The CMS Guide to Management Planning*. The CMS Consortium, Talgarth, Wales.

Alexander, M. (1996). *A Guide to the Production of Management Plans for Nature Conservation and Protected Areas*. Countryside Council for Wales, Bangor, Wales.

Alexander, M. (2000). *Guide to the Production of Management Plans for Protected Areas*. CMS Partnership, Aberystwyth, Wales.

Arizona State Office (1995). *Eagletail Mountains Wilderness Management Plan, Environment Assessment and Decision Record*. US department of the Interior, Bureau of land Management, Arizona, USA.
Birkhead, T. R. and Perrins, C. M. (1997). *Auk studies on Skomer Island*. In: Welsh *Islands: Ecology, Conservation and Land Use*. Rhind, P. M., Blackstock, T. H. and Parr, S. J. (ed.), Countryside Council for Wales, Bangor, Wales, pp. 104–109.
Clarke, R. and Mount, D. (1998). *Site Management Planning*. Countryside Commission, Cheltenham, UK.
Elzinga, C. L., Salzer, D. W., Willoughby, J. W. and Gibbs, J. P. (2001). *Monitoring Plant and Animal Populations: A Handbook for Field Biologists*. Blackwell Science, Inc. Malden, Massachusetts, USA
English Nature (2005). *NNR Management Plans: A Guide*. English Nature, Peterborough, UK.
Eurosite (1999). *Toolkit for Management Planning*. Eurosite, Tilburg, The Netherlands.
Hamel, P.B. (2000). *U.S. Fish & Wildlife Service Cerulean Warbler Status Assessment*, U.S. Forest Service, Southern Research Station, Fort Snelling, MN, USA.
Hatton-Ellis, T. (2011). *Can the Keystone Species and Keystone Habitats concepts be usefully applied in the Natural Environment Framework?* Unpublished report Countryside, Council for Wales, Bangor, UK.
Idle, E. T. and Bines, T. J. H. (2005). *Management Planning for Protected Areas, A Guide for Practitioners and Their Bosses*. Eurosite, English Nature, Peterborough, UK.
JNCC (1998). *A Statement on Common Standards Monitoring*. Joint Nature Conservation Committee, Peterborough, UK.
JNCC (2004). *Handbook for Phase 1 Habitat Survey – A Technique for Environmental Audit, Revised reprint 2003*. Joint Nature Conservation Committee, Peterborough, UK.
Kaljuste, T. (1998). *Guidelines for Development of Management Plans for Protected Areas*. Ministry of Environment of Estonia, Tallinn, Estonia.
Lindenmayer, D. B., Margules, C. B. and Botkin, D. B. (2000). *Indicators of biodiversity for ecologically sustainable forest management*. Conservation Biology, 14(4).
Manning, R. E. (1986). *Studies in Outdoor Recreation*. Oregon State University Press, Cornwallis, Oregon, USA.
Margoluis, R. and Salafsky, N. (1998). *Measures of Success: Designing, Managing, and Monitoring Conservation and Development Projects*. Island Press, Washington DC.
McCool, S. F. (1989). *Limits of acceptable change: some principles towards serving visitors and managing our resources*; Proceedings of the visitor management strategies symposium. University of Waterloo, Ontario, Canada, pp. 195–200.
Ministere charge de l'Environnmement (1991). *Plans de gestion des reserves Naturelles*. Direction de la Protection de la Nature, Paris.
National Parks and Wildlife Service (2005). *Management Plan for the Wicklow Mountain National Park (2005 – 2009)*. The National Park and Wildlife Service – Department of the Environment, Heritage and Local Government, Ireland.
NCC (1981). *A Handbook for the Preparation of Management Plans for National Nature Reserves in Wales*. Nature Conservancy Council, Wales, Bangor, UK.
NCC (1983). *A Handbook for the Preparation of Management Plans*. Nature Conservancy Council, Peterborough, UK.
NCC (1988). *Site Management Plans for Nature Conservation, A Working Guide*. Nature Conservancy Council, Peterborough, UK.
Peterken, G. F. (1993). *Woodland Conservation and Management*, 2nd edn. Chapman and Hall, London.
Peterken, G. F. (1996). *Natural Woodland, Ecology and Conservation in Northern Temperate Regions*. Cambridge University Press, Cambridge, UK.
Pollard, E. and Yates, J. T. (1993). *Monitoring Butterflies for Ecology and Conservation*. Chapman and Hall, London, UK.

Pullin, A. S. (2003). *Support for decision-making in conservation*. English Nature research reports, 493. English Nature, Peterborough, UK.

Ramsar Convention Bureau (2002). *New Guidelines for Management Planning for Ramsar Sites and Other Wetlands, Ramsar resolution VIII.14.* Ramsar Convention Bureau, Gland, Switzerland.

Rowell, T. A. (1993). *Common Standards for Monitoring SSSIs.* JNCC Peterborough. UK.

RSPB (1993). *RSPB Management Plan Guidance Notes,* version 3. Internal publication, The Royal Society for the Protection of Birds, Sandy, UK.

Samways, M. J. (2005). *Insect Diversity Conservation.* Cambridge University Press, Cambridge, UK.

Sinclair, A. R. E., Fryxell, J. M. and Caughley, G. (2006). *Wildlife Ecology, Conservation, and Management, second edition.* Blackwell Publishing, Oxford, UK.

Stankey, G. H., McCool, S. F. and Stokes, G. L. (1984). *Limits of acceptable change: A new framework for managing the Bob Marshall Wilderness.* Western Wildlands, 10(3), 33–37.

Tasmania Parks and Wildlife service (2003) www.parks.tas.gov.au.

Thomas, L. and Middleton, J. (2003). *Guidelines for Management Planning of Protected Areas.* IUCN Gland, Switzerland and Cambridge, UK.

Wilson, S. A. (1992). *Population trends of guillemots Urial aalge and razorbills Alca torda on Skomer Island, Wales, 1963 – 1991.* Unpublished report to Countryside Council for Wales, Bangor, Wales.

Wood, J. B. and Warren, A. (ed.) (1976). *A Handbook for the Preparation of Management Plans – Conservation Course Format, Revision 1.* University College London, Discussion Papers in Conservation, No. 18. London.

Wood, J. B. and Warren, A. (Ed.) (1978). *A Handbook for the Preparation of Management Plans – Conservation Course Format, Revision 2.* University College London, Discussion Papers in Conservation No. 18, London.

Chapter 16
Rationale for Biological and Other Features

Abstract The rationale is the process of identifying, in outline, the most appropriate management for the various site features. The procedure comprises two distinct phases, beginning with the identification of the status of the feature and an assessment of current conservation management. We will have some confidence in current management when the feature is considered to be at Favourable Conservation Status and little confidence when it is not. The second stage considers the relationship between factors and the condition of the feature, and the implications of the factors to management.

Fig. 16.1 Sand dune management

M. Alexander, *Management Planning for Nature Conservation: A Theoretical Basis & Practical Guide*, DOI 10.1007/978-94-007-5116-3_16,

16.1 Background

The 'rationale' is a British innovation. It first appeared in published form in the Nature Conservancy Council's *Site Management Plans for Nature Conservation – a working guide* (NCC 1988). At this time, the rationale was used '… to consider how the ideal objectives of management can be achieved, or if necessary modified'.

By 1996, the purpose of the rationale had been completely revised (Alexander 1996) so that its primary function was to identify the management requirement for each feature. The following list of specific functions was given:

The purpose of the rationale is:

- To consider the implications of the current status of a feature
- To identify the factors which are relevant to the feature and ensure that these are monitored or recorded
- To consider how a feature might be influenced by the impact of any significant factors
- To identify the most appropriate management and management options
- To consider the relationships and in particular any conflicts between the different site features
- To identify any factors that may require operational limits

The *CMS Guide to Management Planning* (Alexander 2005) recommends a modified structure for the management plan. Recognising that factors are used at several different times in the planning process, and to avoid unnecessary repetition, purposes 2 and 6 above were moved to an earlier section in the management plan.

The *Eurosite Management Planning Toolkit* (Eurosite 1999), which is a derivative of the NCC (1988) guide, replaces 'rationale' with 'second evaluation': 'The purpose of the second evaluation is to measure the effects of the identified constraints and modifiers on the ideal objective and to arrive at a set of practically achievable objectives'. At a later stage in the Eurosite plan there is a section called 'management strategies', which identifies management options and management actions.

Margoluis and Salafsky (1998) use 'activities' to describe management. They define 'activities' as: 'specific actions or tasks undertaken by project staff designed to reach each of the project's objectives'. The activities are linked to an objective, and they specify the task to be carried out. Activities are also achievable and 'appropriate to site-specific cultural, social, and biological norms'. This process is similar to the rationale described by Alexander (2005) in that it links management to objectives. Margoluis and Salafsky also consider the relationship or potential conflict between the different features. They differ in that there is no assessment of current management, probably because this planning approach is intended for new conservation and development projects.

16.2 Conservation Status

The use of 'conservation status' in the rationale section of a management plan was introduced by the Countryside Council for Wales (Alexander 1996). There do not appear to be any examples of the use of Favourable Conservation Status, or similar concepts, in the rationale or any other section of management plans outside Europe.

The concept of Favourable Conservation Status (FCS) originates in international and European treaties. The statutory definition as applied by the European Union dates back to 1992. 1993 marked a significant development in UK conservation planning. A report, *Common Standards for Monitoring SSSIs*, was prepared for the UK Joint Nature Conservation Committee (Rowell 1993). Rowell developed the concept of different states to describe the condition of a feature at a particular point in time. These were: optimal-maintained, optimal-recovered, sub-optimal-recovering, sub-optimal-not-changed, sub-optimal-declining, partially-destroyed and totally-destroyed. By 1998, in order to obtain consistency with Europe, 'optimal' was replaced by 'favourable'. The latest EU guidance (2011) recommends that, in order to avoid confusion, FCS should not be used for features at site level. It is extremely difficult to understand why anyone might be confused. FCS as a concept can be used, and is used, at any level, everywhere: from small populations on tiny sites, to describing the status of a feature within a bio-geographical region in a country. The EU now recommends the use of 'degree of conservation' at site level. At best, this is clumsy, uncomfortable language; at worst, it will create unnecessary confusion. This is an EU recommendation and not a directive. My advice would be to ignore it and to continue using FCS at any appropriate level. There are no implications or complications when reporting to the EU. If organisations are concerned, they could simply use 'status': i.e. the status of a feature could be described as favourable or unfavourable. Or, we could always revert to using 'optimal'. Regardless of the language that we choose to use, the concept remains the same. It is the definition of FCS that really matters, and not what we call it.

Status or Condition

In an attempt to establish common standards for monitoring features on domestic and European statutory conservation sites in Britain, the Joint Nature Conservation Committee (JNCC) published *A Statement on Common Standards Monitoring* (1998). The publication contained a section on 'judging the condition of site features' and provided five categories that could be used to report on the *'condition'* of the features. The use of 'condition' and not 'status' for reporting, particularly when reporting on Natura 2000 sites to Europe, is very confusing. This confusion is restricted to the UK: other European countries are content with 'status'.

The difference between status and condition is very important. 'A habitat or population can be at one of several states at any particular point in time.' (Rowell 1993) Thus, the condition of a feature is its state at any given time. Rather like a snapshot, it describes what is present at that time, but no more. The condition that we require for a feature is defined by the objective and specifically by the attributes which are used as performance indicators. The attributes are quantified and, when monitored, they allow us to differentiate between favourable and unfavourable condition. If the feature is monitored on several occasions it is also possible to determine whether change is taking place and the direction of change, i.e. the feature can be recovering or declining.

The JNCC (1998) provide seven categories to describe the 'condition' of a feature:

- **Favourable–maintained**
  An interest feature should be recorded as 'maintained' when its conservation objectives were being met at the previous assessment, and are still being met.
- **Favourable–recovered**
  An interest feature can be recorded as having 'recovered' if it has regained favourable condition, having been recorded as unfavourable on the previous assessment.
- **Unfavourable–recovering**
  An interest feature can be recorded as 'recovering' after damage if it has begun to show, or is continuing to show, a trend towards favourable condition.
- **Unfavourable–no change**
  An interest feature may be retained in a more-or-less steady state by repeated or continuing damage; it is unfavourable but neither declining nor recovering. In rare cases, an interest feature might not be able to regain its original condition following a damaging activity, but a new stable state may be achieved.
- **Unfavourable-declining**
  Decline is another possible consequence of a damaging activity. In this case, recovery is possible and may occur either spontaneously or if suitable management input is made.
- **Partially destroyed**
  It is possible to destroy sections or areas of certain features or to destroy parts of sites with no hope of reinstatement because part of the feature itself, or the habitat or processes essential to support it, has been removed or irretrievably altered.
- **Destroyed**
  The recording of a feature as destroyed will indicate the entire interest feature has been affected to such an extent that there is no hope of recovery, perhaps because its supporting habitat or processes have been removed or irretrievably altered.

'Status' takes things further: the definition of Favourable Conservation Status for a habitat includes the requirement that the 'ecological structure and function of the habitat must be maintained and that the factors that affect the habitat, including its typical species, must be under control' (European Communities 2000). For a species to be at FCS the definition includes: 'the population must be sustainable in the long term'. The status of a feature is, therefore, defined by a combination of its condition and additional evidence that make it possible to assess whether a feature is sustainable. This additional evidence is obtained by monitoring, directly or indirectly, the factors which influence the feature. If the factors are under control, and there is some evidence that they can be kept under control, we can assume that the feature can be maintained in a favourable condition. A feature that is, and can be maintained, in a favourable condition is at Favourable Conservation Status.

The categories that are used to describe 'condition' are not, as they stand, entirely suitable for describing the status of a feature. Whenever an objective for a feature is based on FCS, the following categories are more appropriate. When these are used in the UK, to avoid any confusion, it must be made clear in the text that these define 'status' and not 'condition'.

Categories That Describe the 'Status' of a Feature

- **Favourable–maintained**
  All the attributes of the feature are within the specified limits, and all the attributes of the feature were also within the limits at a previous assessment. (This can be expressed as: The feature is in a favourable condition and was also in a favourable condition at a previous assessment.) The factors are also under control and there is evidence that they can be kept under control, i.e. they are within limits and were within limits at a previous assessment.
- **Favourable–recovered**
  A feature is in favourable condition but was unfavourable at a previous assessment. The factors are also under control and there is evidence that they can be kept under control.
- **Favourable–unknown**
  A feature is in a favourable condition and the factors are within limits. There has been no previous assessment and consequently it is not possible to differentiate between 'maintained' and 'recovered'.
- **Recovering**
  A feature is in an unfavourable condition, but the factors are under control and there is a trend towards favourable condition. (The word 'unfavourable' has been omitted from the description because it conveys an unnecessarily negative message.)
- **Unfavourable-declining**
  The feature is in an unfavourable condition and the factors are not under control.
- **Unfavourable–unknown**
  There is insufficient, or no, evidence on which to come to a safe conclusion about the direction of change. The precautionary principle is applied, and the feature is recorded as unfavourable.
- **Partially destroyed**
  It is possible for sections or areas of features to be destroyed with no chance of recovery. This means that the feature will be unfavourable, because for a feature to be at FCS the size must be stable or increasing. The feature must be reassessed, and if it is considered to be viable at the reduced size the objective is reapplied, albeit on a smaller area.
- **Destroyed**
  The feature is completely destroyed with no potential for recovery or so damaged that complete and permanent loss is inevitable.

Note: 'Unfavourable–no change' has been omitted as this is something that would be extremely difficult or impossible to prove. A pragmatic approach is to record the status as 'unfavourable–unknown' or, when taking a precautionary approach, as 'unfavourable-declining'.

16.3 Rationale – Conservation Status

This section in the management plan is repeated for each feature. The rationale should begin with a statement on the current conservation status of each feature or parts (compartments) of a feature. It should also provide an outline of the evidence that led to the judgement and indicate the level of confidence in the judgement. Ideally, the evidence would be based on monitoring the performance indicators, attributes and factors, following the methodology prescribed earlier for each attribute. Unfortunately, this may not always be possible. However, the status of a feature can be established, albeit provisionally, if an assessment can be made with some level of confidence. By this stage in the process, all relevant site information will have been collated and the management objectives prepared. This suggests that any individual or team engaged in preparing the plan will have acquired a good understanding of the site and the features. It should, therefore, be possible for someone involved in the planning process to make a provisional assessment of the status of the features. Beware: it is never possible to conclude that a feature is changing, increasing or decreasing on the basis of a single round of monitoring.

The status of a habitat will sometimes vary across a site. This can happen on large, multi-ownership sites, mainly as a consequence of differences in past management. The obvious implications are that, although the objective should not vary over the site, the management requirement will. There is a legal obligation to report on the status of the features on statutory sites, and this has led to the development of monitoring systems that describe the overall status of a feature. These may meet a bureaucratic requirement but are less useful for management purposes. The assessment of status must be established at compartment level if the information is intended to guide management. For example, monitoring could reveal that 80% of the area of a site is at FCS and the remainder unfavourable. If we do not know which areas or compartments are unfavourable, we will not be able to target management where it is needed.

The status of a feature is always associated with management. When features are at Favourable Conservation Status, or recovering, it is probably safe to assume that management is appropriate, at least for the time being. Conversely, when a feature is unfavourable and declining, present management must be considered inappropriate. The only complication is that a newly introduced management regime may need to be in place for some time before a change in the status of a feature is detectable.

The following table can be used to provide a structured approach for identifying, in outline, a management response following the assessment of status. Clearly, the performance indicators, attributes and factors should have been monitored, or there should have been some less formal assessment of the performance indicators.

Table 16.1 Assessment of status

Assessment			
Current	**Comparison with previous assessment**	**Conservation status**	**Outline management response**
1. Both attributes and factors are within limits, and …	attributes and factors were within limits at last visit	Favourable maintained	No change to management is required.
	attributes were outside limits at last visit	Favourable recovered	Change in management may be required since management that has been in place to restore condition may not be appropriate for maintaining it.
	there is no previous assessment	Favourable (unknown)	No change to management is required.
2. Attributes are within limits but factors are outside limits, and …	factors were outside limits at last visit	Unfavourable (unknown)	Factors may be OK. The limits should be reviewed. No change of management.
	both attributes and factors were within limits at last visit	Unfavourable declining	We can expect condition to deteriorate and therefore a change of management is required to bring factors within limits.
	there is no previous assessment	Unfavourable declining	
3. Attributes are outside limits but factors are within limits, and …	attributes were outside limits at last visit, but we believe that if the factors are kept within limits for a longer period the feature will recover.	Recovering	We believe that management is appropriate. More time is needed for the condition to recover, so maintain current management.
	attributes were outside limits at last visit and the factors have been within limits for some time. There is no sign of any improvement in condition.	Unfavourable declining	The condition ought to be showing signs of recovery by now. Therefore, the limits should be re-assessed and management should be changed.

(continued)

Table 16.1 (continued)

Assessment			
Current	**Comparison with previous assessment**	**Conservation status**	**Outline management response**
	attributes were within limits at last visit	Unfavourable declining	The condition has deteriorated and changes to management are required (i.e. limits for factors are inappropriate, or new factors have appeared).
	there is no previous assessment	Unfavourable unknown	We are unable to make a decision with confidence. Management remains unchanged.
4. Both attributes and factors are outside limits, and …	recovery is possible if factors can be brought under control	Unfavourable declining	Changes to management are required.
	recovery of part of the feature is possible if factors can be brought under control	Part destroyed Unfavourable declining	The feature must be reassessed. If it is viable at the reduced size then the objective is reapplied to the smaller area.
	there is no prospect of recovery	Destroyed	Abandon the feature.
5. Attributes within limits, factors not assessed		Unfavourable (unknown)[a]	No basis on which to change management.
6. Attributes outside limits, factors not assessed		Unfavourable declining	
7. Attributes not assessed, but factors within limits		Unfavourable (unknown)	
8. Attributes not assessed, factor outside limits		Unfavourable declining	Management is required to bring factors back within limits.
9. Attributes and factors not assessed		Unfavourable (unknown)[a]	No basis on which to change management.

[a]*It could be argued that the conservation status is simply 'unknown'. However, if there is no evidence to demonstrate that a feature is favourable, and we adopt a precautionary approach, we should assume that it is unfavourable*

This table is a development of an original prepared by Dr Adam Cole King, Countryside Council for Wales 2003

16.4 Maintenance and Recovery Management

Management can be divided into two distinct phases: maintenance and recovery. Recovery management is applied when a feature is unfavourable in order to change it to favourable condition. Maintenance management is applied when a feature is in a favourable condition in order to maintain that condition. If recovery management is successful and the feature becomes favourable, there will often be a change from recovery to maintenance management. This usually means that a different intensity or method of management is applied. For example, unmanaged or abandoned grassland will usually become very rank and inundated with scrub. Recovery management to reinstate the grassland could include cutting the scrub, treating stumps with herbicide, and heavy grazing pressure. However, once the grassland is in a favourable condition there will no longer be a need to control scrub, and heavy grazing with one species may be replaced by lighter grazing with another.

Fig. 16.2 Grazing – essential management

16.5 Management Should Be Effective and Efficient

On sites where there is a known history of management and the status of the feature has been established, it is possible to differentiate between successful and unsuccessful management actions. In an adaptable planning system this is done when the rationale is revisited at the end of the first and all subsequent cycles. Even when management is considered successful it should be reviewed. This is because management, in addition to being effective, must also be as efficient as possible. Conservation managers must always seek the most efficient management techniques available. They cannot afford to waste limited resources. The key to this lies in gaining access to up-to-date information about similar projects which have been successful elsewhere. In the UK, the published journal *Conservation Land Management* provides an excellent source of information on conservation management. The advent of evidence-based conservation management (Pullin and Knight 2001; Sutherland 2000; Sutherland et al. 2004) has led to the development of associated web sites. These could become one of the most important sources of information on conservation management.

16.6 Rationale – Factors

There is a difference in the rationale between planning for the first time, when there is *no* record of management, and on subsequent occasions, when there *is* a record of management. An assessment of status is required for both, but the conclusions reached when planning for the first time will be limited by the lack of any previous assessment and records of management. The use of status as a guide to identifying appropriate management will be extremely limited. In these circumstances, an analysis of the factors is the best method for identifying management.

Management is invariably about controlling factors. Control means the removal, maintenance, adjustment or application of factors, either directly or indirectly. For example, grazing is an obvious factor for grassland habitats. Grazing can be removed, reduced, maintained at current levels, increased or introduced.

All the factors that have the potential to influence or change a feature (directly or indirectly) or to affect the way in which a feature is managed, as well as the relationship between the factors and management of the feature, are identified at an earlier stage in the planning process (see Chapter 13). The use of simple models or diagrams that establish the relationship between factors and management was included at the end of Chapter 14. The diagrams are also relevant to this chapter and are repeated here for convenience.

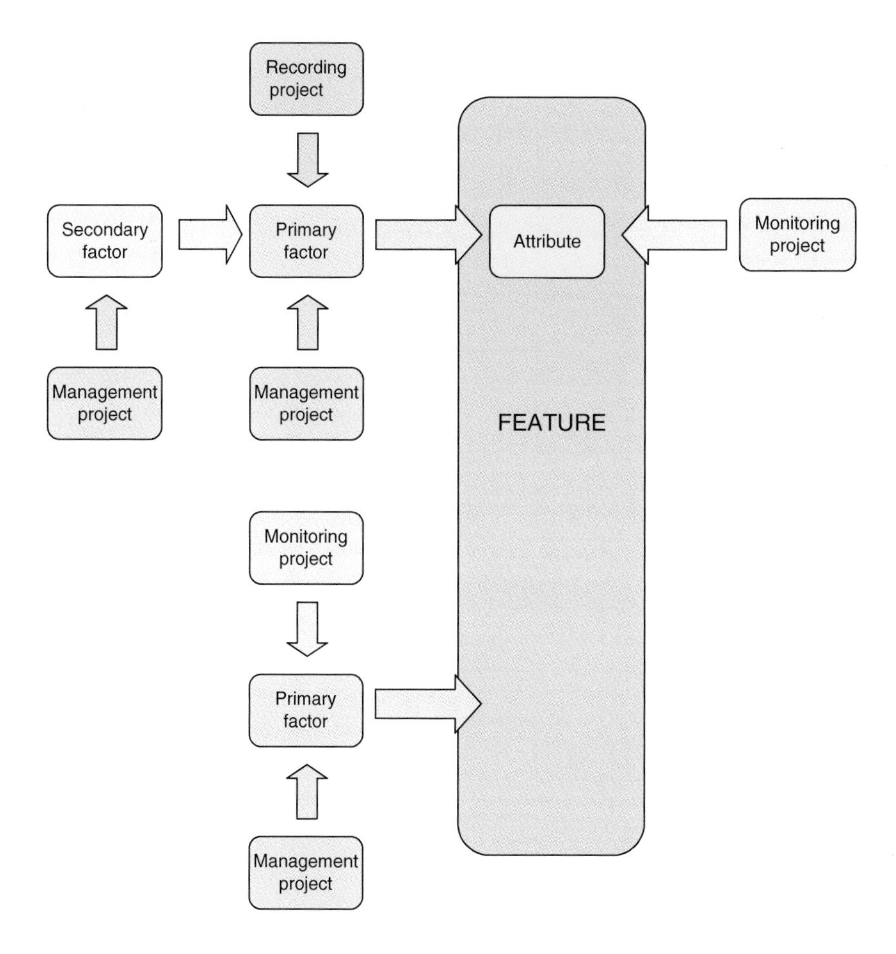

Primary factors will have a direct influence on the feature:as the factor changes so will the condition of the feature.
Secondary factors will not have a direct influence on the feature but will influence primary factors or our ability to manage the primary factors.
An attribute is a characteristic of a feature that can be monitored to provide evidence about the condition of the feature. Attributes change in response to the influence of a factor or factors.
Management projects are the activities or interventions that are employed to control the factors. All management projects are recorded.
Monitoring a feature can be direct, when an attribute is monitored, or indirect, when a factor is or monitored.
Recording projects are generally applied to factors where the range of acceptable levels is unknown, i.e. they cannot be monitored.

Fig. 16.3 The relationship between factors, attributes, management, monitoring, surveillance & recording

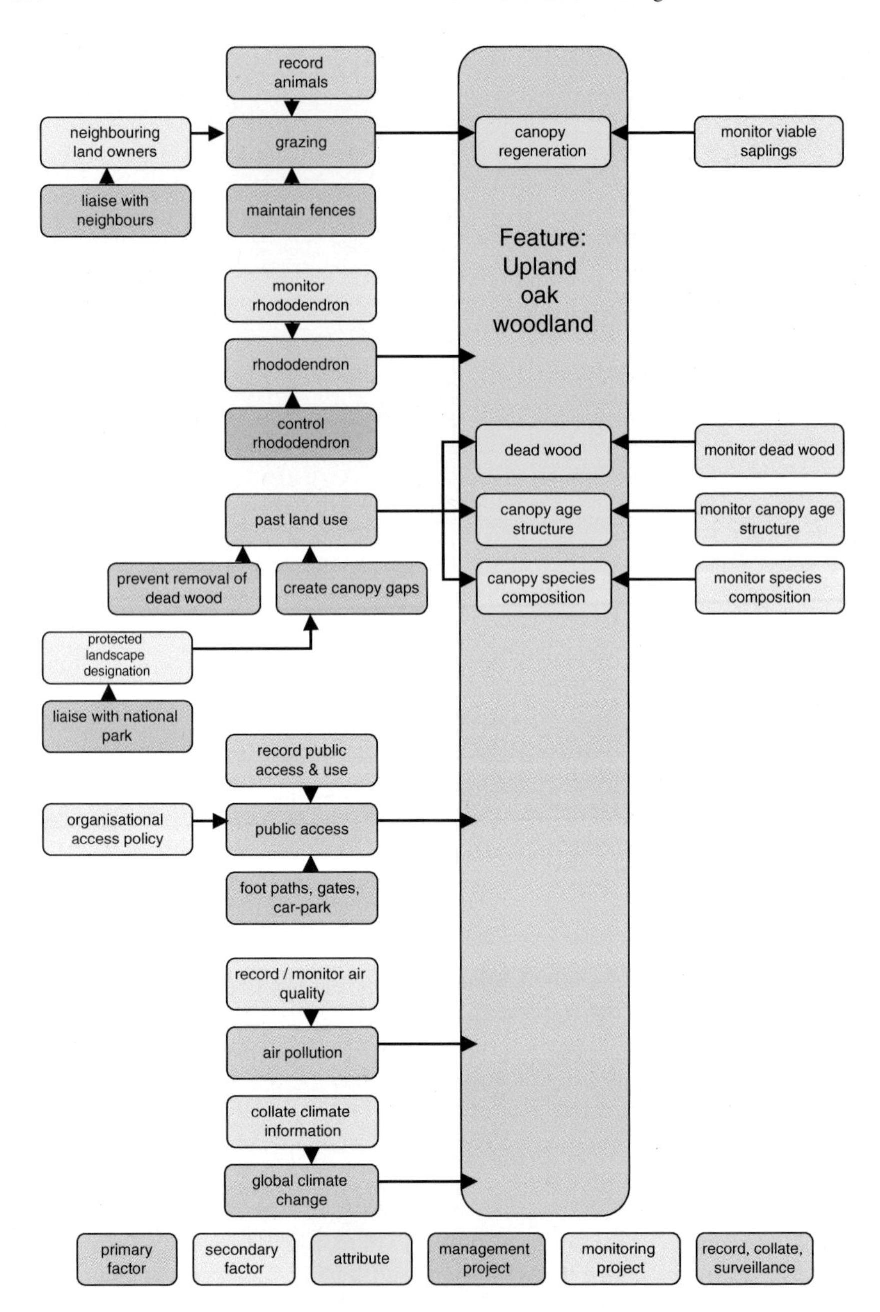

Fig. 16.4 The relationship between factors, attributes, management, monitoring, surveillance & recording in an upland oak woodland

Diagrams of this kind can be very informative, and some people find that information displayed in this way is easier to assimilate. Others are much happier with text. Whether or not a diagram is used, the relationship between management, the factors and the feature is the starting point for the discussion in this section of the plan.

The primary factors provide the subheadings for discussing the factors. Each should be considered in turn, beginning with an assessment of their impact on the feature (i.e. how they have changed, are changing, or could change, the feature). This will have been discussed when factors were used to guide the selection of some of the attributes.

The next step is to consider how the factors should be controlled or managed. Do not forget that factors can be positive and/or negative. The management of the primary factors can be influenced by any number of secondary factors. There is a complication: although individual factors may have a limited impact on a feature, in combination they can become a serious issue. This means that factors should be considered both individually and collectively. This is best achieved by treating the remaining primary factors as secondary factors.

The outcome of this section is an outline of the management required to ensure that the factors are kept under control so that the feature can be restored to, or maintained at, Favourable Conservation Status.

This approach, including the production of the diagram, can be a useful means of confirming or checking the management on sites which have a history of conservation management.

Table 16.2 Vision for an oak woodland with attributes and factors

Definition of Favourable Conservation Status for a habitat	Vision for an upland acidic oak woodland (**attributes** are shown in bold and the factors are underlined)
The area of the habitat must be stable in the long term, or increasing.	The **entire site** is covered by a high forest, broadleaf woodland.
Its quality (including ecological structure and function) must be maintained.	The woodland is **naturally regenerating,** with plenty of **seedlings and saplings** particularly in the canopy gaps. There is a **changing or dynamic pattern of canopy gaps** created naturally by wind throw or as trees die.
Any typical species must also be at FCS.	The woodland has a **canopy and shrub layer** that includes **locally native trees** of **all ages,** with an abundance of standing and fallen **dead wood** to provide habitat for **invertebrates, fungi** and other woodland species. The **field and ground layers** will be a patchwork of the characteristic vegetation communities, developed in response to local soil conditions. These will include areas dominated by **heather, or bilberry**, or a mixture of the two, areas dominated by **tussocks of wavy hair-grass or purple moor-grass,** and others dominated by **brown bent grass** and **sweet vernal grass** with abundant bluebells. There will also be quite heavily grazed areas of more grassy vegetation. Steep rock faces and boulder sides will be adorned with **mosses and liverworts and filmy ferns**. The **lichen flora** will vary naturally depending on the chemical properties of the rock and tree trunks within the woodland. Trees with **lungwort and associated species** will be fairly common, especially on the well-lit woodland margins.
The factors that affect the habitat, including its typical species, must be under control.	The woodland does not contain any rhododendron or any other invasive alien species with the exception of occasional beech and sycamore. There will be periodic light grazing by sheep and very occasionally by cattle. This will help maintain the ground and field layer vegetation but will not prevent tree regeneration.

Discussion of the example given in Table 16.2:

The feature: An upland acidic oak woodland

Rationale

Factor: Past management/land use

The woodland has a long history of human intervention and management. At an early stage, it was utilised for the production of a wide range of woodland products. Later, it became an important source of bark for the leather tanning industry. It was almost completely clear-felled during the First World War and then, most recently, became a nature reserve. There is little detail of earlier management, but the more significant consequences are obvious. The canopy trees are young and even-aged, with no veteran specimens. There are no canopy gaps, and, consequently, there is no oak *Quercus* regeneration. There is insufficient dead wood. Each of these failings led to the selection of attributes which have been monitored. The status of the feature is unfavourable declining.

The woodland will not recover because of the presence of additional primary factors. These are invasive species (rhododendron, sycamore and beech) and grazing by trespassing domestic stock and feral goats. The management of each of these factors will be discussed later. The intended management outcome will be the control or removal of the invasive species and the exclusion of goats and domestic animals.

Once these factors have been controlled, natural processes should, in time, lead to the development of a more robust, sustainable and diverse woodland, and, eventually, veteran trees will be present. There are two potential issues:

The dense, even-aged canopy will suppress the regeneration of oak, but gaps will be produced when the beech and sycamore trees are felled. It is also possible that natural processes, such as wind-throw, will produce additional gaps. In the future, the gap creation rate will be entirely the consequence of natural processes.

The second issue is that neighbours have traditionally collected dead wood for firewood. In the short term this will not be a problem because they can be permitted to take small quantities of felled sycamore and beech. However, in the longer term, regrettably, even this must cease. Consultation and liaison with neighbours will be required to ease the process. The only management work which arises as consequence of this project is liaison with neighbours.

Fig. 16.5 Rhododendron invading an oak wood

Factor: Grazing by domestic animals
The woodland has been overgrazed for many years. This has prevented the natural regeneration of the woodland, since seedlings are given no opportunity to grow into viable trees. If grazing is completely excluded there is potential that a dense field layer, comprising mainly bilberry *Vaccinium myrtillus* and heather *Calluna vulgaris,* will develop and suppress any natural regeneration. All the boundaries will be made stock-proof and, wherever possible, the original stone walls will be repaired and fitted with a wire jump-fence. All other sections will be fenced. Sheep will be kept in the woodland until there is a good crop of acorns. Once the seedlings appear, the sheep will be removed. A short-term grazing licence will be issued to a local farmer. The reintroduction of grazing animals will be considered in the future. The woodland will be regularly checked to ensure that there are no trespassing sheep, and the fences will be inspected and maintained in a stock-proof condition.

The purpose of the rationale is to identify and outline the management requirement. This is followed by the preparation of an action plan. It is at that stage that the detailed methodologies are considered.

16.7 Recording

Reminder: Once management is in place, all the activities should be recorded. This was discussed at the end of Chapter 5.

16.8 Management Options or Strategies

This note is included because, when reviewing older plans, some readers may find that options or management strategies have been used to describe management. The use of options in planning is not recommended.

For many years, UK management planning guides placed considerable emphasis on the use of management options. They first appeared in *A Handbook for the Preparation of Management Plans* (NCC 1983). The handbook describes three management options. They are:

- Non-intervention
- Limited-intervention
- Active-management

Non-intervention, as defined by NCC, was a ‘climax or natural vegetation concept’. In the absence of natural (primeval) habitats in Britain, the aim was to acquire semi-natural habitats and allow them to develop free from as much adverse human influence, direct and indirect, as possible.

> It has sometimes been considered that the aim should be to achieve the type of climax vegetation that is postulated to have existed in the past, but as this cannot be known with certainty and may be difficult to achieve in present day conditions, the selection of this non-intervention option is considered to be the next best alternative. (NCC 1983)

Limited-intervention is rather woolly and suggests that it is an ephemeral option employed when there is doubt that non-intervention will deliver.
Active-management is the approach employed to maintain semi-natural or plagioclimatic habitats and communities.

Eurosite use precisely the same approach, but they rename 'options' and refer to them as 'strategies' (Eurosite 1999).

Options or strategies are best regarded as a statement of the general direction which management will follow in order to achieve an objective. This concept pre-dates the arrival of feature-based objectives which define the condition or status of a feature. In Chapter 6, the concept of adaptable management was discussed. It was emphasised that, although an objective can be valid for a considerable period of time, management will vary much more frequently. This means that the approaches to management options will vary over time. A period of intense, active management could be followed by limited-intervention and eventually non-intervention. The idea that any particular option can be applied regardless of changing conditions or factors has been set aside by most organisations, with the possible exception of Eurosite.

16.9 Nature Conservation Management

It is worth taking a little time to consider the nature of conservation management. Management is a crude, sometimes clumsy, activity: we add or subtract animals, we mow, we burn, we use chainsaws and we dig, clear or block ditches. Much of what we do is primitive, particularly when compared to other areas of land management such as modern agriculture.

Planning the timing of management activities is nearly always hit or miss. Despite any amount of careful consideration, things are done when they can be done: when the opportunity arises, when the rain stops, when nature decides that it is time to mow that meadow, when labour is available or when we can afford to buy that essential piece of equipment.

We often have to adopt an opportunistic approach: we may want a particular breed of hardy cattle to graze a wetland, but when none are available a neighbour may be prepared to put his ponies on the land.

We can rarely predict the type or impact of human activities. A single careless match on a dry, windy day can destroy years of careful rotational patch burning on a heather moor.

And what of global climate change?

Despite all this, some people believe that nature conservation management can, and should, be a precisely defined and accurately timed activity. It is not, and it is extremely unlikely that it ever will be. The only possible exception is highly controlled experimental management that cannot be replicated on a realistic scale. The less energy that we devote to seeking precise management prescriptions and the more that we devote to the science of conservation biology and the development of effective and affordable monitoring methodologies, the greater our chances of conserving wildlife.

References

Alexander, M. (1996). *A Guide to the Production of Management Plans for Nature Conservation and Protected Areas*. Countryside Council for Wales, Bangor, Wales.

Alexander, M. (2005). *The CMS Guide to Management Planning*. The CMS Consortium, Talgarth, Wales.

Eurosite (1999). *Toolkit for Management Planning*. Eurosite, Tilburg, The Netherlands.

JNCC (1998). *A Statement on Common Standards Monitoring*. Joint Nature Conservation Committee, Peterborough, UK.

Margoluis, R. and Salafsky, N. (1998). *Measures of Success: Designing, Managing, and Monitoring Conservation and Development Projects*. Island Press, Washington DC.

NCC (1983). *A Handbook for the Preparation of Management Plans*. Nature Conservancy Council, Peterborough, UK.

NCC (1988). *Site Management Plans for Nature Conservation, A Working Guide*. Nature Conservancy Council, Peterborough. UK.

Pullin, A. S. and Knight, T. M. (2001). *Effectiveness in conservation practice: pointers from medicine and public health*. Conservation Biology, 15, 50–54.

Rowell, T. A. (1993). *Common Standards for Monitoring SSSIs*. JNCC Peterborough. UK.

Sutherland, W. J. (2000). *The Conservation Handbook: Research, Management and Policy*. Blackwell, Oxford.

Sutherland, W. J., Pullin, A. S., Dolman, P. M. and Knight, T. M. (2004). *The need for evidence-based conservation*. Trends in Ecology and Evolution, 19, 305–308.

Chapter 17
Action Plan

Abstract This is a rather 'mechanical' chapter. It is concerned with the process of preparing programmes and schedules, and it requires a methodical or structured approach. The action plan contains descriptions of all the work that needs to be carried out on a site in order to meet the objectives. Each individual task or project is identified and described in sufficient detail to enable the individuals responsible for the project to carry out the work. All the basic information for each project (i.e. when and where the work should be completed, who should do the work, the priority, what it will cost, etc.) is aggregated and used to produce a wide range of work programmes, for example, annual programmes, programmes for a specified period, programmes for an individual, financial programmes, long-term programmes, etc. An action plan is prepared for a specified period, usually 5 years. Action plans also provide a structure, and establish priorities, for recording.

17.1 Background

Most, but not all, guides to management planning recognise the need for an action plan. The earliest UK guides use the word 'prescription' to describe the action plan (Wood and Warren 1976, 1978; NCC 1981, 1983, 1988). In 1992, during the preparation of the first Ramsar guidelines on management planning, several participants felt that 'prescription' held little meaning in languages other than English, and, as a consequence, the term 'action plan' was introduced. Since that time, this has become the most common term used in European planning (Alexander 1996, 2000a, b, 2005; English Nature 2005; Ramsar Convention Bureau 2002). The Eurosite guide (1999) includes a very brief section called 'implementation', which recognises the need to divide work into individual projects and to use these as the basis for generating various work programmes. The action plan is given surprisingly little

M. Alexander, *Management Planning for Nature Conservation: A Theoretical Basis & Practical Guide*, DOI 10.1007/978-94-007-5116-3_17,

attention by Margoluis and Salafsky (1998). They use the term 'activities', but these are not given detailed consideration in their book. The IUCN guide (Thomas and Middleton 2003) pays even less attention to management activities.

17.2 Preparing an Action Plan

17.2.1 Projects

A project is a clearly defined and planned unit of work. These are the cornerstones of action planning. Everything else is derived directly from the information contained in the individual project plan. There is an extremely wide range of different types of projects that can be active on a site. In the mid 1970s, the UK Nature Conservancy Council introduced a system of codes linked to standard descriptions which were intended to provide a common basis for describing work on their National Nature Reserves. This list, significantly modified and updated, is still used by most conservation organisations in the UK and by many organisations in other countries. The list is now maintained by the CMS Consortium. It is organised in a hierarchical structure, beginning with three main divisions: Recording, Management and Administration.

Fig. 17.1 Water level management

Table 17.1 Project codes – The three main divisions

R Recording	**M Management**	**A Administration**
RA Record, fauna	**MA** Manage other land	**AA** Site acquisition/declaration
RB Record, biology general	**MB** Manage habitat, hedgerows	**AE** Employ staff
RC Record cultural heritage	**MC** Manage cultural features	**AF** Financial planning and recording
RF Record, vegetation	**ME** Manage site infrastructure	**AI** Inspections and audits
RH Record, human interaction	**MH** Manage habitat	**AL** Legal matters and payments
RM Record, marine	**MI** Information/education/ interpretation/events	**AN** Site designation
RP Record, physical environment	**ML** Liaison with stakeholders	**AP** Planning, plan preparation and revision
RV Record, archive – general, photos, maps etc.	**MM** Manage machinery and equipment	**AR** Reports and general correspondence
	MN Manage habitat, marine	**AS** Site and species safeguard, law enforcement & admin.
	MP Patrol	**AT** Training and management
	MS Manage species	
	MU Manage earth science	

Table 17.2 Projects for recording

RA Record Fauna, is divided:	Each category is further divided, for example: **RA0** Collect data, mammals
RA0 Collect data, mammals	**RA00** Collect data, mammals, general
RA1 Collect data, birds	**RA01** Collect data, mammals, natural event
RA2 Collect data, herptiles	**RA02** Collect data, mammals, survey
RA3 Collect data, fish	**RA03** Collect data, mammals, monitor
RA4 Collect data, Lepidoptera	**RA04** Collect data, mammals, count/estimate/ measure/census
RA5 Collect data, Odonata	**RA05** Collect data, mammals, research project
RA6 Collect data, Orthoptera	**RA06** Collect data, mammals, list species
RA7 Collect data, other insects	
RA8 Collect data, other invertebrates	
RA9 Collect data, fauna, general	

Table 17.3 Management projects

MH Manage habitat is divided:	Each category is further divided, for example: **MH0** Manage habitat, forest/woodland/scrub
MH0 Manage habitat, forest/woodland/scrub **MH1** Manage habitat, grassland **MH2** Manage habitat, controlling invasive species **MH3** Manage habitat, heath **MH4** Manage habitat, bog/mire/flush **MH5** Manage habitat, swamp/fen/inundation **MH6** Manage habitat, open water/rivers **MH7** Manage habitat, coastal **MH8** Manage habitat, rock **MH9** Manage habitat, upland/montane	**MH00** Manage habitat, forest/woodland/scrub, by coppicing **MH01** Manage habitat, forest/woodland/scrub, by planting/sowing **MH02** Manage habitat, forest/woodland/scrub, by thinning/group felling **MH03** Manage habitat, forest/woodland/scrub, aiding natural regeneration **MH04** Manage habitat, forest/woodland/scrub, maintaining ride/path/glade **MH06** Manage habitat, forest/woodland/scrub, by enclosure/exclosure **MH07** Manage habitat, forest/woodland/scrub, by scrub control **MH08** Manage habitat, forest/woodland/scrub, by managing dead wood **MH09** Manage habitat, forest/woodland/scrub, by other activities

These examples[1] have been included to demonstrate the wide range of projects that can be active on a site. The standard codes and the associated project titles are used, unmodified, by all users. In a site-specific context, each individual project code can be divided up to 99 times, and each division is numbered and accompanied by a 'qualifying phrase'. For example:

RA03 Collect data, mammals, monitor
RA03/01 Collect data, mammals, monitor, grey seals
RA03/02 Collect data, mammals, monitor, bank voles

17.2.2 Relationship Between Projects and Objectives

The list of projects for a site will include all the work that is required to meet all the objectives. The projects must be linked, and relevant, to the management objectives. This is for two reasons: it will enable managers to cost the individual objectives, and, more importantly, it will ensure that there is a purpose for all the

[1] The full list of project codes is available as a free download from: www.software4conservation.com

work planned for a site. When auditing sites, one of the most frequently encountered problems is active projects for which there is no justification. These are things that people do simply because that is the way things have always been done: often they are projects that were initiated at some time in the past by staff who have long since moved on.

There is a slight complication in that an individual project will often have relevance to more than one objective. For example, a project to maintain anti-poaching patrols in an African reserve can be relevant to protecting elephant, large ungulates, crocodiles and even the savannah habitat (poachers burn the vegetation to catch game). Each species and the habitat is a feature which requires an individual objective.

The relationship between the individual monitoring projects and the objectives is established when the objectives and performance indicators are identified. Every objective has associated performance indicators. (Objectives for conservation features and all other sections in a management plan must be quantified and measurable). The performance indicators for conservation objectives are based on attributes and factors: each attribute and measurable factor has an associated monitoring project. (Occasionally, surveillance projects will be used in place of monitoring projects). All management projects are identified, or confirmed, in the rationale (confirmed when there is already a history of conservation management).

17.2.3 Planning Individual Projects

The following information should be included in all individual project plans:

Why the Project Is Necessary

The first consideration when planning every individual project must be: what objective or objectives is the project linked to? This should be followed by: why is the project necessary? (i.e. the intended outcome must be explained). There is a need to provide this information for other people, but the most important function is to ensure that managers pay adequate attention to justifying everything that they do.

Potential Impact on Other Features

When planning management projects, it is essential that the implication of the work for other features is considered. In some cases, this could extend to completing an impact assessment. As an example: A wetland site contains a number of ephemeral lakes that are particularly important because they contain rare communities of aquatic plants. The lakes gradually silt up, and there is a need for occasional dredging. The lakes also contain populations of otter and several important breeding birds (these are also protected features). The management of the lakes is essential, not only for the aquatic flora but also for the otters and birds. An impact assessment will take all of this into account, and a management approach designed to minimise impact on the protected features will be devised. Obviously, the work will be undertaken outside the breeding seasons, and only small sections will be dredged at any time. This will ensure that most of the habitat is always in the required condition.

On some statutory sites, there will be a legal requirement to obtain formal consent before carrying out any management work which has potential to influence any of the features.

When the Project Is Active

Projects can be a one-off activity or something that is repeated annually or several times each year. An action plan is prepared for a specified period, usually 5 years. With occasional exceptions, it is difficult to plan any further ahead. Some organisations, for example the Countryside Council for Wales, maintain their reserve plans on a 5 year programme which rolls forward at the end of each financial year. This means that the plans are always valid for at least 4 years. In addition, expensive capital projects or acquisitions are planned over a 10 year period. For example, it is usually possible to estimate the life expectancy of a major boardwalk or of expensive machinery that will have to be replaced at intervals. The ability to predict financial and staff requirements is a management necessity for many organisations.

Managers will need to decide when, and how many times, during a year a project will be active (sometimes work is seasonal, for example, monitoring nesting birds). There is also a need to identify the year, or years, within the life of a plan that a project will be active. Obviously, some will be active each year or even several times each year.

Where the Work Will Be Carried Out

For many projects, the precise location where the work should be carried out is an important consideration. Maps can be extremely useful to show, for example, the location of surveillance plots or the line that a new fence will follow.

Resources

Resource planning is as important, if not more so, than any other section in the action plan. Whenever possible, the plan should identify the resource requirement (financial and staff) necessary to complete each project. This will allow managers to attribute full costs to a project and enable the preparation of work plans. It is only through assessing the resource requirements for each project and aggregating this data that it becomes possible to make a justified case for resources. This could include the information essential for a successful grant application.

Expenditure

The cost of each project should be calculated and the potential or actual source of funds identified. Some organisations use financial coding systems to organise internal expenditure. The reserve manager will often be obliged to comply with organisational protocols.

Staff

This can include employed staff and, for some organisations, volunteers. Whenever possible, the plan should identify the individuals responsible for carrying out each project and give an indication of the time required of each person. This will enable the preparation of various work plans. Staff shortages are usually one of the major problems faced by conservation managers. There is rarely any purpose in stating the obvious, i.e. we need more staff, unless this can be quantified. The most persuasive argument is to list the work that will not be carried out as a consequence of staff shortages. Demonstrating that there will be a failure to protect the features, to meet health and safety requirements, or to provide safe facilities for visitors will often persuade the most intransigent senior staff.

Priority

It is rarely, if ever, possible to complete all the work on a site, and there can be many different reasons for failure. The completion of any outdoor work will be constrained by seasonal variations, the vagaries of weather and other natural conditions. Resources, or more specifically the lack of adequate resources, will always restrict a manager's ability to do all the work that is necessary to manage a site. Managers can never expect to do everything that they need to do, let alone what they want to do. This is why it is essential that all projects are prioritised: if managers had unlimited resources they would not need priorities.

I have been attempting to devise priority systems for over 15 years, and I have learned that the simpler the system (i.e. the fewer priorities) the more likely it is to be effective. Occasionally, organisations will devise hopelessly complicated, multi-layered priority systems, and managers may have no choice but to comply with this inappropriate system. However, I am convinced that three priorities are adequate in almost all circumstances.

Priority 1

These are the essential projects: work that has to be completed regardless of cost. These will include:

- Projects which carry legal obligations. For example, tenure, vehicle maintenance, public rights of way, Disability Discrimination Act and the communication of legal or safety messages relating to site management.
- Animal welfare – stock husbandry.
- Health & Safety and public liability requirements relating to staff, the public use of the site, access infrastructure and the communication of safety messages.

- Habitat/species management projects which are essential to maintain qualifying features at current status.
- Protection of statutory features through the management of public access or the communication of access messages.
- Monitoring the protected features, but only when they are threatened or in decline.

Priority 2

These projects are essential in the longer term, but the consequences will not be too serious if they are delayed. These include:

- Projects required to meet the long-term objectives of conservation management.
- Projects relating to non-qualifying features and the site fabric.
- Monitoring the protected features, even though they are not threatened or in decline.
- Projects which provide information and interpretation for visitors.

Priority 3

These projects are important in the longer term, but can be deferred to a later date. They can also be regarded as 'if only' projects: if only we had more time or more money; if only a volunteer would appear with specific skills. The identification of priority 3 projects is important when responding to opportunities that may arise. Opportunistic management is generally inappropriate. However, in reality, many conservation organisations are dependent on grants and other donations which are often linked to schemes where the donors dictate terms. The consequence is that, in order to obtain finance, organisations will carry out work that is not entirely appropriate and, more seriously, will neglect essential work as a result. This problem is largely insurmountable in a quick-fix, target-obsessed society, but through identifying work in advance and ensuring that it is relevant, organisations are in a better position to recognise appropriate funding opportunities. This issue can also arise when an organisation discovers an under-spend at certain stages in a financial cycle. A financially astute manager with previously-planned priority 3 projects can use these opportunities to respond rapidly and confidently with a bid for windfall resources.

General Background Information

Quite often, a project will have evolved and been revised over a long period of time. The management methods can have a history of trial and development, both locally and elsewhere. There may be a body of relevant scientific research. This information will provide the background and reasoning behind the selection of any particular management technique or approach to monitoring. It can also be relevant to infrastructure management, for example, an explanation of why an exacting fencing specification or the use of a particular product is necessary.

Methodology

This should contain, or refer to, sufficient guidance to enable anyone required to carry out the work to do so without needing to use any other instructions. When the project is relatively simple or is site-specific, the instructions provided in the project description should be sufficient. However, if a project is based on a standard methodology which is easily accessible, there may be little purpose in repeating the information in the project description. The instructions should be clear and succinct: often a series of points will be more appropriate than large blocks of text.

Organisations that manage several sites will recognise that many of the projects will be common throughout. They may also wish to apply corporate standards for some of their work. In these cases, there is little purpose in each individual site manager independently replicating the same methodology when there are considerable and obvious benefits of sharing. Whenever an individual is planning a project, the first step should be to find out if the organisation has a standard specification or if colleagues responsible for other sites have developed a similar project. Ideally, this information should be available both internally and externally.

It is important that managers seek evidence from other sites, search the scientific and conservation management literature, obtain advice from experts and then follow the adaptable management process. If the outcome is acceptable, continue; if not, modify the management approach or try something different. It is also important to bear in mind that factors and their effect can change with time. Management activities considered appropriate today might be completely inappropriate tomorrow.

Project Work Programme

Often, projects will be phased over a period, for example, a planning phase, a preparation phase and a construction phase, followed by a maintenance phase. Each phase will lead to the completion of the project. When projects are phased within a year, and particularly when they are phased over several years, a work programme should be prepared for each phase of the project.

Table 17.4 Recommended structure for planning a management project

Structure for planning a management project	
Project title:	
Year/s when the project is active:	
Event/s within a year:	
Zones or compartments:	
Expenditure:	Staff Time:
Project priority:	
The feature or objective/s that the project is linked to:	
Justification for the project (i.e. the intended outcome):	
Potential impact on other features:	
General background/bibliography:	
Project methodology:	
Project work programme	

Table 17.5 Recommended structure for planning a monitoring project

Planning a monitoring project
Project title:
Year/s when the project is active:
Event/s within a year:
Zones or compartments:
Expenditure: Staff Time:
Project priority:
The feature or objective/s that the project is linked to:
General background/bibliography:
The performance indicator (attribute or factor)
Project methodology:
Project work programme
(a) Equipment: List all equipment, noting any detailed specifications and location of equipment if appropriate. If you intend to use an obscure piece of equipment, e.g. Borman disc, reference bibliographic material describing equipment in detail.
(b) Location of sample collection: Define the area of sample collection.
(c) Fixed point markers: Describe the type of marker and location of each marker. Describe any programme of maintenance for fixed point markers. Some projects may demand an extensive system of markers.
(d) Sampling technique: Describe the technique used for collecting sample data, referring to use of equipment etc. Include sufficient detail to facilitate repetition by others.
(e) Unit of measurement: Identify units, e.g. individual flowering spikes of orchid spp., cm^2 of lichen spp., cm of rise/fall in water table.
(f) Sample type/specification: Define the sample, e.g.:
Total no. of flowering spikes in 10 quadrats each 1×1 m. Total area in cm^2 of lichen spp. in 4 stands, each 10×10 cm.
(g) Sampling period: State the time period within which the set of sample data is collected. This will usually be a period within 1 calendar year, e.g. May – July.
(h) Frequency of sampling during sampling period: State the interval between sampling during each sampling period.
(i) No. of samples collected during sampling period: State the number of samples to be collected during each sampling period.
(j) Repeat interval: State the interval between sampling periods.
(k) Special considerations: Note any other factors which affect data collection, e.g. limitations imposed by weather conditions.
5. DATA MANAGEMENT
(a) Identification of data format: State the format of stored data (paper report/computerised etc). In the case of computerised data, note the type and name of the software and the version, e.g. Excel 2000
(b) Location of data: Note location of original data. State file ref.
(c) Data security: Monitoring data are irreplaceable. Note the location of all copies of data. State file ref.
(d) Analytical technique: Note the method of data analysis. Refer to statistical techniques etc. State file ref. if relevant

Table 17.6 Example of a management project plan

Project plan

Annual Project Summary

Site: STANNER ROCKS	**Priority:** 3
Project code: MH85	Manage habitat, rock, by scrub control
Project number: 01	Control ivy and bramble on selected outcrops
Year: 2012/13	
Staff: SRM	**Time:** 1.00 day
Expenditure: 0	
Compartments: 00	

Project plan:

1. PURPOSE: To avoid populations of rare plants being smothered by the growth of ivy or bramble.
2. GENERAL BACKGROUND: Current circumstances and those prevalent during the past few decades, possibly centred on a lack of grazing and climatic change, have produced lush growths of ivy and an expansion of bramble. Each of these species can seriously harm the rarer plants and grassland by shading and general competition. Grazing other than by occasional browsing is expected to limit re-growth where available.
3. METHOD: Identify outcrops threatened with major ivy or bramble encroachment using photos taken from a distance and observations of site. Where the rarer plants or habitat are threatened then plan to remove the offending ivy or bramble. In many cases the crags can only be scaled safely by abseiling. Rooting ends of ivy need to be plucked out, stems levered off, and the base cut through. (Note: An attempt to kill off large ivy plants in a dry summer in June by severing the base of the stem did little to weaken the vigour of the plants). Bramble may be uprooted or chemical applications considered. The monitoring of the featured plants will highlight the need for lesser stands of ivy or bramble to be controlled. Minor problems can probably be best approached at the time of monitoring the plants.
4. RISK ASSESSMENT: Refer to the site hazard sheet and all relevant risk assessments.
5. REPORTING REQUIREMENT: Report onto CMS for every year the project is active, giving essential detail and analysis, with recommendations if appropriate. Copy reports to the Regional Reserve Manager and summaries to Senior Conservation Officer for information.
6. PROGRAMME:

Annual for the benefit of the following species:

Scleranthus perennis, Trifolium strictum, Grimmia sp x 3, Bartramia stricta

Plan a more extensive control of ivy and bramble for the benefit of the remainder of features species on a 10-year cycle. This should be done in association with grazing plans.

Table 17.7 Example of a monitoring project plan

Project plan

Annual Project Summary

Site: SKOMER MARINE NATURE RESERVE	**Priority:** 2
Project code: RA03	Collect data, mammals, monitor
Project number: 01	Monitor grey seals
Year: 2007/08	
Staff: AMCO1; AMCO2; SMNRO	**Time:** 28 days
Expenditure: £8,075	

(continued)

Table 17.7 (continued)

Project plan

Annual Project Summary

Compartments: 30

Planned: 01/Jul/2012 –31/Oct/2012

Project plan:

FEATURE: Atlantic Grey Seal population (*Halichoerus grypus*)

ATTRIBUTE: Atlantic grey seals range widely throughout, and beyond, the population's breeding and feeding range; some appear to be quasi-resident, and many return to the same pupping site each year. Therefore the population dependent upon the MNR can only be realistically assessed by monitoring breeding success.

Attribute 1: Total of seal pups born in the MNR:

Upper limit: None set

Lower limit: 190 (170 in any 4 year period, provided numbers recover to over 190 in the following year)

Attribute 2: Survival of Atlantic grey seal pups to first moult:

Upper limit: None set

Lower limit: 70% (67% in any 4 year period, provided survival recovers to over 72% in the following year)

BACKGROUND

Grey seal (Halichoerus grypus) pup production is monitored within the MNR. Pup production on Skomer Island is monitored by contractor. Pup production on the beaches and in the caves of the Marloes Peninsula is monitored by MNR staff.

METHODOLOGY

Equipment: Fluorescent yellow dye in spray gun, torches, binoculars and MNR seal pup proforma.

Site location: All beaches within the MNR are monitored, but seals are usually only born a limited number of beaches.

Caves surveyed: The following caves are known to be used by seals: Martins Haven, Wooltack, Horseshoe, Three Doors.

Sampling technique: Cliff top observations are conducted regularly and beaches/caves are visited as often as possible, subject to weather conditions. Pups are age classed according to E. A. Smith's five-fold system of pup classification. During site visits and inspections, disturbance is kept to a minimum. Where possible, seal pups are given an identity mark using alcohol-based dyes. This is achieved by spraying with yellow dye in a coded system e.g. Single spot on right shoulder, stripe on left side. A list of the different variations is maintained along with the seal pups' class and location on a proforma sheet (kept in Seal (ring) file). It is important that similar dye codes are not given to pups located closely together or neighbouring beaches or caves, as this could cause misidentification if seal pups wander from their birth site. Dead pups are sprayed all over the head area to avoid re-counting. A file system is used to record the progress of each individual pup until the completion of moult.

Units: Numbers of pups. % survival to moult

Sampling period: Breeding season (Sept – Dec)

Frequency of sampling: Annual

Special considerations: To reduce the possibility of interfering with the cow/pup bond, especially at times of high pupping activity on a particular site, cliff top observations are made and beach landings avoided. Pups are not sprayed anywhere in the region of eyes, nose or mouth.

Fig. 17.2 Monitor grey seal pups

17.3 Work Programmes

Site managers will require a range of different work programmes. For the key personnel they will usually need a programme that contains basic information, such as: what they should do; for how long; when or with what frequency; where; the priority of the work. There are many variations and different requirements, but they should all be generated from information contained in the individual project plans.

Unless computer databases or spreadsheets are used this can be an extremely tedious and difficult task. Sites can have many objectives, and each objective can be associated with a range of projects. Often, an individual project will be relevant to more that one objective. Computer databases are the obvious solution, and this has been a justification for the development of a small number of computer systems designed to manage conservation planning systems. The only widely available system is the Conservation Management System (CMS) (see Case Study 5).

The following are examples of a project layout and various work programmes which have been generated by CMS.

Table 17.8 An example of a 'staff time' report

Skomer marine nature reserve – selected projects with planned staff time 2011			
Project	**Qualifier**	**Staff title**	**Estimated staff time days**
ME02/01	Maintain MNR moorings	Assistant marine conservation officer 1	1.00
		Assistant marine conservation officer 2	2.00
		Assistant marine conservation officer 3	1.00
		Skomer marine nature reserve officer	2.00
MM00/01	Operate & service boats	Assistant marine conservation officer 1	5.00
		Assistant marine conservation officer 2	20.00
		Assistant marine conservation officer 3	5.00
		Skomer marine nature reserve officer	10.00
MM00/05	Operate and service road trailers	Assistant marine conservation officer 1	0.50
		Assistant marine conservation officer 2	1.00
		Skomer marine nature reserve officer	0.50
MM00/03	Service road vehicles	Assistant marine conservation officer 1	0.50
		Assistant marine conservation officer 2	0.50
		Skomer marine nature reserve officer	1.50
MM20/05	Maintain/acquire tools & general equipment	Assistant marine conservation officer 2	1.00
		Skomer marine nature reserve officer	0.10
MM20/02	Maintain/acquire diving equipment & air	Assistant marine conservation officer 1	4.00
		Assistant marine conservation officer 2	7.00
		Assistant marine conservation officer 3	1.00
		Skomer marine nature reserve officer	5.00

Table 17.9 A selection of projects planned to occur during a 5 year period

SKOMER MARINE NATURE RESERVE		*04*	*05*	*06*	*07*	*09*
RP04/01	Collect data, climatological, count/estimate/measure/census RECORD METEOROLOGICAL FACTORS	1	1	2	2	2
RP04/02	Collect data, climatological, count/estimate/measure/census RECORD METEOROLOGICAL FACTORS	X	2	X	X	X
RP22/01	Collect data, geological, survey SURVEY SUBTIDAL GEOLOGY	3	3	3	3	1
RP32/02	Collect data, geomorphological, survey SURVEY BENTHIC HABITATS	3	3	3	3	3
RP64/06	Collect data, oceanographic, count/estimate/measure/census RECORD WATER COLUMN CHLOROPHYLL	3	3	3	3	3
RB00/01	Collect data, biological SAC WORK	2	2	2	2	2
RB02/01	Collect data, biological, survey SURVEY DEEP WATER COMMUNITIES	2	2	2	2	2

(continued)

Table 17.9 (continued)

SKOMER MARINE NATURE RESERVE		*04*	*05*	*06*	*07*	*09*
RB02/02	Collect data, biological, survey NON-FEATURE SURVEY SPECIES / COMMUNITIES	1	2	2	2	2
RB03/01	Collect data, biological, monitor MONITOR LITTORAL HABITATS/ COMMUNITIES	1	1	2	2	2
RA03/01	Collect data, mammals, monitor MONITOR GREY SEALS	1	1	2	2	2
RA33/01	Collect data, fish, monitor MONITOR TERRITORIAL FISH POPULATIONS	2	1	2	X	2
RA34/01	Collect data, fish, count/estimate/ measure/census RECORD ANGLING CATCH RECORDS	2	2	2	2	2
RH03/01	Collect data, human impact, count/ estimate/measure/census RECORD BENTHIC IMPACTS: DIVING	2	2	2	2	2
RH07/01	Collect data, human impact, pollution RECORD POLLUTIONS	1	1	1	1	1
RM73/01	Collect data, Echinodermata, monitor MONITOR ECHINUS POPULATION	0	X	X	2	1
ML60/01	Liaise, emergency services LIAISE COASTGUARD	1	1	1	1	1
ML70/01	Liaise, media LIAISE MEDIA	1	1	2	2	2
ML80/01	Liaise, others LIAISE ADVISORY COMMITTEE	1	1	1	1	1
ML80/02	Liaise, others LIAISE COMMERCIAL FISHERMEN	2	1	2	2	2
ML80/03	Liaise, others LIAISE RECREATIONAL BODIES	2	1	2	2	2
E01/01	Boundary structures MAINTAIN MARKER BUOY SYSTEM	1	1	1	1	1
ME02/01	Other structures MAINTAIN MNR MOORINGS	1	1	1	1	1
ME20/01	Comply with legal obligations COMPLY WITH LEGAL OBLIGATIONS	1	1	1	1	1
MM00/01	Acquire/service vehicles/boats OPERATE & SERVICE BOATS	1	1	1	1	1
MM20/02	Acquire/maintain tools/equipment MAINTAIN/ACQUIRE DIVING EQUIPMENT & AIR	1	1	1	1	1
AI40/01	Implement inspection, other DIVING MEDICALS	1	1	1	1	1

Note: The total number of projects on this site is 126

Table 17.10 An example of a finance plan

Skomer marine nature reserve – projects requiring finance			
Project code	**Qualifier**	**Priority**	**Expend**
ME02/01	Maintain MNR moorings	1	750
AI40/01	Diving medicals	1	360
MM20/01	Maintain/acquire personal & protective equipment	1	1,300
MM20/07	Maintain/develop it systems	1	100
MM00/01	Operate & service boats	1	9,500
ME12/01	Maintain MNR office building	1	3,300
ME04/01	Remove debris	1	5
AT30/02	Training: marine operations safety	1	20
MM20/02	Maintain/acquire diving equipment & air	1	2,500
ME01/01	Maintain marker buoy system	1	150
AT50/01	Honorary wardens	1	800
AR30/01	Correspondence/admin/filing	1	900
AP80/01	Convene advisory committee meetings	1	100
MM00/05	Operate and service road trailers	1	450
MM20/05	Maintain/acquire tools & general equipment	1	550
AL00/01	Maintain leases	1	1,600
MM20/06	Maintain/acquire marine electronic equipment	1	920
ME02/02	Maintain visitor moorings	1	200
RP04/01	Record meteorological factors	2	300
RB03/01	Monitor littoral habitats/communities	2	2,530
AS00/01	Promulgate byelaws, codes of conduct	2	2,000
RA03/01	Monitor grey seals	2	8,075
RM03/01	Monitor epibenthic rock communities	2	80
RM13/01	Monitor sponge populations	2	10,080
RM23/01	Monitor eunicella	2	30
RM73/01	Monitor echinus population	2	3,000
RP64/01	Record seawater temperature	2	50
RV10/01	Maintain photographic data	2	100
RV51/01	Record media coverage	2	60
ME02/03	Maintain monitoring site structures	2	300
MM20/03	Maintain/acquire photo & optical equipment	2	1,500

Table 17.11 Skomer marine reserve – staff time

Selected records of staff time 2012			
Project	**Qualifier**	**Staff time days**	**Staff title**
RP64/01	Record seawater temperature	0.40	Assistant marine conservation officer 1
		2.50	Assistant marine conservation officer 2
		1.30	Assistant marine conservation officer 3
		1.30	Skomer marine nature reserve officer
RA03/01	Monitor grey seals	8.20	Assistant marine conservation officer 1
		1.90	Assistant marine conservation officer 2
		6.70	Assistant marine conservation officer 3
		2.20	Skomer marine nature reserve officer
		0.40	Volunteer
RA33/01	Monitor territorial fish populations	15.20	Assistant marine conservation officer 1
		10.00	Assistant marine conservation officer 2
		3.10	Assistant marine conservation officer 3
		3.00	Skomer marine nature reserve officer
		37.00	Volunteer
RM23/04	Monitor cup coral populations	1.50	Assistant marine conservation officer 1
		2.30	Assistant marine conservation officer 2
		0.70	Skomer marine nature reserve officer
		0.70	Volunteer
RM23/05	Monitor parazoanthus populations	2.80	Assistant marine conservation officer 1
		1.70	Assistant marine conservation officer 2
		0.90	Assistant marine conservation officer 3
		0.10	Senior marine conservation officer
		1.20	Skomer marine nature reserve officer
		0.30	Volunteer

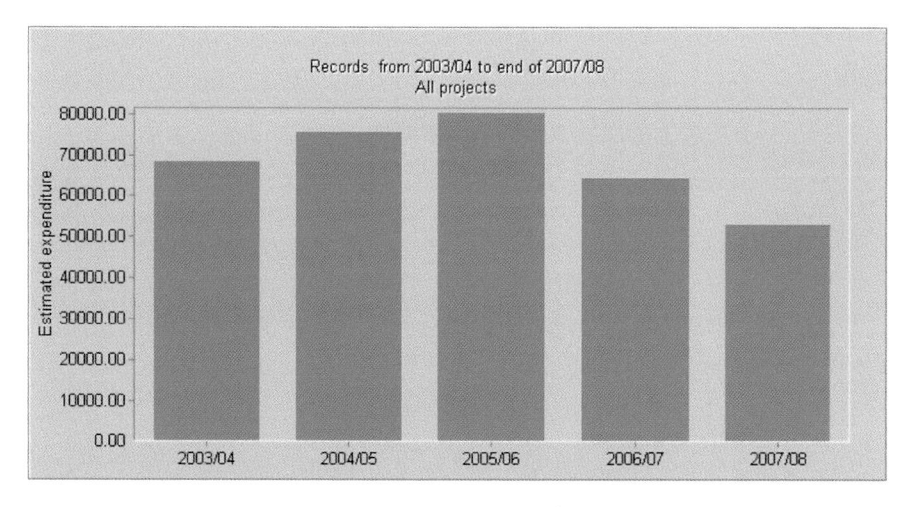

Fig. 17.3 Skomer marine nature reserve expenditure 2003/2004 to 2007/2008

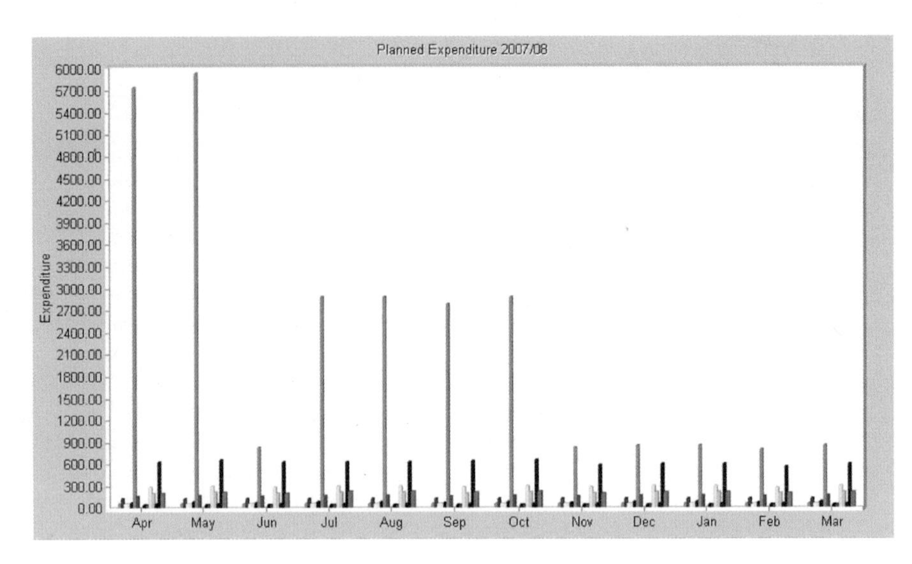

Fig. 17.4 Planned expenditure for different categories of projects 2011/2012

Fig. 17.5 Skomer Marine Reserve

17.4 Operational Objectives

On all sites, irrespective of other considerations arising from the management objectives, there will be a requirement to meet legal, and other similar, obligations. On some larger sites, there may be a need to manage very large operations that are carried out to support many of the different management objectives. For many sites, it would be difficult, and extremely cumbersome, to associate these operations with the individual feature objectives.

The best way to deal with these operations is to treat them almost as an objective, but recognising that the rules applied to the other objectives are not relevant to 'operational objectives'. The structure can be very simple: it begins with a brief introduction which links the operational objective to the main management objectives and provides an explanation of why the operation is necessary. This is followed by an 'operational aim' which is, in many respects, an extremely simplified version of the vision section of a management objective. One or more projects are identified in order to ensure that there is compliance with the objective. These are 'compliance monitoring' projects and will follow a pattern similar to all other monitoring projects. The final stage is to identify the range of individual projects which will describe all the work that will be required to undertake a successful operation.

Some operational objectives are used to bring together a wide range of activities which stem from legal or other obligations. On all sites, there will be a range of obligations that are defined by law, for example, a requirement to comply with any health, safety and public liability legislation. Organisations will also impose procedures and specify certain operational requirements for reserve managers, for example, requirements for reporting and record keeping. Where these requirements are elaborate or complex, and are not easily accommodated elsewhere in the plan, they can be grouped together as major operations. The following is an example of an operational objective to ensure compliance with health and safety obligations:

Operational aim:

- To meet all health and safety regulations and ensure a safe working environment for all reserve staff and volunteers

Projects:

- Prepare and maintain risk assessments for all activities on the reserve.
- Prepare and maintain a hazard risk assessment for the site.
- Provide protective clothing and equipment for all reserve staff and volunteers.
- Provide appropriate safety training for all staff and volunteers.
- Ensure that procedures are in place for handling dangerous substances.
- Undertake regular safety audits and record the level of compliance with all projects in this group.

Operational objectives can, alternatively, be a single major operation. For example, in the Kidepo National Park in Uganda one of the most important activities is maintaining an anti-poaching operation. This is a very large-scale operation with a wide range of individual projects, including maintenance of an armed ranger service, ranger posts, an armoury, vehicles, patrols, etc. The poachers target several different species, or groups of species, each of which is a feature of the site. In order to find and trap game they damage the savannah vegetation by burning. Thus, the operational objective to control poaching is relevant to many different park objectives.

A further example of an operational objective, taken from a management plan for a British wetland site, is the provision of a large and complex footpath system that, in addition to providing access for visitors, is also essential for management works carried out in respect of several habitat features. The outline description would be to: Develop and maintain the network of footpaths and tracks required to meet all legal access requirements, to provide access for management operations and to provide access for visitors to the site.

This could contain the following list of individual projects:

- Ensure that the public rights of way are open at all times.
- Carry out a disabled access audit in compliance with the Disability Discrimination Act.
- Provide and maintain the network of footpaths (refer to the map).
- Construct and maintain 1 k of boardwalk on the raised bog.
- Maintain the private boardwalk that provides access to the moth trap.
- Maintain the 5 footbridges (refer to map for locations).
- Construct and maintain 16 gates and 6 styles.
- Maintain the private tracks and bridges (refer to project map for locations).
- Undertake weekly inspections of the entire access infrastructure.

References

Alexander, M. (1996). *A Guide to the Production of Management Plans for Nature Conservation and Protected Areas*. Countryside Council for Wales, Bangor, Wales.

Alexander, M. (2000a). *A Management Planning Handbook*. Uganda Wildlife Authority, Kampala, Uganda.

Alexander, M. (2000b). *Guide to the Production of Management Plans for Protected Areas*. CMS Partnership, Aberystwyth, Wales.

Alexander, M. (2003). *CCW Guide to Management Planning for SSSIs, Natura 2000 Sites and Ramsar Sites*. Countryside Council for Wales, Bangor, Wales.

Alexander, M. (2005). *The CMS Guide to Management Planning*. The CMS Consortium, Talgarth, Wales.

Eurosite (1999). *Toolkit for Management Planning*. Eurosite, Tilburg, The Netherlands.

English Nature (2005). *NNR Management Plans: A Guide*. English Nature, Peterborough, UK.

Margoluis, R. and Salafsky, N. (1998). *Measures of Success: Designing, Managing, and Monitoring Conservation and Development Projects*. Island Press, Washington DC.

NCC (1981). *A Handbook for the Preparation of Management Plans for National Nature Reserves in Wales*. Nature Conservancy Council, Wales, Bangor, UK.

NCC (1983). *A Handbook for the Preparation of Management Plans*. Nature Conservancy Council, Peterborough, UK.

NCC (1986). *A review of event record system and proposed project recoding system*. Unpublished report, Nature Conservancy Council, Peterborough, UK.

NCC (1988). *Site Management Plans for Nature Conservation, A Working Guide.* Nature Conservancy Council, Peterborough, UK.

Ramsar Convention Bureau (2002). *New Guidelines for Management Planning for Ramsar Sites and Other Wetlands, Ramsar Resolution VIII.14.* Ramsar Convention Bureau, Gland, Switzerland.

Thomas, L. and Middleton, J. (2003). *Guidelines for Management Planning of Protected Areas.* IUCN Gland, Switzerland and Cambridge, UK.

Wood, J. B. and Warren, A. (ed.) (1976). *A Handbook for the Preparation of Management Plans – Conservation Course Format, Revision 1.* University College London, Discussion Papers in Conservation, No. 18. London.

Wood, J. B. and Warren, A. (ed.) (1978). *A Handbook for the Preparation of Management Plans – Conservation Course Format, Revision 2.* University College London, Discussion Papers in Conservation, No. 18. London.

Chapter 18
Access, Tourism and Recreation – Definition and Background

But all conservation of wilderness is self-defeating, for to cherish we must see and fondle, and when enough have seen and fondled, there is no wilderness left to cherish.

(Leopold 1949)

Abstract Access planning is concerned with all the provisions made for people who visit or use a site for any reason other than official, business or management purposes. 'Tourism' is included in the chapter heading because this term is used in place of 'access' in many countries. People visit protected areas to pursue a wide range of recreational activities. These activities can include everything from a quiet walk or bird watching to quite extreme sports. The provision of access for local visitors and tourism, and opportunities for recreational use, is an important, if not essential, function of most nature reserves and protected areas. For some sites it will be the most important function. Access and tourism undoubtedly offer many opportunities, but there is a cost: all activities can have a direct or indirect impact on wildlife, landscape and people. Wildlife can be displaced, habituation can make populations vulnerable, populations of species that are not attractive to visitors can be forgotten and habitat can be damaged. The values of landscape and wilderness areas can be diminished through excessive or inappropriate use, for example, the proliferation of unsightly trails and camp sites, or simply by the presence of too many people. Integration is essential: the management plan must establish a direct and clear link between the protection of wildlife and the management of people.

18.1 Definition

The first and obvious question is: what do we mean by public access, tourism and recreation? *Access planning is concerned with all the provisions made for people who visit or use a site for any reason other than official, business or management purposes.* 'Tourism' is included in the chapter heading because this term is used in

M. Alexander, *Management Planning for Nature Conservation: A Theoretical Basis & Practical Guide*, DOI 10.1007/978-94-007-5116-3_18,

place of 'access' in many countries, particularly those where ecotourism is important. In developing countries there will often be a significant and easily recognised distinction between local visitors and tourists, particularly foreign tourists. However, by contrast, in countries such as Britain it is usually difficult to differentiate between local visitors, British holidaymakers and foreign tourists. With the exception of occasional multilingual signage, there are very few examples where different provisions are made for the different groups.

People visit protected areas to pursue a wide range of recreational activities. These activities can include everything from a quiet walk or bird watching to quite extreme sports. Hunting for sport, fishing and other forms of harvesting (such as collecting edible fungi) can also be included, but activities that support an individual's livelihood, or provide some form of income, are best excluded.

18.2 Introduction

The provision of access for local visitors and tourism, and opportunities for recreational use, are important, if not essential, functions of most nature reserves and protected areas. For some sites it will be the most important function. Nature-based tourism and the public use of the countryside are growing, and there has been a move from the more passive pursuits, such as walking and bird watching, to activities including climbing, mountain-biking, rafting and canoeing (Eagles et al. 2002). This growing interest and use may represent an increased awareness of, and concern for, our natural environment. In this case, there will be opportunities, through sharing the natural experience, to instil values and help to develop a collective sense of responsibility towards the natural world. A more cynical view, however, and one that is difficult to avoid when witnessing the reckless use of jet-skis and powerboats in marine protected areas or off-road vehicles tearing up fragile blanket bogs, is that, for many, the countryside is not much more than an expendable playground.

IUCN's *Sustainable Tourism in Protected Areas – Guidelines for Planning and Management* (Eagles et al. 2002) lists 26 potential benefits from tourism in protected areas. The benefits are divided into three groups:

- Enhancing economic opportunity
- Protecting natural and cultural heritage
- Enhancing quality of life

Although the IUCN guide is clearly intended for larger national parks, and it is unlikely that many of the 26 benefits will be obtained on small nature reserves, all sites will have something to gain under each of the three headings.

The scale of the benefit will vary enormously. A small nature reserve in Britain may employ one or two staff, provide limited opportunities for a few local retailers and contractors, and attract a handful of tourists. In contrast, at peak summer levels in Yellowstone National Park 3,500 employees work for park concessionaires and 800 work for the park.

Although the scale of individual sites may vary, many organisations, particularly NGOs, depend to a very large extent on entry charges to their sites or on providing free entry as a membership incentive. For example, The Royal Society for the Protection of Birds, a UK organisation, finances the management of over 100 reserves in this way.

Access and tourism undoubtedly offer many opportunities, but there is a cost: all activities can have a direct, or indirect, impact on wildlife, landscape and people. Wildlife can be displaced, habituation can make populations vulnerable, populations of species that are not attractive to visitors can be forgotten and habitat can be damaged. The values of landscape and wilderness areas can be diminished through excessive or inappropriate use, for example, the proliferation of unsightly trails and camp sites, or simply by the presence of too many people. A few local people may obtain financial benefit from tourism, for example, hoteliers, retailers, local transport providers and guides, but the majority can sometimes be disadvantaged. This may range from mild irritation, as a consequence of congested roads and loss of opportunities for the quiet enjoyment of the countryside, to much more serious problems such as loss of traditional rights, displacement of people and unsustainable pressure on the local infrastructure, including water, sewage and transport. Clearly, through appropriate management it is possible to ensure that the benefits far outweigh the cost. There should be a strong and positive presumption in favour of providing access and appropriate facilities for visitors to all sites but only in so far as these activities are compatible with protecting the important wildlife features, the interests of local human communities and, in particular, indigenous people, landscape qualities and site fabric. The challenge when planning for access and public use is to optimise opportunities for visitors and tourists, while ensuring that our obligation to protect wildlife is not compromised through the inappropriate management of access.

18.3 Background

18.3.1 The European Perspective

Until very recently, management planning for protected areas in Europe had been mainly concerned with habitat and species protection and had, in general, placed significantly less emphasis on providing for access and people.

The first UK planning guide, *A Handbook for the Preparation of Management Plans for Nature Reserves - Revisions 1 & 2* (Woods and Warren 1976, 1978), which paved the way for most of the later UK and European systems, barely mentions access. A few years later, the UK Nature Conservancy Council published *A Handbook for the Preparation of Management Plans* (NCC 1983). It contains the following statement:

> On most reserves, however, including National Nature Reserves, managers should assume a responsibility to provide for access and enjoyment of visitors whenever possible; but since this could often affect the conservation status of the site, it is necessary to consider access control, by selecting one of the following options: Closed, Restricted Access, Partially Open Access, Open Access.

Apart from identifying the range of options (given above) that could define the levels of access on a site, the 1983 guide gives very little attention to access planning. The emphasis throughout is on controlling people and their activities. A 'popular' version of the 1983 guide was published (NCC 1988), but still the only reference to planning for access was the inclusion of the 1983 options.

In 1986, the Countryside Commission in the UK published *Management plans, a guide to their preparation and use*. The Countryside Commission was an organisation established in 1968 with specific responsibilities for the National Parks and there was a particular emphasis on providing public access to, and use of, the wider countryside. Quite remarkably, given their responsibilities, their guide to planning barely mentions access. Even as late as 1998, when the Commission published a second guide (Clarke and Mount 1998), they did not explicitly recognise a need for access objectives or anything similar. However, this later guide clearly acknowledges that provisions for people and access are important management considerations.

The most widely available European planning guides from this period are the Eurosite *European Guide for the Preparation of Management Plans for Protected and Managed Natural and Semi-Natural Areas* (Eurosite 1992 revised 1996) and the *Eurosite Toolkit for Management Planning* (Eurosite 1999). Both guides, with the exception of the descriptive sections, fail to pay any significant attention to access or people.

Elsewhere, by the 1990s many management planning guides and management plans were beginning to contain detailed prescriptions or action plans which identified the work required to provide access for visitors. One of the first was a Royal Society for the Protection of Birds internal document, *RSPB Management Plan Guidance Notes* (RSPB 1993). Although objectives are not featured in this guide, there is a section on 'Management Policy' where policies are 'the broad goals that we would like to achieve at the site'. Their first policy covers habitat and species management, but the second is 'visitors, interpretation and education'. It is not surprising that it should be the RSPB, an organisation with such an outstanding reputation for visitor management, that took a lead in developing access planning for nature reserves in the UK.

In 1996 the Countryside Council for Wales published *A Guide for the Production of Management Plans for Nature Reserves and Protected Areas* (Alexander 1996). The guide is mainly concerned with planning for wildlife features, but it also introduces the idea of including an objective for access in a nature conservation management plan. The guide contains the following example of an access objective: 'To enable the appropriate and sustainable public use of the reserve providing this does not compromise the nature conservation features.' The objective is accompanied by a number of performance indicators that quantify the access provisions, but the guide does not recognise a need to measure the quality of visitors' experience.

It was not until 2000, when the Countryside Management System Consortium (CMSC) published *The CMS Management Planning Guide for Nature Reserves and Protected Areas* (Alexander 2000b), that a UK guide gave access objectives appropriate attention. This change of emphasis was a direct consequence of the experience that the author gained while working with staff of the Uganda Wildlife Authority during the preparation of a management plan for the Kidepo Valley National Park in Uganda (Alexander et al. 2000) and the production of a management planning guide for use in tropical Africa (Alexander 2000a).

18.3.2 USA and the New World

The European experience contrasts sharply with the requirements for managing visitors in the wilderness, or semi-wilderness, areas of the New World, including Australia, Canada and the USA. There is a wealth of documentation dealing with the management frameworks used in the New World, particularly in the USA.

The most frequently encountered management frameworks include:

ROS	Recreation Opportunities Spectrum	Clark and Stankey (1979)
LAC	Limits of Acceptable Change	Stankey et al. (1984)
VAMP	Visitor Activities Management Planning	Graham et al. (1988)
VIM	Visitor Impact Management	Graefe et al. (1990)
TOS	Tourism Opportunities Spectrum	Butler and Waldbrook (1991)
VERP	Visitor Experience and Resource Protection	Manning et al. (1995)
ECOS	Ecotourism Opportunity Spectrum	Boyd and Butler (1996)
PAVIM	Protected Area Visitor Impact Management	Farrell and Marion (2002)

18.3.2.1 ROS Recreation Opportunities Spectrum

ROS was one of the first frameworks; it was developed for the Forest Service and Bureau of Land Management in the USA. It defines a range of landscape zones, called 'settings', which vary from pristine wilderness to high-density urban areas. Six specific attributes define the types of opportunities for recreation that are considered suitable within each setting. These are: access, management, social interaction with other users, non-recreational resource uses, acceptability of impacts from visitor use and acceptable levels of control of users. In short, this system identifies the range of recreational opportunities that can be accommodated in a zone, while taking account of visitor preferences. Sensitive areas are protected, and robust areas are used for recreation.

18.3.2.2 LAC Limits of Acceptable Change

The LAC system, the most widely used approach in the USA, was also created by the Forest Service and Bureau of Land Management. This system appears to be a development of its precursor ROS and utilises the same classes of opportunities for recreation. It takes the process further, based on the recognition that all recreational use of wilderness causes some impact. It considers the impact that visitors and recreational use have on the environment and on other visitors. LACs define the carrying capacity of an area by placing limits on the amount of change that can be tolerated. The process, in its original form, is driven by issues (factors) (Stankey et al. 1984). It does not define objectives (goals or targets) and gives no indication of what the desired outcomes might be. This is a planning process that aims to prevent or control use and activities, and to avoid the need for expensive or difficult remedial management. It has been criticised because there is no guarantee that management values and decisions will meet visitor preferences (Boyd and Butler 1996). The paper that introduced LACs claimed a shift in focus from 'how much use' to 'how much change' (Stankey et al. 1984). However, the actual examples of limits provided by Stankey appear to be concerned with 'how much use':

> Number of other parties met per day while travelling.
> Number of other parties camped within sight or sound per day.
> Percentage of trail systems with severe erosion miles with multiple trails.

18.3.2.3 VAMP Visitor Activities Management Planning

This is a Parks Canada system and was designed specifically for their planning programme. It incorporates elements of the ROS framework. The system focuses on the provision of opportunities. The impact of visitors and activities are dealt with in other sections of Park Canada's Natural Resource Management Planning System.

18.3.2.4 VIM Visitor Impact Management

VIM is a US National Parks system designed to control the negative impact that visitors and their activities can have on a park. It does not use a zoning system. VIM identifies the problems, or factors, that will result in damage or degradation and develops management strategies for restoration, or for the prevention of further damage, but it does not assess potential impacts.

18.3.2.5 TOS Tourism Opportunities Spectrum

This is a variation of ROS, an adaptation designed to meet the need to provide a system for developing tourism in the Canadian Arctic. It provides a framework for collating the data required before decisions can be made about the activities that can be permitted in an area. The emphasis is on seeking opportunities for tourism, but it also considers the impact of tourism.

18.3.2.6 VERP Visitor Experience and Resource Protection

The authors of this system claim a conceptual shift from 'issue driven' planning, epitomised in the LAC process, to 'goal driven' planning. Their philosophy is that, 'issues are nothing more than obstacles that lie between existing conditions and future desired conditions'. This implies that you must know what your desired state is (goals) before you can really understand 'issues' (Hof and Lime 1997). (We can assume that 'goals' are very similar to 'objectives', as defined elsewhere in this book.) Management zones are central to this process, and appropriate levels of use are identified for the zones. This is a framework where carrying capacity is defined by the quality of the resources and the quality of the visitor experience.

18.3.2.7 ECOS Ecotourism Opportunity Spectrum

This is a framework developed specifically for managing ecotourism; it is based on ROS and TOS. In particular, it provides a framework for identifying and promoting opportunities for ecotourism within an area.

18.3.2.8 PAVIM Protected Area Visitor Impact Management

This is the most recent addition. It was developed as an alternative to carrying capacity frameworks, such as LAC and VERP, because these are often too costly to implement. The system places considerable emphasis on public participation, and one of the most unusual components is that the views of an expert panel replace the use of indicators, monitoring and standards. The 'public' (stakeholders) identify area values, purpose and management zones. They also define specific objectives and identify and prioritise impact problems.

The paper that introduces the framework (Farrell and Marion 2002) contains an interesting description of management objectives: they must be specific, realistic, achievable and time-bounded. The omission of 'measurable' from an otherwise complete definition of 'SMART'[1] is probably because they rely on expert opinion in place of monitoring. However, in reality, they have replaced an approach to measuring objectives which is not affordable in many circumstances with a pragmatic and affordable approach. The paper provides three different examples of objectives:

- Social condition objectives: 'that define desired visitor experiences for zones and may be characterised in terms of interaction with park staff, amount or type of visitor use, contact with or proximity to natural environments and level of knowledge, effort or risk needed to experience the area.'
- Resource objectives: 'specifying desired trail tread conditions can be assessed for degree of compliance.'

[1] **S**pecific, **M**easurable, **A**chievable, **R**elevant, **T**ime-based.

– Managerial condition objectives: 'regarding level or type of facility development can guide selection of corrective actions.'

The authors state that:

> The PAVIM framework provides a professional impact identification and evaluation process, represents cost effective and timely means of managing visitor impacts, and may also better integrate local resource needs and management capabilities and constraints into decision-making.

18.3.2.9 Discussion

Overall, this is an extremely impressive body of literature, and there are many useful ideas, concepts and approaches that can be applied to other types of protected areas. It has not been possible here to do more than identify, and briefly outline, the various processes. For further information read *A Comparative Analysis of Protected Area Planning and Management Frameworks* (Nilsen and Tayler 1998) and the IUCN publication *Sustainable Tourism in Protected Areas Guidelines for Planning and Management* (Eagles et al. 2002). The latter contains an appendix that provides a useful comparison of five visitor management frameworks.

The definition of access and tourism used in this book is discussed at the beginning of this chapter: 'Access planning is concerned with all the provisions that are made for people who visit or use a site for any reason other than for official, business or management purposes.'

The only detectable difference between this definition and that implied by the USA frameworks is the length of time that visitors spend on a site. There is probably significant overlap, but the difference is emphasised by the authors of the ECOS framework. They state that: 'Ecotourism in this context does *not* include most of the short term visits to natural and semi-natural areas, especially those in developed countries where the emphasis is on participation in an activity rather than experiencing nature' (Boyd and Butler 1996). The focus of most of these frameworks appears to be the management of long-stay visits. This is clearly different to the situation experienced at most European sites where the majority of visits are short-stay.

The second part of the statement, 'emphasis is on participation in an activity rather than experiencing nature', is very difficult to understand. The paper does not offer any evidence to support the claim, and there is contradictory evidence available from the UK. In England and Wales, 291 National Nature Reserves are visited by over 18 million people each year. These are people who visit reserves to experience or enjoy nature and not to participate in activities other than viewing wildlife (Natural England, internal report 2006; Countryside Council for Wales, internal report 2005). There is certainly greater use by local people, who generally visit for shorter periods, but tourism is also a very important function of many sites. Regardless of continent, New or Old World, visitors to protected areas are most often there for a wilderness or wildlife experience in some form or other.

It is difficult to see how any of the USA frameworks could be applied elsewhere, particularly to sites managed for specific wildlife features that are protected by legislation, without some modification. The most significant reason is that these frameworks do not give sufficient attention to the biological or wildlife features on a site. In New World conservation, the priority has been, and continues to be, the protection of wilderness areas, with an emphasis on managing people to minimise their impact on the wilderness. In the Old World, Europe, for example, the cultural landscape, comprising mainly semi-natural habitats, is not only the priority for conservation but, in most places, it is all that remains to be protected. These mainly semi-natural or plagioclimatic habitats require active management if they are to survive. In Europe we are mainly tied to a features approach to management planning. For the most important European sites (Natura 2000 sites) there is an obligation to protect the features for which the site is designated. This obligation has precedence over almost everything else and compels managers of these sites to define the carrying capacity in relation to the implications for the features. In other words, access and recreational activities may be encouraged but only in so far as they are consistent with protecting the wildlife features. This may help to explain why, until recently, some countries in Europe have placed so much emphasis on habitat and species management.

A second reason for rejecting the USA frameworks, unless they are substantially modified, is that most of these systems are driven by negative issues or factors and are not objective led. The main problem with any issue- or factor-driven approach is that some factors may be missed, dismissed or not recognised as a threat until there has been damage. Some of the systems, for example PAVIM, claim a move to an objective- or goal-driven approach. However, the examples that they provide are very general or broadly defined. They are not sufficiently specific, measurable or detailed for feature-based management, but this is not to suggest that they are unsuitable for wilderness management.

Excluding VERP (Manning et al. 1995), which is a component of a general management plan, it would appear that these frameworks, which are almost entirely focused on recreational use, are the only systems used to identify and control anthropogenic influences. Unfortunately, in addition to visitor impacts, there are many other human influences, often originating outside the protected area, that can have a much greater impact on the site and its wildlife. These will include invasive species, aerial pollution, water-borne pollution, loss of connectivity with other areas, unregulated hunting or harvesting offsite and global environment change. The management of any feature is obtained through controlling factors. Consequently, planning should take adequate account of all the factors that have changed, are changing, or may change, a feature.

Although the frameworks may not be entirely appropriate for use outside wilderness areas, some of the ideas, components and features contained in the frameworks will be very useful elsewhere. The lessons learned from the frameworks will be carried forward in the following section, which focuses on access planning for protected areas that are mainly semi-natural and where the emphasis is on managing wildlife features.

18.3.2.10 USA Frameworks and IUCN Protected Areas

The International Union for the Conservation of Nature and Natural Resources, IUCN, is the world's largest and most important conservation network. They describe a protected area as a clearly defined geographical space, recognised, dedicated and managed, through legal or other effective means, to achieve the long term conservation of nature with associated ecosystem services and cultural values.

IUCN Protected Areas Categories System

IUCN protected area management categories classify protected areas according to their management objectives. The categories are recognised by international bodies such as the United Nations and by many national governments as the global standard for defining and recording protected areas and, as such, are increasingly being incorporated into government legislation.

Category Ia: Strict Nature Reserve
Category Ia are strictly protected areas set aside to protect biodiversity and also possibly geological/geomorphological features, where human visitation, use and impacts are strictly controlled and limited to ensure protection of the conservation values. Such protected areas can serve as indispensable reference areas for scientific research and monitoring.
Primary Objective

- To conserve regionally, nationally or globally outstanding ecosystems, species (occurrences or aggregations) and/or geodiversity features: these attributes will have been formed mostly or entirely by non-human forces and will be degraded or destroyed when subjected to all but very light human impact.

Category Ib: Wilderness Area
Category Ib protected areas are usually large, unmodified or slightly modified areas, retaining their natural character and influence without permanent or significant human habitation, which are protected and managed so as to preserve their natural condition.
Primary Objective

- To protect the long-term ecological integrity of natural areas that are undisturbed by significant human activity, free of modern infrastructure and where natural forces and processes predominate, so that current and future generations have the opportunity to experience such areas.

Category II: National Park
Category II protected areas are large natural or near natural areas set aside to protect large-scale ecological processes, along with the complement of species and ecosystems characteristic of the area, which also provide a foundation for environmentally

and culturally compatible, spiritual, scientific, educational, recreational and visitor opportunities.

Primary Objective

– To protect natural biodiversity along with its underlying ecological structure and supporting environmental processes, and to promote education and recreation.

Category III: Natural Monument or Feature

Category III protected areas are set aside to protect a specific natural monument, which can be a landform, sea mount, submarine cavern, geological feature, such as a cave, or even a living feature, such as an ancient grove. They are generally quite small.

Primary Objective

– To protect specific outstanding natural features and their associated biodiversity and habitats.

Category IV: Habitat/Species Management Area

Category IV protected areas aim to protect particular species or habitats, and management reflects this priority. Many Category IV protected areas will need regular, active interventions to address the requirements of particular species or to maintain habitats, but this is not a requirement of the category.

Primary Objective

– To maintain, conserve and restore species and habitats.

Category V: Protected Landscape/Seascape

A protected area where the interaction of people and nature over time has produced an area of distinct character with significant, ecological, biological, cultural and scenic value, and where safeguarding the integrity of this interaction is vital to protecting and sustaining the area and its associated nature conservation and other values.

Primary Objective

– To protect and sustain important landscapes/seascapes and the associated nature conservation and other values created by interactions with humans through traditional management practices.

Category VI: Protected Area with Sustainable Use of Natural Resources

Category VI protected areas conserve ecosystems and habitats, together with associated cultural values and traditional natural resource management systems. They are generally large, with most of the area in a natural condition, where a proportion is under sustainable natural resource management and where low-level, non-industrial use of natural resources, compatible with nature conservation, is seen as one of the main aims of the area.

Primary Objective

– To protect natural ecosystems and use natural resources sustainably, when conservation and sustainable use can be mutually beneficial. (www.iucn.org)

The IUCN protected areas are categorised according to their primary management objective. The categories imply a gradation of public access, ranging from effectively none at all in Category Ia areas to quite high levels of access in Category V areas. I am not sure about Category VI. Clearly these are areas where significant levels of human impact are acceptable, and priority is given to cultural values and sustainable use, but there is no explicit mention of recreation or tourism. Category Ia is easily understood. Access is strictly controlled and probably not promoted. The USA wilderness frameworks certainly relate to category Ib areas. Although wilderness is not specifically mentioned in the definition of category II areas, the definition clearly states that these areas are designated mainly for ecosystem protection, education and recreation. The USA frameworks are applicable in these areas, but their relevance will diminish if the sites are managed for specific conservation features and are also, for example, Natura 2000 sites. The IUCN category III areas are concerned with natural monuments. The frameworks are probably not suitable for use in areas established to protect specified features without significant modification. The same clearly applies to category IV areas. Areas V and VI require a different approach to planning. This is more than adequately covered in the WCPA publication, *Management Guidelines for IUCN Category V Protected Areas, Protected Landscapes/Seascapes* (Phillips 2002).

The IUCN publication, *Sustainable Tourism in Protected Areas* (Eagles et al. 2002), contains a list of guidelines for the development of tourism policies. The first two entries on the list are:

> The natural and cultural environment within the protected area should form the basis for all other uses and values affecting the park and its management. These fundamental assets must not be put at risk;
>
> Protected area tourism depends on maintaining a high quality environment and cultural conditions within the area. This is essential to sustaining the economic and quality of life benefits brought by tourism;

Most of the USA planning frameworks use issues or factors as the focus for planning, and these factors are the various forms of human impact resulting from tourism and recreation. Bearing in mind that the frameworks are mainly intended for IUCN Category Ib or II sites, where tourism and recreation are the primary or secondary objectives of management, there is some compatibility between the frameworks and the IUCN guidelines.

The Relationship Between Access and Wildlife Conservation

When managing Natura 2000 sites there is no doubt that the features which were the basis of designation come first. There may be a presumption on these sites which is strongly in favour of encouraging access, but access will only be permitted in so far as it is compatible with the primary objective of management, which is the protection of wildlife features.

The priority given to the wildlife features on non-statutory sites will be determined by the policies of the organisation which manages the site or as a consequence of the legal status of the feature. For example, many species are protected by law. For all areas managed primarily, or in part, to protect nature conservation features, it would make little sense if the quality of the wildlife suffered in order to accommodate people. As a consequence, the principle of allocating secondary status to objectives for access should be central to any planning approach applied to these sites.

There are amenity sites, which are managed primarily to provide recreational and other opportunities for people, where the wildlife component, although less important, is the main attraction. Even on these sites, given that in the absence of wildlife there would be little purpose in people visiting the site, the maintenance of the wildlife features should have at least the same priority as access.

For all of these reasons, the planning process described in the next chapter (Chapter 19) gives the protection or management of wildlife priority over access provisions. The first, and most significant, consideration in the access section of a management plan is the carrying capacity of the wildlife features; this will be covered in detail later. The implication is that objectives for the wildlife features must be completed before embarking on the access section of the plan. The objectives for wildlife will define the condition that is required for the habitat and species features. For a feature to be at Favourable Conservation Status the factors must be under control. Access should be regarded as a factor that is changing, or could change, the wildlife features. 'Change is a natural, inevitable consequence of recreational use' (Stankey et al. 1984).

Integration is essential: the management plan must establish a direct and clear link between the protection of wildlife and the management of people. There are so many planning systems that compartmentalise the plan and fail to provide the essential links between sections. Worse still, some organisations divide the responsibility for producing a plan between different sections or teams within the organisation: one prepares a plan for wildlife while another prepares an access plan.

Recommended Further Reading

Eagles, P. F. J., McCool, S. F. and Haynes, C. D. A. (2002). *Sustainable Tourism in Protected Areas: Guidelines for Planning and Management.* IUCN, Gland, Switzerland and Cambridge, UK.

Nilsen, P. and Tayler, G. (1998). *A Comparative Analysis of Protected Area Planning and Management Frameworks.* Pages 49–57 in McCool, S. F. and Cole, D. N. (Compilers), Limits of Acceptable Change and Related Processes: Programs and Future Directions. Proceedings of conference, 1997 May 20–22, Missoula, MN. Gen. Tech. Rep. INT-GTR-371. U.S. Department of Agriculture, Forest Service, Rocky Mountains Research Station, Ogden, UT, USA.

References

Alexander, M. (1996). *A Guide to the Production of Management Plans for Nature Conservation and Protected Areas.* Countryside Council for Wales, Bangor, Wales.

Alexander, M. (2000b). *Guide to the Production of Management Plans for Protected Areas.* CMS Partnership, Aberystwyth, Wales.

Alexander, M. (2000a). *A Management Planning Handbook*. Uganda Wildlife Authority, Kampala Uganda.

Alexander, M., Hellawell, T. C., Tillotson, I. and Wheeler, D. (2000). *Kidepo Valley National Park Management Plan*. Kampala, Uganda, Uganda Wildlife Authority.

Boyd, S. W. and Butler, R. W. (1996). *Managing ecotourism: an opportunity spectrum approach*. Tourist Management, 17(8) UK.

Butler, R. W. and Waldbrook, I. A. (1991). *A new planning tool: the Tourism Opportunity Spectrum*. Journal of Tourism Studies, 1991 2(I), 1–14.

Clarke, R. and Mount, D. (1998). *Site Management Planning*. Countryside Commission, Cheltenham, UK.

Clark, R. N. and Stankey, G. (1979). *The recreation opportunity spectrum: a framework for planning, management and research*. Gen. Tech. Rep. GTR-PNW-98. Portland, Oregon, USA.

Eagles, P. F. J., McCool, S. F. and Haynes, C. D. A. (2002). *Sustainable Tourism in Protected Areas: Guidelines for Planning and Management*. IUCN, Gland, Switzerland and Cambridge, UK.

Eurosite (1992). *European Guide for the Preparation of Management Plans*. Eurosite, Tilburg, The Netherlands.

Eurosite (1996). *Management Plans for Protected and Managed Natural and Semi-natural Areas*. Eurosite, Tilburg, The Netherlands.

Eurosite (1999). *Toolkit for Management Planning*. Eurosite, Tilburg, The Netherlands.

Farrell, T. A. and Marion, J. L. (2002). *The Protected Area Visitor Impact Management (PAVIM) Framework: A Simplified Process for Making Management Decisions*. Journal of Sustainable Tourism, 10(1). Portland Press Ltd. London, UK.

Graefe, A., Kuss, F. R. and Vaske, J. J. (1990). *Visitor Impact Management: The Planning Framework*. National Parks and Conservation Association, Washington, DC, USA.

Graham, R., Nilsen, P. and Payne, R. J. (1998). *Visitor Management in Canadian National Parks*. Tourism Management, 1988, 9(1), 44–62.

Hof, M., and Lime, D. W. (1997). *Visitor Experience and Resource Protection Framework in the National Park System: Rationale, current status, and future direction*. In *Proceedings–Limits of Acceptable Change and related planning processes: Progress and future direction*, comps. McCool, S. F., and D. N. Cole, 29–36. 1997, May 20–22; Missoula, MT. General Technical Report, UT: USDA.

Leopold, A. (1949). *A Sand County Almanac, and sketches here and there*. Oxford University Press, New York.

Manning, R. E., Lime, D. W., Hof, M. and Freimund, W. A. (1995). *The Visitor Experience and Resource Protection (VERP) Process*. The George Wright Forum, 12(3).

Phillips, A. (2002). *Management Guidelines for IUCN Category V Protected Areas, Protected Landscapes/Seascapes*. IUCN Gland, Switzerland and Cambridge, UK.

RSPB (1993). *RSPB Management Plan Guidance Notes, version 3*. Internal publication, The Royal Society for the Protection of Birds, Sandy, UK.

Stankey, G. H., McCool, S. F. and Stokes, G. L. (1984). *Limits of Acceptable Change: A new framework for managing the Bob Marshall Wilderness*. Western Wildlands, 10(3), 33–37.

NCC (1983). *A Handbook for the Preparation of Management Plans*. Nature Conservancy Council, Peterborough, UK.

NCC (1988). *Site Management Plans for Nature Conservation, A Working Guide*. Nature Conservancy Council, Peterborough, UK.

Nilsen, P. and Tayler, G. (1998). *A Comparative Analysis of Protected Area Planning and Management Frameworks*. Pages 49–57 in McCool, S. F. and Cole, D. N. (Compilers), Limits of Acceptable Change and Related Processes: Programs and Future Directions. Proceedings of conference, 1997 May 20–22, Missoula, MN. Gen. Tech. Rep. INT-GTR-371. U. S. Department of Agriculture, Forest Service, Rocky Mountains Research Station, Ogden, UT, USA.

Wood, J. B. and Warren, A. (ed.) (1976). *A Handbook for the Preparation of Management Plans – Conservation Course Format, Revision 1*. University College London, Discussion Papers in Conservation, No. 18. London.
Wood, J. B. and Warren, A. (ed.) (1978). *A Handbook for the Preparation of Management Plans – Conservation Course Format, Revision 2*. University College London, Discussion Papers in Conservation, No. 18. London.

Chapter 19
Preparing an Integrated Plan for Access and Recreation

Abstract The section on preparing a plan for access and recreation should be regarded as an integral component of a full management plan and not as an independent document. There may be occasions when there is a need to present the section as a stand-alone document, but its relationship with the remainder of the site plan must not be forgotten. One of the key components of the access section is the identification of the carrying capacity of the site. This is the level of access that can be accommodated without detracting from the quality of the experience that visitors enjoy on the site. There will be two main areas of impact: Visitors can have a direct impact on the infrastructure, landscape and wilderness qualities of a site, for example, paths may become over-wide and unsightly. People can also visit sites in such large numbers that they become a distraction to others. This is particularly important in areas of high landscape or wilderness value. An access objective with performance indicators is required for most sites. Performance indicators for access must be measurable and quantified (i.e. so that they can be monitored), and the data should be easy to collect. The number of indicators should be kept to a minimum, but there should be sufficient to provide the evidence necessary to ensure that the quality of the access provisions can be measured.

19.1 Introduction

This section on planning for access is intended for nature reserves and areas managed primarily, or in part, to protect nature conservation features. It is possibly less relevant to vast wilderness or semi-wilderness areas, for example, the USA national parks or the large European parks established to protect cultural landscapes, which include a high proportion of inhabited areas.

The section on access should be regarded as an integral component of a full management plan and not as an independent document. There may be occasions when there is a need to present the section as a stand-alone document, but its relationship with the remainder of the site plan must not be forgotten.

M. Alexander, *Management Planning for Nature Conservation: A Theoretical Basis & Practical Guide*, DOI 10.1007/978-94-007-5116-3_19,

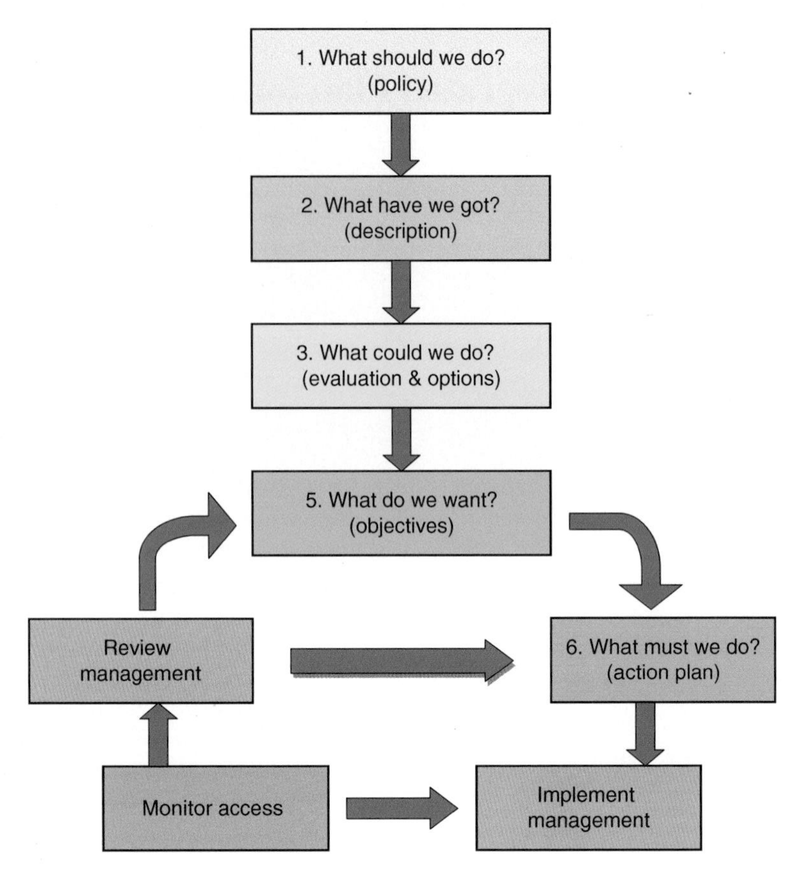

Fig. 19.1 The structure of the access section in relation to the entire plan

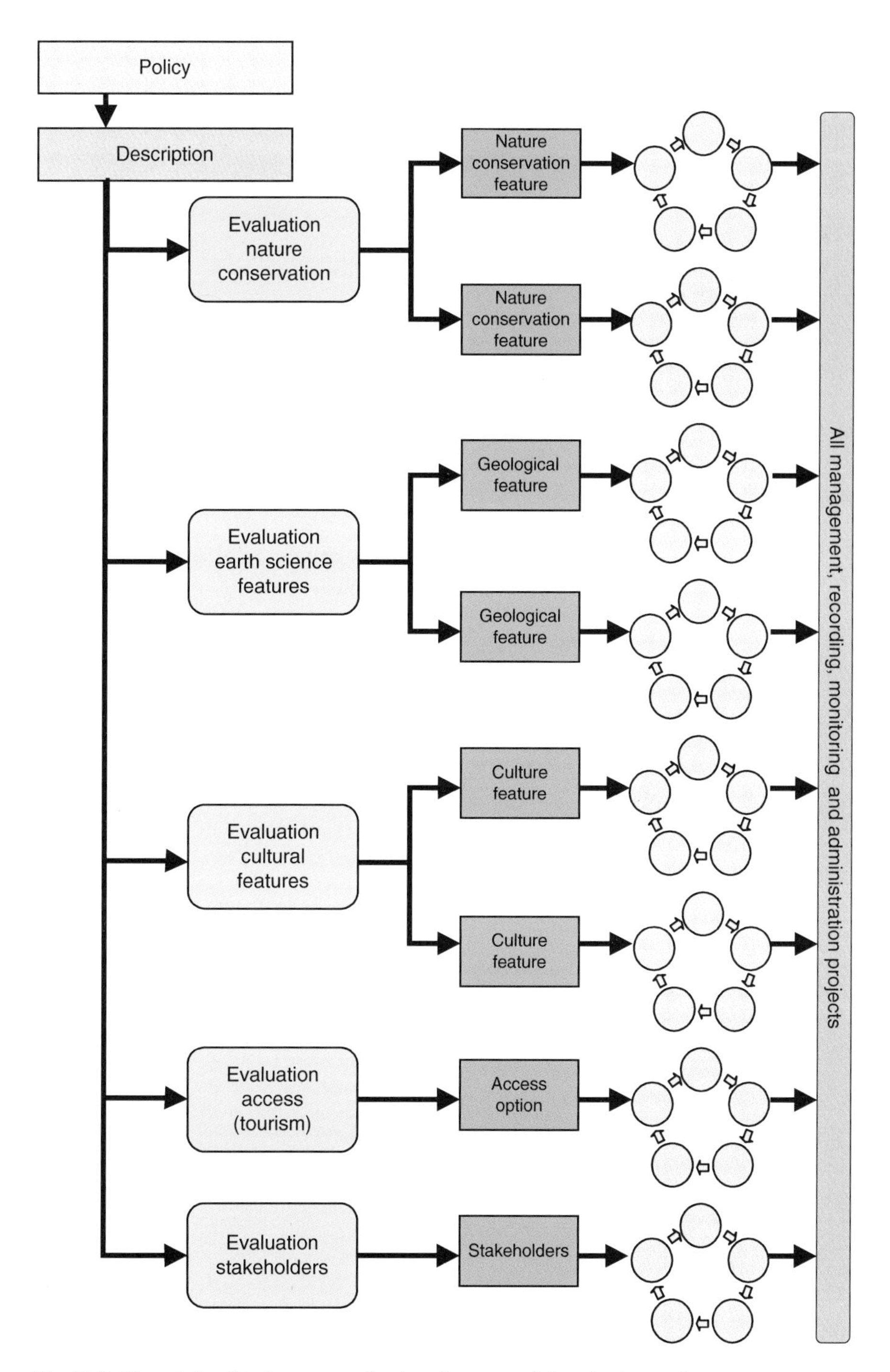

Fig. 19.2 The relationship between evaluation, features and the adaptive cycle

19.2 The Contents of an Access Section

The sections in a management plan which contain information relevant to access

(The sections in bold are general sections with some access information. The remainder are specific access sections.)

1 Plan Summary
2 Legislation & Policy
3 Description

3.4 People – Stakeholders, Access, Etc. (section in the main description)
3.4.2 Access
3.4.2.1 Visitor Numbers
3.4.2.2 Visitor Characteristics
3.4.2.3 Visit Characteristics
3.4.2.4 Access to the Site
3.4.2.5 Access Within the Site
3.4.2.6 Visitor Facilities and Infrastructure
3.4.2.7 The Reasons Why People Visit the Site
3.6.2.7.1 Wildlife Attractions
3.4.2.7.2 Other Features That Attract People
3.4.2.7.3 Recreational Activities
3.4.2.8 Current and Past Concessions
3.4.2.9 Stakeholder Interests
3.4.2.10 The Site in a Wider Context
3.4.3 Interpretation and Information
3.4.4 Educational Use

5 Access Section

5.1 Evaluation
5.1.1 Actual or Potential Demand
5.1.2 Accessibility of the Site
5.1.3 Access Within the Site
5.1.4 Site Safety
5.1.5 Implications of Stakeholder Interests
5.1.6 Carrying Capacity of the Features
5.1.7 Carrying Capacity of the Site
5.1.8 Availability of Resources
5.1.9 Summary of the Evaluation
5.2 Access Option
5.3 Access Objective
5.3.1 Vision
5.3.2 Performance Indicators & Monitoring
5.4 Status & Rationale
5.4.1 Status

5.4.2 Rationale
- 5.4.2.1 Legislation
- 5.4.2.2 Access to the Site
- 5.4.2.3 Access Within the Site
- 5.4.2.4 Visitor safety
- 5.4.2.5 Seasonal Constraints
- 5.4.2.6 Public access
- 5.4.2.7 Excessive demand
- 5.4.2.8 Visitor infrastructure
- 5.4.2.9 Information
- 5.4.2.10 Interpretation
- 5.4.2.11 Education

6 Action Plan

Important: The remainder of this chapter follows the numbering system given above as used in a management plan, and not the sequence used elsewhere in this book.

1 Plan Summary – Access Section

When there is good reason for preparing a large section on access in a management plan, and particularly if there is a need to use the section as a stand-alone document, a summary should be included. The summary contains a succinct outline of all the main subsections in the access plan. This should be sufficient to provide readers with a rapid overview and understanding of the main provisions in the plan.

2 Legislation & Policy – Access Section

This section can be included within a subsection of the management plan that deals entirely with access, or, alternatively, it can be placed in the general policy section. Please refer to Chapter 11, which provides full guidance on dealing with legislation and policy.

2.1 Legislation

The planning and management of all sites will be influenced by legislation. This will include Health & Safety, Public Liability and other general legislation which relates to a duty of care for all visitors. For example, in England and Wales, as a consequence of the Disability Discrimination Act 1995, it is unlawful for a *provider* of services to discriminate against a disabled person. Reserves that allow public access are 'providers'. This means that site management must make reasonable provisions for disabled people.

In addition, there will usually be some specific legislation in respect of access to the countryside. The following are examples of British countryside legislation:

- *Rights of way.* These are usually minor routes that exist for the benefit of the community at large. Historically, they were an integral part of the country's transport system, but they have long since evolved into a recreational web which enables people to explore the countryside. Where these rights of way pass through a site, with few exceptions, they must be kept open at all times.
- *Access to open countryside.* The Countryside and Rights of Way Act 2000 provides a statutory right of public access on foot for informal recreation over mountain, moor, heath, down and all registered common land. Maps show where rights over open country and registered common land under Part 1 of the CRoW Act apply. The implications of a site falling within an open access area are obvious: the pubic has right of access. Limited restrictions can be applied to protect sensitive wildlife.

This section of the plan must contain reference to all legislation that has implications for site access and particularly for the safety of visitors. This is not the place to discuss the implications of the law.

2.2 Policy

The development of the access, public use and tourism section of the plan is entirely guided by policy, and policies must reflect legislation. This section should describe all organisational, and any other, policies that have relevance to access provisions on the site. Much of the following evaluation is concerned with assessing the extent to which organisational policies can be met on individual sites. Local conditions, for example, dangerous features, fragile wildlife, inaccessibility, etc., can significantly influence the ability to meet policies.

The following is an example of the access policies from a UK government organisation which has a responsibility for managing a large number of nature reserves:

The sustainable public use of the National Nature Reserves will be encouraged in so far as such use:

- Is consistent with our duty to maintain or restore the nature conservation and geological features to Favourable Conservation Status
- Does not expose visitors or staff, including contractors, to any significant hazards

All legitimate and lawful activities will be permitted in so far as these activities:

- Are consistent with our duty to maintain or restore the nature conservation and geological features to Favourable Conservation Status

- Do not expose visitors or staff, including contractors, to any significant hazards
- Do not diminish the enjoyment of other visitors to the site

This next example is the general policy which was used in a management plan for a Ugandan national park. It demonstrates the relationship between an access policy and the other park policies (Alexander et al. 2000).

- The Kidepo Valley National Park will be managed by the Uganda Wildlife Authority with the prime purpose of sustaining viable habitats and their associated wildlife populations in the long term. All other functions, including tourism, are secondary.
- All parts of the Park, including both the Narus and Kidepo Valleys and those mountainous areas within the boundaries, will be managed to contribute towards sustaining viable habitats and their associated wildlife populations in the long term.
- The Park will, in the future, serve as a reservoir of wildlife resources for Uganda and specifically for re-colonisation of other suitable protected areas.
- The outstanding landscape and wilderness qualities will not be compromised. All management operations, including the provision of facilities for tourists, will be designed to minimise their impact on the Park.
- Sustainable and appropriate levels of tourism will be encouraged in so far as this provision is compatible with maintaining the landscape and wilderness qualities and the associated wildlife.

All relevant policies should be included in this section of the plan. The areas which are of particular relevance to the site and the plan should be highlighted.

Access policy (strategy) for a suite of sites

There are many reasons for attempting to prioritise access provisions over a suite of sites. The most usual is insufficient resources. If an organisation attempts to spread resources too thinly over a number of sites it is unlikely that anything of real value will be achieved anywhere. A strategic approach is essential when an organisation aims to ensure that each individual site within a suite of sites is managed in the most appropriate way to meet organisational policies. There is no need, or justification, to do everything everywhere.

A full analysis of the access potential of all sites should be completed. Each site should be examined against a range of criteria; these will be more or less the same as those used to evaluate individual sites. They will include, for example, accessibility of the site, accessibility within the site, safety, fragility of features and site fabric, features of public interest, current public use, facilities and provisions, and suitability to meet other organisational objectives. Clearly, the ideal way of obtaining an overview would be to prepare access plans for each site. Unfortunately, even the cost of preparing the simplest access plan can be prohibitive for some organisations. An initial assessment, based on the above evaluation criteria, would identify the priorities for the subsequent preparation of management plans.

The following is an example of a strategy identifying the priorities for access provisions applied to a suite of National Nature Reserves in Wales. The priorities were developed by considering each site against a list of criteria similar to those mentioned above (CCW 2004).

The NNRs in Wales can be divided into three categories:

(a) **Sites where access is a major issue**
This group comprises some of the most important tourist attractions in Wales. These sites, with few exceptions, attract very large numbers of people regardless of their status as nature reserves. Unfortunately, with the exception of several key well-known sites, the access potential of some sites in this group has not been realised. These sites provide ideal opportunities to promote the value of the National Nature Reserves, nature conservation and the countryside.
Resources will be made available to optimise public use of these sites and to ensure that an appropriate infrastructure is in place. In many cases, the carrying capacity of this group of sites will have been reached, and there will be no justification for increasing the number of visitors. However, there should be scope for improving the quality of the visit on some sites. The completion of the access sections of the management plan for these sites is given the highest priority.

(b) **Sites where access is important but where, for a variety of reasons, there are relatively few visitors**
These may not be the most important tourist sites, but they are very important sites for local people and those who are particularly interested in the countryside and wildlife. Many have an underdeveloped infrastructure, and few have any significant provisions for visitors. The importance of this group of sites must not be underestimated as they provide obvious opportunities for development.
Wherever possible, access to these sites will be improved. A full and fresh appraisal of these sites is required and will be undertaken as part of the site management planning process. Plans for these sites will be prepared once plans for the category a sites have been completed.

(c) **Sites where there is little public interest and where there is a range of factors that severely restricts, or prevents, access**
These sites can be important to local people and individuals with specialist interests. They should be regarded as areas with potential for improvement. However, it is important to recognise that there will be a small number of sites in this group that are not suitable for public access.
These sites must not be neglected. Although not currently important for access, unless there are compelling reasons for doing otherwise, steps should be taken to facilitate at least limited access. Planning should commence once plans for the two preceding tiers are complete.

3 Description

Access Zones (access compartments)

Note: this information is best held at the beginning of the full management plan description and not in the access section.

The USA ROS system (Clark and Stankey 1979) was possibly the first management framework to introduce the concept of 'settings'. Settings are more or less the same as compartments or zones. International designations, for example, Biosphere Sites, tend to favour the word 'zone'. A 'settings' approach is used in most of the American frameworks. The range and intensity of activities that can be encouraged, or tolerated, within a protected area will depend on a number of biological, physical and social factors. The frameworks suggest that there are considerable advantages in recognising that an area can be divided into settings which provide different levels of opportunity for access and recreation.

On many sites, the conclusions of the following access evaluation will vary from place to place within the site and, consequently, the level of access will vary across a site. Some parts may be suitable for access, while others are unsafe or fragile. Many other factors can also influence the selection of visitor zones, for example, the distribution of features of interest to visitors, the availability of access routes, and the protection of wildlife, landscape or wilderness qualities.

As the evaluation progresses, consideration should be given at each stage to the need to divide the site into access zones. A range of different levels of access may be identified for the whole site or for the zones within a site. For example, it would be quite reasonable to include total exclusion zones, controlled access zones and open access zones within an individual site. It is also sometimes necessary to establish activity zones, i.e. areas where specified activities are permitted or prohibited. Zones can be seasonal. For example, exclusions can be imposed during the breeding season of vulnerable nesting birds.

It is important that zonation systems are regarded as flexible management tools that can be introduced, removed or modified according to need. They can be used for a very wide range of different purposes. The only important rules are:

- They should be clearly shown on a map.
- Maps must be made available to all interested parties.
- The boundaries of the zones must be marked, or otherwise easily located, on the ground.

A site may have been previously divided into zones for a variety of different reasons, but these may not necessarily be relevant to access. This is not an issue, since it is perfectly acceptable to have several different overlapping zoning systems on an individual site. When it is necessary to divide a site into visitor zones, the delineation and description of the zones, along with an explanation outlining the basis for their selection, is required.

Access Section – Description

The USA frameworks recognise the need for an *inventory* system, and this, more or less, equates to the structured description contained in many European plans. However, they do not appear to integrate procedures for maintaining or improving their inventories within the structure of the frameworks.

The section of the plan that contains the general site description was dealt with in Chapter 12. All the basic site information, much of it relevant to this section, will have been included there, for example, location, tenure and descriptions of the features. Some additional information is required when preparing a management plan for access. Ideally, this should also be included in the general description. Occasionally, organisations may wish to present the access section of a full management plan as a stand-alone document. In these cases, the parts of the description that specifically relate to access can be held in the access section of the management plan.

The following is a recommended list of contents for the access section of the description:

Contents:

- 3.4.2 Access and Tourism
  - 3.4.2.1 Visitor Numbers
  - 3.4.2.2 Visitor Characteristics
  - 3.4.2.3 Visit Characteristics
  - 3.4.2.4 Access to the Site
  - 3.4.2.5 Access Within the Site
  - 3.4.2.6 Visitor Facilities and Infrastructure
  - 3.4.2.7 The Reasons Why People Visit the Site
    - 3.5.2.7.1 Wildlife Attractions
    - 3.5.2.7.2 Other Features That Attract People
  - 3.4.2.8 Recreational Activities
  - 3.4.2.9 Current and Past Concessions
  - 3.4.2.10 Stakeholder Interests
  - 3.4.2.11 The Site in a Wider Context

Full guidance on the preparation of the sections on description which are directly relevant to access is given in Chapter 12.

5 Access Section

5.1 Evaluation

The outcome of this section is a clear statement of the level of access, including recreational activities, that is appropriate for a site, or parts of a site. In other words, to what extent can an organisation's access policy be applied to the site?

Organisational access policies were discussed at the beginning of this section. They provide the basis for developing the site-specific access objectives. The level at which access can be provided on individual sites, or areas within a site, will be dependent on a range of local factors and on any strategic plans.

The evaluation can be based on the following list of criteria. This list is offered as guidance and should not be regarded as definitive. Some of these criteria will not be relevant to some sites and will not be used. Conversely, additional criteria, not mentioned below, may be useful in some situations. The criteria are interrelated and cannot be dealt with in isolation; each is dealt with in turn but considered within the context of the whole.

The order in which the questions are presented is quite important. It is intended to avoid unnecessary effort on sites where access is either not possible or extremely low-key. Although an organisation may have an access policy that is aimed at encouraging appropriate levels of access and recreational use on sites, there can be occasions when, for good reasons, access cannot be provided or there is no need to provide access. For example, there are sites where, with the exception of a few scientists, people have no interest, and there is no potential to encourage interest. Sites can be completely inaccessible, possibly as a consequence of legal or physical barriers. Some sites are so dangerous that access cannot be permitted. If it becomes apparent, at any stage in the evaluation, that there are good reasons for not providing access then the process can come to an end. There is no point in considering carrying capacity for a site where there will be no, or very few, visitors. Later, in the rationale, there may be opportunities for some form of virtual access. This could be anything from interactive computer programmes to remote viewing. For example, remote cameras are an excellent means of providing the public with opportunities to view rare birds at their nests.

Evaluation is always about asking questions and providing answers. It must not become a rambling, inconclusive discussion. All the information needed for the evaluation should be contained in the description.

The criteria provide a logical sequence of questions:

1. What is the actual or potential demand?
2. Is the site accessible?
3. Is access possible within the site?
4. Is the site safe?
5. What are the implications of stakeholder interests?
6. What is the carrying capacity of the features?
7. What is the carrying capacity of the site?
8. Are sufficient resources available?

5.1.1 Actual or Potential Demand

The first step in the evaluation is an assessment of the demand, or requirement, for access. When trying to assess potential public demand, one of the more important questions is: how popular is the site with visitors, and could promotion or publicity increase interest and demand? There is no need to consider detail at this stage; an outline will suffice. Projects to promote or publicise the site can be developed at a later stage. The current information on visitor numbers and profile should be held in the general site description, so there is no point in simply repeating that information.

The reason for beginning with this question is that if, for any reason, there is no actual, or potential, demand there is little purpose in continuing with the evaluation. Whether or not a site has visitors it is important to ask why this is the case. The features of interest have been described in the preceding description. At this stage, the features are assessed to establish the actual, or potential, interest. Many features mentioned in the description may, for a variety of reasons, fail to interest visitors, while others may attract thousands of people each year.

The seasonal nature of features should be considered. For example, the feature that attracts the majority of visitors to an offshore island nature reserve is the breeding population of seabirds, particularly the puffins, but these birds are absent in the winter.

What are the recreational opportunities; what do people do on the site, or what do they want to do; how many of them want to do it? This is an expression of interest and not a replay of the description. The actual carrying capacity, or tolerance to recreational activities, will be considered in the sections of the management plan that deal with the wildlife features. This will be summarised later in the subsection of this evaluation which deals with carrying capacity.

5.1.2 Accessibility of the Site

The obvious question is: how accessible is the site, or parts of the site, and can people get there? Is it close to, or easily accessible from, major highway networks? If it is remote, are there any roads, trails or footpaths, and are these in a condition that can be used, for example, by vehicles, bicycles or on foot? If people travel by private vehicle, can they park? The legal rights of access are also important; in some countries access through privately owned land is not permitted. There may be seasonal aspects; some sites are not accessible in winter. If sites are not accessible, and there is little or no potential for improvement, there will be no point in making access provisions.

5.1.3 Accessibility Within the Site

How easily can visitors gain access within the site? What is the capacity of the current infrastructure? What are the limitations, if any? For example, footpaths may be extremely severe and only suitable for fit, active people, or, alternatively, the site may contain a network of level, wide, well-surfaced tracks that are suitable for everyone. In some countries, the UK, for example, land can be designated as open access, where there is no restriction on access, but there may be restrictions on activities such as the use of vehicles.

Is access controlled? For example, are vehicles, bicycles or horses permitted? Are there periods of the year when parts of the site may have to be closed to protect wildlife or for any other reason?

When a site cannot meet its potential carrying capacity because of problems of access within the site, there may be opportunities for remedial action. At this stage in the plan there is no need for detail. An indication will suffice, as the detail will be included in the management rationale. The conclusion of this section will be the extent to which access within the site will have an influence on the potential for the site to provide for visitors.

5.1.4 Site Safety

Access to any site may be restricted by the presence of hazards. In extreme circumstances, there may be an obligation to close parts of sites, or even entire sites. The first step when completing this section is to ensure compliance with all statutory and organisational health and safety procedures. For example, in the UK all organisations that employ staff on sites, or provide public access to sites, must complete a detailed risk assessment or audit of the site. All potential dangers or threats on the site must be identified. All the implications for the health and safety of visitors are considered, and then limits, if necessary, are established and applied. For example, a section of a site may have to be closed to public access. Of course, in some instances, it will be possible to take remedial action to remove or isolate the risk and ensure visitor safety. The conclusion of this section is an assessment of the extent to which safety considerations limit public use of the site.

5.1.5 Stakeholder Interests

Are there any stakeholder interests, rights and expectations that will influence access to the site, and will they influence access provisions on the site? This extends the evaluation to consider the concerns, expectations and aspirations of stakeholders. Some stakeholders may benefit, but others will be adversely affected as a consequence of visitor activity. Is there any potential for conflict with other local provisions, or opportunities for working with other providers? It is important that opportunities for working with others are considered. It may be possible to work with stakeholders to provide improved opportunities for visitors, thereby enhancing their experience and providing income or other benefits to the stakeholders.

Carrying Capacity

Recreation carrying capacity was the focus of most of the research into wilderness management in the USA. Between 1970 and 1990, over 2,000 papers were published (Stankey and McCool 1990). Unfortunately, despite the extraordinary attention given to this subject, there does not appear to be any consensus. In fact, the literature can be very confusing. The USA frameworks recognise that there are two divisions: a biological and a social component. Superficially, the biological component, that is the extent to which the wildlife features can tolerate human presence and activities, would appear to be reasonably easy to understand. However, in the context of wilderness planning, nothing could be further from the truth. There is no static, unchanging state that can easily be described. Wilderness is dynamic; plant communities and their associated fauna are constantly changing.

Ideally, managers need to differentiate between changes that result from natural processes and which, in the context of wilderness management, are considered acceptable, and changes resulting from the impact of human activities, which are not acceptable. In reality, it may not be possible to differentiate, with any certainty, between the effects of natural processes and the effects of anthropogenic activity, since change is often the consequence of the combined impact of several factors, both natural and anthropogenic.

Biological carrying capacity may be complicated, but it pales into insignificance when carrying capacity is used to define the quality of the human experience. The values are subjective and, consequently, difficult to defend. Limits on the level of access, i.e. the total number of visitors, can be meaningless. Activities also need to be limited, and some activities will be intolerable to all but a very small minority. In addition to the type of activity, the size of groups, the behaviour of individuals, and the time that they spend in an area will all have implications for the enjoyment of others.

One way forward, which is recognised by most systems and is highly developed in the PAVIM approach (Farrell and Marion 2002), is to ensure public participation in this process. But for sites that offer opportunities for a multitude of competing or incompatible leisure activities it will not be possible to please everyone. Different people have different interests and expectations. They will judge their experience of a site from an almost infinite range of personal perspectives. At one extreme, some individuals will seek solitude, and even a single encounter with other people will diminish the quality of their experience. At the other extreme, many people feel very uncomfortable when they are away from the crowd; wilderness, wild places and nature can be threatening to those who see them as unfamiliar territory.

5.1.6 Carrying Capacity of the Features

The relationship between access and the wildlife features of a site is an extremely important consideration in a plan. It is essential that public access does not put the wildlife features at risk.

Establishing the carrying capacity of individual features, i.e. their tolerance of human activities, is quite different to dealing with the carrying capacity of a wilderness. In many, if not most, cases, it is possible at least to identify the activities that could damage features. However, defining acceptable levels is more difficult. The process that identifies the carrying capacity of the important wildlife features will be contained in the section of the plan that establishes the limits for the factors which have implications for the wildlife features (Chapter 15). Consequently, the sections on wildlife objectives *must* be completed before dealing with access.

Except in rare circumstances, people will have some level of impact on the site features; in other words, they, and in particular their activities, are *factors.* The key role of nature reserves, and most protected areas, is to ensure that wildlife is safeguarded against the excesses of uncontrolled, illegal or destructive human behaviour. The consequence is that human activity must be controlled. There may, occasionally, be justification for some compromise, and areas of habitat may have to be sacrificed to provide the infrastructure necessary to accommodate people (for example, paths, roads, parking facilities, etc.). Also, be aware that the landscape and wilderness qualities of a site can easily be compromised by the construction of inappropriate boardwalks, footpaths or other management infrastructure.

Some aspects of public use can have very serious and obvious consequences for wildlife features, for example, climbing on cliffs used by seabirds, dog walking (emptying) in sensitive botanical sites, wildfowling where the feature is a wintering population of wildfowl. Where an activity is changing, or has obvious potential to change, a feature, these activities should be regarded as *factors* which must be kept under control. When our tolerance of the factor can be defined by specified limits this will provide a performance indicator. The following examples demonstrate the use of limits:

(a) An example where the *factor* is a recreational activity, i.e. climbing, and where the features are populations of cliff-nesting birds. The limit is:

> 'Climbing is prohibited in the marked areas between 1st February and 1st August.'

This limit will be monitored and used as a performance indicator to determine the status of the feature (cliff-nesting birds). If monitoring indicates that climbing is not under control, the assumption will be that there are adverse implications for the birds, and, consequently, the population will be considered to be at an unfavourable conservation status. The limit is linked to management actions, for example, the production and distribution of a leaflet, liaison to gain the cooperation of the climbers, and regular patrols on the cliffs.

(b) An example for a raised bog where the factor is: 'people want to walk on the bog'. The access policy for most organisations will be to provide access for people whenever this is possible. Clearly, in the case of a bog which is extremely fragile and very dangerous, limits cannot be set for access levels, apart from the obvious, which would be to prohibit all public access. However, it is possible to compromise by sacrificing a very small area of the bog to construct a boardwalk. The limit could be:

> 'Public access to the bog is restricted to the boardwalk.'

This can be monitored. The management implications are obvious, i.e. construct and maintain a boardwalk. This example illustrates the need to seek management solutions to accommodate public access rather than simply denying access.

This process, or analysis, is *not* part of the access section. These limits are established when preparing the objectives for the important wildlife features and landscape (if the site has a landscape objective). They can then be copied into this section of the plan. In some circumstances, there may be a need to return to the wildlife objectives to make sure that the access factors have been given adequate attention. The impact of all current activities on each of the wildlife features, on the landscape and on other important features, particularly those that have legal protection (for example, archaeological features), must be considered. Examples of activities include climbing, cycling, canoeing, fishing, wildfowling and, in fact, any activity that could change a feature.

Note: The *management* required to control human activities should be identified and described in the 'rationale' section for each of the objectives for the wildlife features. The same projects will also be linked to the access objective.

The Precautionary Principle

Establishing the carrying capacity of features where access and recreational use does not have any easily measured impact on the important features is more complicated. From an ethical, and sometimes legal, position, it would be extremely difficult to defend a situation where an area is declared a nature reserve and the consequence is that subsequent public use damages the wildlife. Managers must avoid this situation, and they must not leave anything to chance. A precautionary approach should be adopted; this was established in Principle 15 of the 1992 Rio Declaration on Environment and Development adopted by the United Nations Conference on Environment and Development (UNCED), which affirms that:

> In order to protect the environment, the precautionary approach shall be widely applied by States according to their capabilities. Where there are threats of serious or irreversible damage, lack of full scientific certainty shall not be used as a reason for postponing cost-effective measures to prevent environmental degradation.

If the precautionary principle is applied, there is no need for scientific proof in order to restrict human use or any specific activities when there is a reason to believe that they are a threat. In essence, the precautionary principle is about not taking chances with our environment. So, logically, when applying the principle to the carrying capacity of a site, or a feature, there should be an obligation to prove, with full scientific certainty, that an activity will *not* cause any damage before that activity, or level of activity, is permitted.

Turning to Europe, the legal situation for Natura 2000 sites is clearly set out in an official European Commission document, *Managing Natura 2000 sites: The provisions of Article 6 of the 'Habitats' Directive 92/43/EEC* (EC 2000). Although the legal implications of Article 6 are only relevant to Natura sites, the concept that gave rise to the article is relevant to all sites. The article states:

> Member States shall take appropriate steps to avoid, in the special areas of conservation, the deterioration of natural habitats and the habitats of species as well as disturbances of the species for which the areas have been designated, in so far as such disturbance could be significant in relation to the objectives of this directive.

In addition to the article, the document contains two particularly appropriate passages:

> This article should be interpreted as requiring Member States to take all the appropriate actions which it may reasonably be expected to take, to ensure that no significant deterioration or disturbance occurs.
>
> In addition, it is not necessary to prove that there will be a real significant effect, but the likelihood alone ('could be') is enough to justify corrective measures. This can be considered consistent with the prevention and precautionary principles.

Clearly, there is an obligation to limit or manage access and activities, rather than expecting wildlife to adapt to the presence of people. The precise impact that public use will have on a feature is rarely understood, so the potential for establishing evidence-based limits is low. The current level of use is a good starting point. An obvious question would be: has the past, or current, level of use had a detrimental impact on any of the wildlife features? If the answer is 'no', but there is reason to believe that an increase in public use would put features or the site at risk, then limits can be set at the current level. If there are no reasons to believe that an increase in public use will have any detrimental impact, limits need not be applied. If the answer is 'yes', the activities must be controlled or reduced. The factors with limits (access and recreational activities are factors) that are used as performance indicators for the wildlife features must be set below a level which threatens a feature.

Even when public access and recreation is not considered a threat, it must be recorded. If in the future any damage to the features is detected, or there is concern that there is potential for damage, and this is linked to an increase in public use, access limits can be established at that time.

When people seek permission to engage in new recreational activities or to significantly increase a current activity, a full impact assessment must be completed before consent is given. Logically, the applicant, and not the site manager, should finance this assessment.

5.1.7 Carrying Capacity of the Site

This is the level of access that can be accommodated without detracting from the quality of the experience that visitors enjoy on the site. There will be two main areas of impact:

Direct Impact on Infrastructure or Landscape Qualities
Visitors can have a direct impact on the infrastructure, landscape and wilderness qualities of a site, for example, paths may become over-wide and unsightly. It is relatively easy to deal with the direct physical impact; the consequences of over use are tangible and measurable. Carrying capacity can be defined by the condition required of, for example, trails and viewing areas. Trails should not become too wide or develop into multiple tracks, and viewing areas should not become hopelessly eroded areas of mud or dust. In short, carrying capacity in this context is an

Fig. 19.3 Visitors can have a direct impact

expression of the how many people a site can accommodate without showing unacceptable signs of wear.

This issue is complicated because footpaths and trails of all sorts can be constructed to withstand pressure. For example, in the UK, upland National Park footpaths have been constructed using heavy stone slabs. Some people consider that these diminish the 'natural' qualities of sites, while others appreciate the opportunity to gain relatively easy access to these upland sites. This section on access planning begins with policy because it is so important. Organisations have choices, and these choices are expressed through their policies: they can provide access opportunities regardless of impact on the intrinsic values of a site, or they can restrict access and maintain some of the natural wilderness character. There must be room for both, and a strategic approach that identifies different levels of use for different sites would be ideal. Where site mangers are not blessed with guidance from a wider strategy, they should at least look to other providers in the local area, and perhaps seek to offer alternative experiences.

On very large sites it will be possible to delineate different zones for different levels of use within the site. For example, footpaths close to parking areas could be provided with robust, even surfaces while, at the opposite extreme, the footpaths in remote areas could be left unmade. The decisions to open up or restrict access to areas by managing the condition of the footpaths will, of course, be influenced by

Fig. 19.4 A mountain path, Yr Wyddfa

the carrying capacity of the wildlife features and by most of the other criteria included in this evaluation.

Impact on the Quality of the Experience Available for Visitors
People can visit sites in such large numbers that they become a distraction to others. This is particularly important in areas of high landscape or wilderness value. There is also a problem with some recreational activities which, although perfectly legal, may be considered intrusive and antisocial by other visitors. Assessing the quality of experience is never easy; it will always be an entirely subjective analysis. Different people will have very different views: for some a visit to a beach is about being in a crowd, and they obviously enjoy that experience, but others deliberately seek out wild and lonely landscapes, where they have few encounters with others.

There is a view that activities of all kinds, in so far as they are compatible with protecting the important wildlife features, should be promoted on nature reserves. Often, in our rush to apply scientific reasoning, i.e. to establish a scientific basis for approving or encouraging an activity, the very special character of these sites is forgotten. Protected areas provide some of the few remaining places where people can find opportunities to enjoy nature. These connections with the wild provide the intangible, almost spiritual, experiences that can enhance our sense of well-being. Regardless of our inability to obtain empirical evidence to support this, there must be an obligation to protect and, if at all possible, improve opportunities for these experiences.

Perhaps we should question the need for noisy or intrusive activities in these very special areas. Would it not be more constructive to promote a sense of respect for the unique atmosphere of these places? There are now so few remaining opportunities, particularly in the developed world, to experience a true closeness with nature, that it may be better to decide that the more disruptive activities should be accommodated elsewhere. There are many places that provide ideal conditions for off-road driving or jet skiing, but a landscape with wilderness qualities, that gives a sense of peace and solitude, is something rare and precious that is too easily destroyed. This is not a call for exclusivity or some form of elitism. Nature reserves should fulfil the role of providing opportunities for *anyone* who wishes to enjoy the unique experiences that they can provide.

Clearly, there is an obligation at least to attempt to provide opportunities to suit everyone somewhere. Ideally, a strategic approach should be adopted that is not limited to any individual nature reserves, but would take into account all the access opportunities in a given area. A strategy should recognise the need to provide the widest possible range of recreational activities, including opportunities for people to enjoy the tranquillity of a nature reserve that is largely undisturbed.

In the absence of a wider strategy, the management planning process, and particularly this section on carrying capacity, can, and should, consider the site within a wider geographical context. A good example of this is the Meirionnydd coast in North Wales, where the beaches at Barmouth, a busy seaside resort, are crowded with holiday makers while, in contrast, except for the immediate vicinity of the entrance area, the beach at Morfa Harlech National Nature Reserve, just a few miles to the north, is almost empty. The management plan for this site recognises that there are plenty of opportunities in the vicinity for holiday makers to enjoy a traditional, bucket and spade, ice-cream, donkey riding, amusement arcade experience. In contrast, the nature reserve provides for those who seek spectacular coastal scenery, wildlife and solitude.

All existing, and potential, recreational activities should be considered, and a decision should be made at this point whether to encourage, permit, control or prohibit them. As with all other sections in this evaluation, whenever decisions are made reasons must be provided. This is particularly important when making decisions that cannot be based on objective analysis. There is no scientific or magic formula that can be used to calculate a carrying capacity that is consistent with maintaining a high quality experience, because, in this context, 'quality' has an infinite range of values.

5.1.8 Availability of Resources

The level of resources available, or anticipated, will almost always be a consideration that has significant implications for the level of access that can be provided on a site. One of the functions of any management plan is to provide the justification for resources. If the resource level has not been previously specified, it is best not to allow planning to be constrained by a perceived lack of resources at this stage. The level of access should be determined by the other evaluation criteria, and this should provide a site-specific interpretation of organisational access policy. However, some organisations may specify resource levels prior to the preparation of a plan. In these cases, the availability of resources becomes a very important factor, which will have implications for the site-specific implementation the organisation's general access policy. For example, at an organisational level there may be a strong presumption in favour of providing access, but the lack of resources to provide a safe infrastructure on a site could severely restrict access.

5.1.9 Summary of the Evaluation

This is the final stage in the evaluation. A succinct summary, based on the preceding evaluation, is prepared. It describes the extent to which organisational policies for access can be met when taking account of the prevailing circumstances on an individual site. An indication is given of the number of people and what activities the site, or zones within a site, will accommodate.

5.2 Access Options

Access options are a simple means of indicating the level of access that is considered appropriate for the site, or for zones within the site, following the evaluation. (If a site is divided into different zones a map should be included.)

Options are, in fact, site-specific access *policies*. They are best defined at an organisational level and should, ideally, be applied in a standard way to all the sites managed by an organisation. This would contribute to the development of a strategic access approach, described in an earlier section of this chapter. The following are examples of access options:

(a) Access is encouraged with no limits applied to any legal recreational activity.
(b) Access is encouraged and recreational activities are controlled within specified limits.
(c) Access levels and recreational activities are encouraged but controlled within specified limits.
(d) Access is permitted, but only unobtrusive or passive activities are allowed.
(e) Access is limited to legal rights of way, courtesy paths and other facilities.
(f) Access is limited to legal rights of way.
(g) No access.

5.3 Access Objective

It is a very small step to move from an option (site-specific policy) to an objective. A policy is a broad or general statement of intent, while an objective is, or should be, *SMART*. The concept of SMART objectives was covered in detail in the chapter on wildlife objectives. As a reminder:

Specific

Measurable
Achievable
Relevant
Time-based

An access objective will be specific to the provision of access on an individual site. It will be measurable because there will be associated performance indicators. Certainly, in the long term, it will be achievable. Objectives will always be relevant to an organisation's policies. 'Time-based' requires a significantly different definition to that used in the version of SMART for wildlife objectives. Access objectives should be time-based and written for specified periods. The period can range from as little as a year to, in exceptional circumstances, as long as 10 years. There is little purpose in trying to predict what will be relevant beyond that time scale.

For sites that are robust and resilient, and where organisational access policies can be applied without significant modification, an access objective could be:

To encourage the sustainable and inclusive public use of site X in so far as such use is consistent with maintaining the nature conservation features at Favourable Conservation Status and provided that visitors are not exposed to any hazards.

For a fragile and dangerous site, where there is very little public interest, the access objective could be:

To enable limited public access to the site. This will be mainly a facility intended for local people. For their own safety, all visitors will be restricted to the system of public rights of way and courtesy footpaths.

These examples are specific, achievable and relevant. Performance indicators will be used to quantify and measure achievement of the objective. The time-based component will be considered in the management rationale and specified in the individual management projects. This means that each project can have a different time scale and that work can be scheduled in a logical order. For example: year 1, complete the construction of a boardwalk; year 2, construct a public hide at the end of the boardwalk and car parking facilities at the side of the public road; year 3, open the boardwalk, organise publicity and begin patrolling. Thereafter, implement an ongoing inspection and maintenance project.

5.3.1 Vision for Access

This is the point in the access section of the plan where the levels of access have been identified and a simple objective has been prepared, but this can be developed much further. Providing opportunities for people to gain access to sites is not simply about enabling them to enter and wander around the site. There is an obvious need to provide visitors with a very positive experience, and it is possible to describe the experience that they should gain when visiting a site. Strictly, this section is not essential; a plan can function without it. However, if we are able to describe what we are trying to provide and share this with others, there is a greater chance that we will find support and will, consequently, be successful in achieving our aims.

The vision for access is based on the preceding evaluation and the general site description. A vision must be easily understood by the intended audience. Management plans are about communicating our intentions, sometimes to a very wide audience, many of whom will not be professional countryside managers. The vision should, therefore, be written in plain language; it is a portrait in words. There is no point in writing a vision that simply describes the facilities or infrastructure. These will be dealt with later, and facilities will probably vary over time in order to meet the access vision.

A vision is best regarded as an aspiration. Perhaps it should be achievable in the long term, if resources are available, but the vision should not be constrained by resource considerations since these can change over time.

Visions Are Best Written in the Present Tense

A vision for access is a description of an outcome. It can specify a range of different conditions and facilities. At any given time, some of these conditions and facilities may be in an acceptable condition and others will not. For this reason, a vision is best written in the present and not future tense. Otherwise, it could contain a mixture of tenses, *future* for conditions that are currently unfavourable and *present* for conditions that are favourable, and these will change and potentially alternate.

The following is a very simple example for a small woodland site. It is a site where the provision of access is regarded as important by the manager but where, for a variety of reasons, there are relatively few visitors. Access to the site is poor, being three miles from the nearest village on small country lanes. It is not in a tourist area and, with the exception of local people, there is very limited interest in the site. There are few features of interest to the general visitor; most people visit to enjoy a quiet walk and to obtain views of the surrounding area. The reserve manger maintains close contact with local residents.

Coedydd Maentwrog (Fig. 19.5)

Visitors, and particularly local people, feel encouraged to visit Coedydd Maentwrog NNR, and enjoy the opportunity to walk through all parts of the site. However, the nature of the terrain is severe, steep and rocky. This means that access for people with mobility problems is extremely limited and is restricted to viewing the site from the Ffestiniog Railway.

When arriving by car, visitors discover limited, but adequate, parking opportunities. They can also enter the site on foot from the Ffestiniog Railway stations. Some regular visitors already know that this is a special place, and all others discover that the area is a National Nature Reserve as they enter the site. They are aware of all the site hazards, particularly the difficult terrain and the danger presented by the Ffestiniog Railway. They have access to a network of well-maintained paths which are easy to follow and are marked on the ground with distinctive wooden pegs. The footpaths are shown on the site map, which is included on the NNR signs provided at all entrances. Visitors can use this information to plan their walk through the site. They have the choice of several circular routes through the woodlands, or they may decide to walk from one railway station to the next.

Fig. 19.5 Coedydd Maentwrog in winter

Visitors discover spectacular woodland and ravine scenery, including high waterfalls, moss carpets and veteran trees. In spring and early summer, people are delighted by the volume and variety of bird song, and evening walkers may have close encounters with patrolling bats. They can also enjoy the spectacular woodland scenery and views across the Vale of Ffestinog from many safe points within the site.
The following example is for a large coastal dune and estuary nature reserve.

Morfa Harlech
Visitors, local people and educational groups are encouraged to visit Morfa Harlech National Nature Reserve, an extensive coastal dune and estuarine system offering safe access opportunities across most of the site. People visiting the reserve can choose from four main access points, all of which have car parking facilities suitable for disabled people. The parking areas are linked to the wider countryside by a network of minor roads and footpaths. However, to the north, on the estuarine flats of Afon Dwyryd, the nature of the terrain and the limited number of formal access routes means that people with mobility problems can only appreciate the site from

the safety of the car-parking area at Ynys, and from a small number of paths at neighbouring Portmeirion.

People visiting Morfa Harlech are made aware that the area is a National Nature Reserve as they enter the site; there is clear and accessible information on access possibilities and descriptions of the wildlife they might see on the site and its environs. They realise that this is a very special place offering the opportunity to enjoy a range of stunning costal environments, including estuarine salt marshes, sand flats, dunes and wooded coastal cliffs, all set within the majestic Snowdonia National Park. Some people visit to see the shifting flocks of wintering birds, the evocative display of summer breeding lapwing or the riot of colour from the summer dune flowers. Others simply come to enjoy the sea air and the long, golden arc of Harlech beach or the solitude of a mid-winter visit.

People who wish to visit the site without using personal transport can easily make their way to the main access points by either the regular bus services on the A496 and A487 or from train stations on the Porthmadog to Harlech Cambrian Coast line.

All visitors understand that a number of natural hazards exist on the site and are aware of areas where access is not recommended for safety reasons.

Access Vision – Alternative Version

It is possible to improve access statements. The vision can, and perhaps should, be used to describe some of the deeper, less tangible, experiences that these very special places can provide. An access vision written in this way will help everyone involved in managing a site to understand what they are trying to provide for visitors. The value of the visit is expressed in terms of the quality of experience and is not simply about the availability of infrastructure and facilities.

An example of an access vision is included in Case Study 2. It demonstrates this approach within the context of the complete access section of a management plan.

The following example is a vision, or description of an experience, which may catch the readers' imagination and help them to recognise why these places are so important to all of us.

Skomer Island (Figs. 19.6, 19.7 and 19.8)

Skomer Island is a destination that captivates visitors, almost regardless of age or interest; it attracts those who already have a deep understanding of natural history as well as those who simply want to experience the uniqueness of an island and its spectacularly accessible wildlife. At Martins Haven, the departure point for the boat, there is a visitor centre providing all the necessary information about the island and its facilities to help people plan their excursion.

Almost as soon as the mainland is left behind visitors begin to sense that they are heading for somewhere special. In spring and early summer the sea is strewn with seabirds. Puffins, guillemots and razorbills scatter as the boat approaches, spreading trails of footprints across the water or diving deep, leaving nothing but a plume of bubbles. As the island comes closer, an exuberant cacophony of birdcalls echoes from the cliffs. The boat pulls gently alongside the rocks in North Haven, giving access to a sturdy flight of concrete steps that takes visitors to the cliff top. On arrival, all visitors are welcomed by island staff, who can advise on the best places to see wildlife and also explain the care that must be taken to avoid causing damage to some of the island's more vulnerable areas. Visitors are also made aware of issues relating to their personal safety while on the island. Guidebooks and other information are available.

From the landing point, the well-marked paths offer a choice of routes, varying in length, that are designed to give the best and safest access to the island's most spectacular scenery and wildlife. Visitor numbers are regulated so that, with the miles of footpaths and variety of things to see, people become dispersed around the island and it never loses its quality of unspoilt tranquillity. Though the wildness of an island is what draws many people, few want to experience the full force a cloudburst, so shelters are available if the weather decides to do its worst.

In early spring, when winter storms have left the vegetation crushed and faded to a rustling blanket of pale ochre, the deeper layers of the island's history show through most clearly. Because it has been largely undisturbed, Skomer is almost unique in the completeness of its prehistoric landscape, and visitors can see the lives of these farming communities laid out in detail beneath their feet. An archaeological trail gives access to some of the most interesting features and, while the prehistoric hut circles will be obvious to many, a closer inspection reveals the patchwork of enclosures and field systems unfolding around them.

Spring comes slowly to such a windswept island but, when it finally arrives, the bluebells flood across the island inundating the drab remnants of winter with lakes of milky indigo. Footpaths run through drifts of bluebells that appear to stretch out and touch the sea giving the impression of endless blue. To be able to walk immersed in the scent of so many flowers is a highlight of Skomer's year.

One of the most dramatic sights is The Wick, an inlet on the south coast of the island where the ribbon of enclosed water is polished deep, glassy green. The black basalt cliff that forms one face of the inlet is carved with ledges that are ideal for the thousands of nesting seabirds. Guillemots, razorbills, fulmars and kittiwakes all crowd onto the cliffs, so that the air is hazed with the shimmer of birds. For anyone who has never experienced the sound of a seabird colony at such close quarters it is truly a revelation, something that no picture or guidebook could ever convey. The strident growling of guillemots and razorbills blends with plaintive mewing of kittiwakes while, above it all, the chuckling of fulmars soars cheerfully free. And yet, despite this breathtaking display, visitors may find themselves distracted by the puffins parading on the grassy banks at their feet.

For many day visitors the worst part of their visit is leaving: watching from the departing boat as puffins skim above their heads, carrying ashore the iridescent rainbows of fish for their young. For those who would like to know how it feels to stay on the island, there are a few rooms available for overnight guests. It is an unforgettable experience because darkness on Skomer brings one of the most stunning encounters with birds to be found in Britain. When the light has faded completely the first shearwaters arrive, tumbling and crashing out of the blackened sky. As tens of thousands of these nocturnal seabirds return to their underground burrows they call to each other with loud, tuneless cries, filling the air with noise.

Fig. 19.6 South Haven, Skomer

Fig. 19.7 Skomer, south towards Skokholm

Fig. 19.8 Puffins

Even when the seabirds have left and the early autumn storms have singed the vegetation with salt, a new season is beginning which has a special appeal for visitors. Seals start to breed at the end of summer and, though the white-coated pups may be well hidden, the increasing numbers of adults are easy to see from the boat and from the cliff-top paths, while their mournful howls echo against sea and stone. After the crowds of seabirds and swathes of spring flowers, it is a chance to see the island in a quieter mood before the boat stops running for the winter.

5.3.2 Performance Indicators & Monitoring

Background

All the USA frameworks recognise the need for monitoring strategies, but there may be a problem with language. Monitoring, according to common use in the UK, seems to mean almost any kind of measurement, including survey, census and even research. In fact, it has such a broad range of meanings that it can be very misleading. The USA frameworks do not even define 'monitoring'. Their approach is to develop strategies that seem to be an almost independent component of the framework. Their emphasis appears to be on monitoring the impact of activities, and this is mainly concerned with the direct and obvious impacts on the site fabric. Erosion scars or multiple trails are often given as examples, and these can be reasonably easy to measure. Monitoring the environmental impacts of visitors on a site is mentioned (Eagles et al. 2002). The approach is mainly one of surveillance, where changes in the size or distribution of populations of key, important or sensitive species are used to assess human impact. This approach is probably adopted because it is so difficult to identify or define the conditions required in a natural ecosystem.

One of the most innovative advances, which is at the core of the PAVIM framework (Farrell and Marion 2002), is the use of expert views as a replacement for monitoring. This will be controversial and possibly difficult to promote, but the very common alternative is no monitoring. This is because monitoring protocols are often derived from research protocols. They are very sophisticated, and, although they might provide accurate and reliable results, they are far too expensive. The world of protected area management is littered with the good intentions of planned, but never implemented, monitoring projects.

Performance Indicators

Monitoring is central to any management planning process. If there is no monitoring, it is not possible to know that an objective is being met, and there is no means of knowing that management is appropriate. Monitoring requires a focus. The planning process identifies the objective with performance indicators, which are, in fact, the formulated standard referred to in the definition of monitoring.[1] Without a standard there can be no monitoring.

Performance indicators for access need to be selected with care. They must be measurable and quantified (i.e. so that they can be monitored), and the data should be easy to collect. The number of indicators should be kept to a minimum, but there should be sufficient to provide the evidence necessary to ensure that the quality of the access provisions can be measured.

There is no need to include performance indicators for the condition of the infrastructure at this stage. The access infrastructure (for example, the roads, footpaths, trails, boardwalks, hides, etc.) will be described later in the action plan. The condition of all infrastructure provisions must be monitored to ensure that it meets prescribed standards and, more specifically, legal safety standards. However, this is best regarded as *compliance* monitoring (i.e. compliance with the plan). So, whenever a plan identifies the need to construct or maintain a structure in order to meet an access objective (for example, a boardwalk over a raised bog), there will be an associated project for inspecting or monitoring the condition of the structure.

Monitoring visitor attendance and activities on a site (how many, how often, when and where) will provide a useful range of performance indicators. Examples can include:

- The total annual number of visitors, or a representative sample, for the whole, or part, of the site. (This can be used to measure trends.)
- The spatial distribution of use within a site.
- The seasonal distribution of visits.
- The number of different tour operators, or the total annual number of organised tours, on a site.
- The number of educational groups.

[1] 'Monitoring is surveillance undertaken to ensure that formulated standards are being maintained' (JNCC 1998).

As with all performance indicators, the relationship between the number of visitors to a site and the quality of the access provisions can be very tenuous. There are many examples of nature reserves where high levels of public use are not related to the status, the features of interest, or the provisions on a site. Visitors may be passing through a site for some unrelated purpose. For example, disproportionately high numbers of people are recorded on a footpath that leads through a nature reserve in South Wales; this is because the footpath provides a short cut between a residential area and a main bus stop. Visitors sometimes visit a site for reasons that have nothing to do with the wildlife or the fact that it is a nature reserve. For example, there are many coastal sand dune nature reserves with beaches that attract people not because of the wildlife but because they want to sunbathe or swim in the sea.

A second area with potential for misinterpretation is that people will often visit sites regardless of the quality or quantity of access provisions. The features of interest, wildlife or otherwise, can be so special that visitor numbers will exceed expectation and capacity even in the absence of any provisions intended to attract them.

Taken alone, the total number of people visiting or using a site has little meaning unless their reasons for visiting are understood and there is some relationship with the quality of the services provided. In other words, quantity as a performance indicator should be accompanied by indicators of quality. Quality is not easily measured. The following are a few examples of approaches that have been used with some success:

Quality

- The number of repeat visits by individuals, or by a particular tour operator.
- Level of satisfaction measured informally by, for example, visitor books.
- Level of satisfaction measured formally by, for example, structured questionnaires or visitor surveys.
- The number of complaints or compliments.

Limits

All performance indicators must be quantified. Specified limits define the degree to which the value of a performance indicator is allowed to fluctuate without creating any cause for concern. In ideal circumstances, two values are required: an upper limit and a lower limit. Limits can be used for access performance indicators in much the same way that they are used for wildlife objectives. Limits can provide a warning that there are too many or too few visitors to a site. Too many can damage the features of interest or the fabric of a site; too few can, in some circumstances, lead to a loss of direct income, or indirect contributions to a local community.

It is important to remember that the identification of specified limits will always require a degree of judgement. The best that can be done, in many cases, is to set limits using expert judgement, backed up by some form of peer review, corporate ownership and consultation with stakeholders. The planning process is cyclical and iterative; limits will be tested and, if they fail, changed. An adaptable approach to planning allows us to learn from our mistakes.

Reminder: Access objectives are about what we *want* to provide on a site, and this is not necessarily what we *currently have*.

5.4 Status & Rationale

5.4.1 Status

The status of access provisions is the difference between the current state and the required state, as defined by the objective and the performance indicators. Terms such as favourable and unfavourable can be used to describe the status of access provisions. However, the definition of Favourable Conservation Status, as applied to wildlife features, is not relevant and cannot be applied to the status of access provisions.

If the status is unfavourable, reasons should be provided. Consideration is given to the quality of the provisions currently on offer. Are they adequate? The shortfall, if any, is noted. Whenever possible, reasons should be given for the failure, but remedial management actions are not considered at this stage. They will be identified in the next section.

One of the failings may be that insufficient numbers of people are visiting the site. It may be that, as a consequence of the lack of public transport, access is only available to car owners. If the site objective is to ensure that, for whatever reason, people are not, or do not feel, excluded, does the actual visitor profile match the aspirations for the site?

In some cases, the problem may be that more people want access to a site, or parts of a site, than can be safely accommodated without putting themselves or the integrity of the site at risk. These problems will be identified in this section and carried through to the rationale, where management projects will be identified.

5.4.2 Rationale

Having decided what is to be achieved and the extent to which the access objective is being met, this is the stage at which all the management actions that are required to meet the access objective are identified and outlined. Many activities or projects will be derived from the preceding assessment of status. If, for example, the conclusion is that access is already at appropriate levels, all the work that is currently being undertaken to maintain the provision should be continued. However, if there is a shortfall (for example, very few people visiting a site that should attract, and could easily accommodate, larger numbers) the reasons or *factors* that contribute to the shortfall must be identified. Management projects can then be introduced to manage, reduce or remove the influence of the factor.

There is a difference in the rationale between planning for the first time, when there is *no* record of management, and on subsequent occasions, when there is a record of management. An assessment of status is required for both, but the conclusions reached when planning for the first time will be limited by the lack of any previous assessment and records of management. The use of status as a guide to identifying appropriate management will be extremely limited. In these circumstances, an analysis of the factors is the best method for identifying management.

The rationale should identify projects that include all construction and maintenance work, liaison, people management, provision of information and, in some circumstances, interpretation and education. The projects will not be described in any detail at this stage; that comes later. The range of projects will vary enormously from site to site.

The following is a list of factors, along with related questions, that may help to structure the rationale. This is *not* a definitive list, and some of the factors will not be relevant to some sites. For many large and complex sites there will be many more factors. In these cases, begin by including the following examples and then list all the additional factors that may influence access provisions. The factors should be treated as a series of questions; the answers will be the work or projects required to provide access for visitors. For example, one factor that will nearly always influence the number of visitors is the accessibility of a site. If visitor numbers are low ask the obvious question: is accessibility a contributing factor? If the answer is yes, identify a project or projects to improve accessibility.

5.4.2.1 Legislation

Legislation, or the need to comply with legislation, must be given adequate attention. All the legislation that is relevant to access management should have been included in the description. This is the stage at which the implications are considered. Some of the most important legislation will be Health & Safety and Public Liability. The plan must recognise the duty of care to all visitors, and this will, of course, include staff. Other areas of legislation will place an obligation on site management to maintain routes or areas of the site open to unimpeded public access. There is also legislation to protect the rights of disabled visitors. All of these will have implications for management and will give rise to specific projects, for example, safety signs and information, safety barriers, exclusion zones, safety inspections and audits, and rights of way management. There will also be legal implications for many of the projects identified when considering the following factors. For example, when considering 'access within the site' the conclusion could be that it is inadequate because there are no safe routes across a raised bog. The solution is to provide a boardwalk. However, as a consequence of the Disability Discrimination Act, the boardwalk must be constructed to a standard suitable for disabled people.

5.4.2.2 Access to the Site

If this is inadequate, can any projects, for example, liaison with the local Highways Department, be identified to improve the situation? One of the issues that can exclude potential visitors is the lack of public transport. In these circumstances, a project could be identified to assess the potential for public transport, and this could lead to liaison with local providers. Are there adequate and safe parking facilities? Projects can be identified to ensure that car parks are constructed to an appropriate standard and that they are maintained.

5.4.2.3 Access Within the Site

Are there adequate roads, bridleways, paths and boardwalks? Do all existing routes meet legal and any other specified standards? Projects must be identified for the construction and maintenance of all existing and planned routes.

5.4.2.4 Visitor Safety

The most obvious and important question has to be: is there an up to date site hazard assessment? The health and safety of visitors must always be a prime concern. Consideration must be given to what steps should be taken to minimise the risks to visitors. Depending on the nature of the hazard, there are a number of management actions that can be employed. The most obvious is to prevent access to dangerous areas or objects. In all cases, there is a requirement to ensure that visitors are aware of the hazards and of any steps that they must take to avoid risk. A management project must be included to ensure that a formal risk assessment is implemented and recorded. In addition, projects that cover the implementation and maintenance of all safety provisions must be identified.

5.4.2.5 Seasonal Constraints

Are there any seasonal constraints? For example, some sites are only accessible at certain times of the year. This could be the consequence of seasonal weather or because the site contains species which are easily disturbed during the breeding season. Seasonality can also be a problem when too many people want to visit the site at the same time, usually because the wildlife interest is only present or accessible at certain times. Are there any projects that can be identified to help resolve these problems?

5.4.2.6 Public Awareness

Are potential visitors aware that the site exists and what it has to offer? This question is only relevant when visitor numbers fall below potential. If numbers are low,v identify projects (for example, publicity, liaison or open days) which raise the public awareness of the site.

5.4.2.7 Excessive Demand

Some sites, or parts of sites, can be extremely popular, and demand will far exceed carrying capacity. In these situations, consideration should be given to finding opportunities to discourage excessive use or to improve the distribution of numbers on the site. For example: close, reduce or move car parks, close footpaths or establish new routes, reconsider signage, restrict publicity.v

5.4.2.8 Visitor Infrastructure

In addition to footpaths and other routes, what sort of visitor infrastructure is required? This should be based on the current visitor facilities, but will also consider the facilities required to meet the access potential of the site as defined by the objective. Outlines of the work required should be provided, and the projects for both construction and maintenance should be identified.

5.4.2.9 Information

What information, signs, leaflets, etc. are required to help visitors find their way both to and around the site, locate the various areas or features of interest, and avoid any dangers? Some of these projects will arise elsewhere. The need for safety signs should be considered as part of the section on 'visitor safety'.

5.4.2.10 Interpretation

What level of interpretation would be appropriate or necessary? This is not the place for an interpretation plan: that is a separate exercise. The intention here is to give an indication of the scale of interpretive facilities that would be relevant to the site.

5.4.2.11 Education

Is there potential and demand for the provision of educational opportunities or facilities on the site? If this is at a very low level, for example, providing opportu-

Fig. 19.9 Students visiting a nature reserve

nities for a local primary school, this can be dealt with in this section. However, once it is recognised that that there is a significant demand, and therefore the justification for providing facilities, the preparation of an education plan should be considered.

In some cases, it may not be possible to conclude with any certainty what the appropriate level of facilities should be. The only approach is to rely on experience, seek the best available advice and run a trial. If the outcome is acceptable, continue; if not, modify the approach or try something different. To some extent, most management is trial and error; we learn through experience what the most effective and efficient management may be at any given time.

Fig. 19.10 The visitor centre at Newport Wetlands

6 Action Plan

Management Projects

The rationale is complete when all the management projects required to meet the objective have been identified and described. Management projects can include, for example, provision of site infrastructure, paths, car parks, bridges, etc. The final stage in planning for access is to provide, in sufficient detail, all the information that the individuals who will be required to carry out the work will need to ensure the successful completion of the project.

Occasionally, access will be a major operation on a site and can include the construction and maintenance of substantial buildings, for example, information centres, shops and restaurants. When confronted with a very large-scale project, consider the need for operational objectives. These are described in Chapter 17.

References

Alexander, M. Hellawell, T. C., Tillotson, I. and Wheeler, D. (2000). *Kidepo Valley National Park Management Plan*. Kampala, Uganda, Uganda Wildlife Authority.

CCW (2004). *Review of the National Nature Reserves*. Internal publication, Bangor, Wales.

Clark, R. N. and Stankey, G. (1979). *The Recreation Opportunity Spectrum: A Framework for Planning, Management and Research*. Gen. Tech. Rep. GTR-PNW-98. Portland, Oregon, USA.

Eagles, P. F. J., McCool, S. F. and Haynes, C. D. A. (2002). *Sustainable Tourism in Protected Areas: Guidelines for Planning and Management.* IUCN, Gland, Switzerland and Cambridge, UK.

Farrell, T. A. and Marion, J. L. (2002). *The Protected Area Visitor Impact Management (PAVIM) Framework: A Simplified Process for Making Management Decisions*. Journal of Sustainable Tourism, 10(1), Portland Press Ltd. London, UK.

JNCC (1998). *A Statement on Common Standards Monitoring.* Joint Nature Conservation Committee, Peterborough, UK.

Stankey, G. H. and McCool, S. F. (1990). *Managing for Appropriate Wilderness Conditions: The Carrying Capacity Issue.* Hendee, J. C., Stankey, G. H. and Lucas, R. C. (ed.), Wilderness Management, North American Press, Golden, CO, pp. 215–240.

Case Study 1
Extracts from a Conservation Management Plan

Doug Oliver[1] **and Mike Alexander**[2]

Abstract This case study is included to demonstrate the management planning process for a habitat and a species feature. The study is based on a management plan for a small, uncomplicated site. It contains all the sections, extracted from the full site plan, which deal with the biological site features. The contents have been edited to remove any irrelevant information. The name of the site, and other information which provided a location for the site, have been changed.

Trawsgoed Nature Reserve – Management Plan

Fig. CS1.1 Oak in flower

[1] Countryside Council for Wales.

[2] CMS Consortium Wales.

M. Alexander, *Management Planning for Nature Conservation: A Theoretical Basis & Practical Guide*, DOI 10.1007/978-94-007-5116-3,

1 Plan Summary

Trawsgoed is a small nature reserve, 67 ha in size, which has been designated a Site of Special Scientific Interest and is managed by the Meirionnydd Wildlife Trust. The site comprises three rocky ridges which are covered by broadleaved, oak-dominated woodland. There are occasional, small, bracken-dominated glades and a few acidic flushes. The canopy is overwhelmingly dominated by sessile oak *Quercus petraea*, except for the base of the various inclines where ash *Fraxinus excelsior* is also an important species. In addition to the woodland habitat, the only other important biological feature is a population of lesser horseshoe bats. The reserve provides limited access opportunities for local people.

2 Legislation and Policy

2.1 Legislation

2.1.1 Legal Status of the Site and Features

SSSI

Trawsgoed was notified as an SSSI in 1960 and re-notified in 1986.
The SSSI features are:

1. Oak woodland

   W17 – *Quercus petraea/Betula pubescens/Dicranum majus* woodland
   W17b – Typical sub-community
   W17c – *Anthoxanthum odoratum/Agrostis capillaris* sub-community

2. Lesser horseshoe bats *Rhinolophus hipposideros*

Oak Woodland

The oak woodland is a Natura 2000 Annex 1 habitat: 91A0 Old sessile oak woods with *Ilex* and *Blechnum* in the British Isles

Description and ecological characteristics
This habitat type comprises a range of woodland types dominated by mixtures of oak *Quercus robur* and/or *Q. petraea* and birch *Betula pendula* and/or *B. pubescens*. It is characteristic of base-poor soils in areas of at least moderately high rainfall in northern and western parts of the UK.

Lesser Horseshoe Bats

This species is included in Appendix II of the Bonn Convention (and its Agreement on the Conservation of Bats in Europe) and Appendix II of the Bern Convention (and Recommendation 36 on the Conservation of Underground Habitats). It is also listed on Annexes II and IV of the EC Habitats and Species Directive. It is protected under Schedule 2 of the Conservation (Natural Habitats, etc.) Regulations 1994 (Regulation 38) and Schedule 5 of the Wildlife and Countryside Act 1981. The 1996 IUCN Red List of Threatened Animals classifies this species as *Vulnerable* (VU A2c) (UK Biodiversity Action Plan).

2.1.2 Other Relevant Legislation

Occupier's Liability Act

This Act imposes on the Trust an obligation to ensure that every reasonable care is taken to remove any risk both to legitimate visitors and to trespassers. A safety audit must be carried out on this site and updated as required. The Trust must ensure that the site hazard assessment is available to people using the site, for example, for research purposes, and that all relevant personnel are issued with a licence.

Health and Safety at Work Act

The Trust has a duty under this Act to ensure the health and safety of its staff when engaged on official business. All operations carried out on site must be undertaken by trained personnel using methods and equipment which comply with Approved Codes of Practice arising from the Health and Safety at Work Act 1974, and also in compliance with both national and local safety procedures (i.e. the Trust safety manual). The need for an up-to-date hazard plan and for regular safety inspections also applies here.

Public Rights of Access

The access road to Trawsgoed is a public highway, but no public footpaths cross this reserve and there are no permanent rights of access for the general public.

2.2 *Policy*

The Meirionnydd Wildlife Trust policy for the management of nature reserves:

The management of our nature reserves which have statutory protection is guided by the relevant legislation. This policy, whenever applicable, will also be applied to all other Trust reserves. The primary land use of all the Trust reserves will be nature conservation. The Trust recognises a responsibility to maintain or restore the nature conservation features to Favourable Conservation Status. All Trust reserves will be managed to the highest possible standard.

The Trust will encourage the sustainable public use of its reserves in so far as such use is consistent with our responsibility to protect wildlife and does not put any visitors or staff at risk. Visitors will be permitted to engage in activities providing that they are legal, that they are not a danger to other visitors or staff and that the activities do not detract from the ability of other visitors to enjoy the site.

3 Description

3.1 General Information

3.1.1 Location & Site Boundaries

National Grid Reference: The centre of reserve is SH444216.

Trawsgoed Nature Reserve lies approximately 2 miles south of the village of Llanilar and 2 miles north of the village of Cantre'r Gwaelod. The reserve is situated within the district of Meirionnydd in the county of Gwynedd and is within the Snowdonia National Park.

3.1.2 Zones

For the purposes of management the site has been divided into 8 zones. These are based on a combination of the tenancy boundaries, habitat types and management requirements. Zones are indicated on the attached map and include the following:

1. Cadair Rhys Goch
2. Bryn Twr
3. Bryn Twr exclosure
4. Parc Mawr
5. Hafod-y-llyn-isaf
6. Coed Hafod-y-llyn
7. Llyn Hafod-y-llyn
8. Trawsgoed

3.1.3 Tenure

The reserve is owned by the Trawsgoed Estate and has been leased by the Meirionnydd Wildlife Trust since 1992.

3.1.4 Past Status of the Site

Trawsgoed is a Nature Conservation Review Grade 1 site.

3.1.5 Management/Organisational Infrastructure

The site is managed as a nature reserve by the Meirionnydd Wildlife Trust. The Trust employs 6 full-time staff and is also supported by a considerable contribution from volunteers. The responsibility for managing Trawsgoed is currently delegated to the senior reserves manager, who manages a team of two estate workers and is also responsible for seven other woodland reserves. Contract labour is employed to carry out capital works, such as fence-line replacement and rhododendron control.

3.1.6 Site Infrastructure

Boundaries

Since 2002, the entire external boundary and all internal compartment boundaries have been fenced. Gates and stiles are provided wherever necessary. All fences are in good condition, having mostly been replaced or erected relatively recently. There is a planned programme of replacement.

Buildings

There is a large barn and a smaller building adjacent to each other near Trawsgoed House. The 'studio', known as Ysgubor, is used as a working base for the estate worker, and the upper floor of the barn is used for storing tools and materials for use on site.

Roads, Tracks and Paths

There is an unclassified road which runs the length of the site along the eastern boundary. A small car park at the southern extremity of this road has space for four cars. Five hundred meters west of this, on the opposite side of the river, a parallel track provides vehicular access to the buildings at the centre of the reserve. There are no public footpaths on the nature reserve, but there are several paths and tracks within the site.

3.1.7 Map Coverage

The following maps, or copies of maps, are held in the Reserve files.

Contour maps

1:50,000 Ordnance Survey Landranger No. 124 Dolgellau & surrounding area 1991
1:25,000 Ordnance Survey Outdoor Leisure No. 18 Snowdonia – Harlech & Bala areas 1990

Edition of 1916, 2nd Series Ordnance Survey map, Meirionethshire sheet III.13
Soil Survey of England and Wales, Sheet 2 – Wales 1983.
1 in. Geological survey, Sheet 65 NE. 1851 and 1854. Uses 1840 sheet as base.
Land Utilisation Survey 1931–2. Copy in National Library of Wales, Aberystwyth.
Land Utilisation Survey 1971. G. Sinclair. Copy held at the trust office.

3.1.8 Photographic Coverage

Fixed Point Photographs: A series of rotational pans was taken in March/April 1989 on 35 mm black and white print film from 24 different fixed points. These cover much of the southern half of the reserve, including Bryn Twr, Parc Mawr and the area to the west of Parc Mawr. Details of the location of each fixed-point and the contact prints are held in the photo-monitoring file. The negatives are held in the Trust office.

General photographs: Various 35 mm colour slides of the site are available. These need to be collated and evaluated.

Aerial photographs: A selection of aerial photographs is held in the reserve files. Dates: May 1946, September 1971, October/November 1986, June 1993.

3.2 Environmental Information

3.2.1 Physical

3.2.1.1 Climate

The following information on climate is based on meteorological information recorded at Cantre'r Gwaelod, which is approximately 1 km north of the site.

Temperature: Temperatures may vary between an absolute minimum of around –8°C and a maximum of around 30 °C. February is the coldest month, with a mean daily temperature of about 4–5 °C. July is the warmest month, with a mean daily temperature of 14–15 °C. The frost-free period generally extends from mid-April until November.

Rainfall and snow: The mean annual rainfall is 2,121 mm. The wettest period is from August through until March (though February is generally drier), and the driest months are May, June and July. Snow falls on an average of 15–17 days a year, with January, February and March being the most likely months.

Sunshine: The average daily duration of bright sunshine is about 4 hours, and the annual average percentage of the possible sunshine is about 30%.

Wind: The strongest winds tend to occur between September and March, and these are likely to be contained within an arc between south and northwest, southwest being the strongest. The maximum gust speed recorded is between 70 knots and 80 knots. The site is afforded some protection by the mountains to the north and west, but is fairly open to winds blowing up the estuary from the south and southwest.

3.2.1.2 Geology

Trawsgoed lies at the western edge of the geological area known as the 'Harlech Dome'. This is an area of uplifted and eroded geology exposing the largest area of Cambrian sedimentary rocks in Wales. It has been surveyed by numerous geologists and is well documented, including a report by the British Geological Survey (Allen and Jackson 1985).

It is thought that the initiation of this present day structure occurred during the Caledonian Orogeny. Pressure from the southeast against the ancient rocks to the north (now Anglesey) caused the uplift of the younger rocks in the Meirionnydd region in the form of a dome extending roughly from the River Mawddach to the River Dwyryd. Subsequent erosion has exposed the older Cambrian rocks in the central area of the dome, with the younger Ordovician igneous rocks remaining around the periphery. These later rock types form the bulk of the mountains surrounding the Rhinogau range, e.g. Snowdon and Cader Idris.

Trawsgoed is situated in the western section of the Cambrian rocks of the dome. The rocks are all slates, grits and shales of the Harlech Grits group, consisting mainly of shales and mudstones of the Maentwrog formation. These are relatively fine-grained rocks, quartz-rich and in flaggy layers. There are also basic Doleritic intrusions, providing a range of pH conditions. The Cambrian rock formations are base-poor, hence the generally acidic nature of this site. There are, however, some localised areas of base-flushing, probably caused by ground-water passing through certain of the beds that contain more basic minerals, manganese beds in particular.

3.2.1.3 Soil and Substrates

There has been no detailed survey and analysis of soils on this site. The following account is based on the rather general details provided by the Soil Survey of England and Wales (1983). The main types are podzolic soils and raw peat soils.

The greater part of the site is overlain by podzolic soils: this is the soil type covering the three main rock ridges. These soils result from the pedogenic accumulation of iron and aluminium or organic matter or a combination of these. Those found here are typical brown podzolic soils of the Manod Association. They have developed over Ordovician shales and have a dark brown or ochreous subsoil with no overlying "bleached" layer. They are well-drained, fine, loamy or silty soils, often shallow with exposed bare rock. While there have been no detailed studies, where larger trees have been uprooted by wind, soil depths have been observed to be little more than 30 cm, with deeper roots being entwined with rock fragments. Soil depths on the tops of the ridges are probably less, perhaps 15–20 cm.

3.2.1.4 Hydrology

The reserve essentially comprises three roughly parallel rock ridges separated by two shallow troughs. The ridges are largely wooded and mainly free draining. However, there are some waterlogged areas in the occasional hollows and some areas of flush, notably at the northern tip of the site. The more southerly and deeper trough between Parc Mawr and Bryn Twr is, in fact, the lower end of the Gwaelod valley. A fairly major stream flows through this valley, fed by a relatively large catchment area to the north. This stream bisects the southern half of the site, eventually flowing into the river. Prior to the construction of the Cantre'r Gwaelod embankment (1815) and the reclamation of the estuary, the southern periphery of this site would have verged on the intertidal zones, with the three wooded ridges representing small peninsulas. This is no longer the case: most of the land immediately seaward of the site is now improved grassland.

3.2.2 Biological

3.2.2.1 Flora

3.2.2.1.1 Flora – Habitats/Communities

The three main rocky ridges are covered by broadleaved, oak-dominated woodland, with occasional bracken-dominated glades and a few acidic flushes. The 'dry' broadleaved, oak-dominated woodland covering the rocky ridges amounts to a total area of c.66 ha. Most of this is the NVC type W17 *Quercus petraea -Betula pubescens – Dicranum majus* woodland, including both the typical sub-community b, and the *Anthoxanthum odoratum – Agrostis capillaris* sub-community type c.

The canopy is overwhelmingly dominated by sessile oak *Quercus petraea* (some hybrids, *petraea* × *robur,* also occur), except at the base of the various inclines where ash *Fraxinus excelsior* is an important component. The build-up of minerals through run-off is the probable reason for the difference in these localities; the NVC type is not changed. Downy birch *Betula pubescens* and silver birch *Betula pendula* are also important components of the canopy. Rowan *Sorbus aucuparia* is a frequent addition, while crab apple *Malus sylvestris* and holly *Ilex aquifolium* occur occasionally, with some particularly large, old specimens of the latter.

There is no detailed information on the age of the woodland stand. Subjective observations indicate that most of the growth dates from around the beginning of the twentieth century. There are evidently some areas where the stands are younger, but also several trees throughout the site which are clearly much older (c.200 years).

The drought in 1976 led to the die-back in the crowns of several trees and the eventual death of some. The worst affected areas were along the tops of the various ridges, especially on south-facing locations, where the soils are thinnest and most free draining. The effects of this drought are still apparent through die-back of branches, but most trees have since recovered.

The under-storey is much influenced by past grazing and, as with most western oak woods, it is generally sparse. A scattering of saplings/young trees of fairly even age (c. 15–25 years) is present across most of the woodland, suggesting a period of relaxed grazing pressure at some time in the past. The main component of the understorey is rowan *Sorbus aucuparia*, while sessile oak *Quercus petraea*, birch *Betula* spp., hazel *Corylus avellana* and holly *Ilex aquifolium* occur frequently. There is occasional beech *Fagus sylvatica* and crab apple *Malus sylvestris*.

The field layer is short and generally dominated by grasses. The main species are common bent *Agrostis capillaris*, velvet bent *A. canina*, sweet vernal grass *Anthoxanthum odoratum*, sheep's fescue *Festuca ovina*, purple moor-grass *Molinia caerulea* and wavy hair-grass *Deschampsia flexuosa*. In compartment 3, where grazing has been effectively removed since 1990, much of the field layer is currently dominated by purple moor-grass *Molinia caerulea*. The other main components of the field layer are wood-sorrel *Oxalis acetosella*, tormentil *Potentilla erecta*, heath bedstraw *Galium saxatile*, bramble *Rubus fruticosus agg.* and bluebells *Hycacinthoides non-scripta*. There are also a number of small glades dominated by bracken *Pteridium aquilinum*. The ericaceous species normally associated with this habitat type, i.e. heather *Calluna vulgaris* and bilberry *Vaccinium myrtillus*, have been largely suppressed by the prolonged period of grazing pressure. Tree seedlings are evident throughout the woodland at ground level.

The ground layer comprises mainly a carpet of bryophytes, though there are some extensive areas of litter. The more abundant species are *Thuidium tamariscinum, Polytrichum formosum, Rhytidiadelphus loreus, Dicranum majus* and *Isothecium myosuroides* var. *myosuroides*.

Fig. CS1.2 Trawsgoed, a distant view

3.2.2.1.2 Flora – Species

Vascular Plants

The flowering plant records for the site were compiled in 1977 (Howells & Ward) and 1980 (Blackstock). The site manager has added to the reserve lists. The list is contained in Appendix 1. It is certainly not complete, and more species are likely to occur. At present, the list amounts to a total of 82 flowering plants, which include 14 tree species, 9 shrubs, 9 grasses (*Gramineae*), 11 sedges (*Cyperaceae*) and 4 rushes (*Juncaceae*). There are no rare or notable flowering plants recorded on the site.

Ferns

The fern records for this site were compiled in 1977 (Howells & Ward) and 1980 (Blackstock). A total of 12 fern species has been recorded on the site (a complete list is contained in Appendix 1). Of particular note is the presence of lanceolate spleenwort *Asplenium obovatum*. A total of ten plants was recorded at two localities in 1980, both being rock crevices on steep, west-facing rocks under moderate tree cover. This is a nationally scarce species, with its distribution in Britain being confined to coastal areas of Wales and southwest England.

Bryophytes

D.A. Ratcliffe compiled a list of bryophytes for this site during a visit in 1977. Some additional species have since been recorded by the site manager (Oliver 1992, 1994). A complete list of all mosses and liverworts recorded to date is contained in Appendix 1. It is certainly not complete, and many more species are likely to occur.

Much of the site is grazed by sheep and/or cattle and, in some areas, by horses/ponies. As a consequence, bryophytes are a major component of the woodland ground flora throughout, forming a luxuriant bryophyte carpet. To date, 53 species have been recorded, of which 35 are mosses and 18 are liverworts. Two of the mosses, *Rhabdoweisia crenulata* and *Leucobryum juniperoideum,* are nationally scarce oceanic species. A liverwort, *Cryptothallus mirabilis,* is also nationally scarce (after Hodgetts 1992). Trawsgoed is one of the few sites in Britain where *Leucobryum glaucum* has been seen in fruit. The woodland communities are characterised by a consistently rich bryophyte ground layer, in which six of the eight constants of the field and ground layer are mosses: *Dicranum majus, Plagiothecium undulatum, Polytrichum formosum, Rhytidiadelphus loreus, Hylocomium splendens* and *Pleurozium schreberi*. All of these, together with *Thuidium tamariscinum* and *Isothecium myosuroides* var. *myosuroides,* make up the more abundant mosses in the Trawsgoed woodlands.

Certain moderately Atlantic liverworts occur in abundance, notably *Bazzania trilobata, Plagiochila spinulosa, Saccogyna viticulosa, Scapania gracilis* and *Lepidozia cupressina*. This indicates that conditions in the woodland are currently favourable for oceanic species which require a shaded and moist environment. During his 1977 visit, D.A. Ratcliffe noted the apparent absence of species such as *Plagiochila punctata, Adelanthus decipiens* and *Lepidozia pinnata.* This suggests a period of forest clearance in the past, during which many of the sensitive shade- and moisture-loving species were eliminated. When tree cover was subsequently re-established, only those species with a strong capacity for spreading were able to re-colonise. Species with a poor capacity for spreading have not managed to return (Ratcliffe 1977).

Lichens

A survey of the lichen flora was made in November 1974 and June 1978 (Pentecost). A full account, with a species list, is included in Appendix 1. A total of 99 species was recorded: of these, 87 were of the corticolous type, i.e. growing epiphytically on bark. The saxicolous lichens, i.e. those growing on rocks, were generally typical of species found in acidic conditions. Two of the species recorded rank as nationally scarce (after Hodgetts 1992): *Lecidea phaeops* and *Parmelia horrescens*. The list also includes several species which are strongly Atlantic in distribution, *Lobaria laetevirens, L. scrobiculata, Parmelia taylorensis, Sticta limbata* and *S. sylvatica.*

Only 16 of a possible 70 lichens 'indicative of ecological continuity' occur (after Hodgetts 1992), possibly suggesting a period of forest clearance in the past.

3.2.2.3 Fauna

Mammals

There has been no systematic survey or recording of mammals on this site. Grey squirrel are abundant. Fox and badger frequent the reserve and there are several badger sets. Otter spraint is found regularly on a rock in the outflow from Llynddu, indicating a good population in the valley. There is no information on small mammals.

There is a large and important breeding roost of lesser horseshoe bat in the upper room of the barn within the farmyard complex of Trawsgoed House. There are at least seven other known breeding roosts within the valley for this species, and it is possible that this may qualify the valley for SAC designation for its population of lesser horseshoe bat.

Birds

A total of 38 species has been recorded on the site. There is no information on breeding populations. A full species list is held in Appendix 2. The list includes a number of Candidate Red Data Book species: buzzard, snipe, swallow, dipper, redstart, whitethroat and raven.

Invertebrates

This site supports an impressive invertebrate fauna. This has been comprehensively surveyed, with over 400 species recorded to date. Most of these records were compiled by H. N. Michaelis and M. J. Morgan between 1978 and 1980.

Butterfly species of local interest include green hairstreak *Callophrys rubi*, purple hairstreak *Quercusia quercus* and small pearl-bordered fritillary *Boloria selene*.

Several noteworthy moths occur, including three nationally scarce species (Notable A): ringed carpet *Cleora cinctaria*, large red-belted clearwing *Synanthedon culiciformis* and Ashworth's rustic *Xestia ashworthii*. A further seven nationally scarce (un-graded) moths have been recorded. These are marsh oblique-barred *Hypenodes turfosalis*, silver hook *Eustrotia uncula,* dotted carpet *Alcis jubata, Rheumaptera hastata, Hypenodes humidialis, Tetheella fluctuosa and Alcis jubata.*

3.3 Cultural

3.3.1 Archaeology

Trawsgoed is of considerable historic interest. A 'mansion' stood here in the late medieval period when it was owned by the poet Rhys Goch Eryri. He supported Owain Glyndwr in his rebellion of 1400–1404 and, according to local tradition, it was while Glyndwr was a guest of Rhys Goch that they were betrayed by a local supporter of the king and escaped in servants' clothes, with the English soldiers in close pursuit, to a cave on Moel Hebog. Some of Rhys Goch's work (31 poems survive) was said to have been composed in a tower on the hill above the house which became known as *Gadair Rhys Goch*. This had become ruinous by the last century, and the stones were re-used in the 1970s to make a seat, which is now overgrown. On one hill within the woods, Cadair Rhys Goch, there are several very small enclosures marked out by lines of fallen stones representing the remains of low walls. They are certainly ancient and possibly prehistoric.

Parc Mawr was described in the last century as, 'coedwigfa unig ac anhygyrch… le am… feudwyaeth a llonyddwch…hen goedydd yn gwyro o henaint teg' (a lonely and inaccessible plantation – a place for hermitage and tranquillity with old trees bowed over with fair age).

3.3.2 Past Land Use

The old wall system suggests a long association with pastoralism. There are several level areas of ground, apparently built up on the lower sides, which have very distinctive ridges and furrows. These possibly indicate early or peasant agricultural use in these locations.

Historical maps provide some information on past land use. The 1st edition OS map (1839) shows the boundary of the Parc Mawr woodland, but there do not appear to be any tree symbols (although this may, in fact, be due to deterioration of the map in question). The 1899–1914 1:25,000 OS map also shows the boundary of Parc Mawr woodland, and in this instance there are tree symbols, indicating that the area was wooded at the time. More recent 1:25,000 OS maps, 1912–14 and 1949, both show the site as largely wooded and/or with areas of rough pasture.

3.3.3 Present Land Use

The site is managed as a nature reserve.

3.3.4 Past Management for Nature Conservation

Tree planting – Cadair Rhys Goch

Several oak seedlings and saplings (30–50) have been planted in the woodland on Cadair Rhys Goch by the owner. These have been grown from native seed and vary in age, the oldest being 20 years. In Feb/Mar 1995, the Trust planted a further 60 trees in compartments 1 and 2.

Woodland exclosure

The south-eastern quarter of Bryn Twr (Compartment 3) was made stockproof in 1989/1990, and stock have since been excluded. The rest of the woods were made stockproof, and most sheep excluded, in early 2002.

Rhododendron control

Rhododendron *Rhododendron ponticum* became a serious problem in various parts of the broad-leaved woodland and is most abundant along the course of the stream. It has been systematically removed from the reserve woodlands, wetlands, and ditch and stream banks, and from adjacent land, since 1994. Control has been successful, although there are still some seedlings and small bushes which have been missed.

3.4 People – Stakeholders, Local Communities and Access

3.4.1 Stakeholders

The owner of the nature reserve

The owner of the site is also the owner of the Trawsgoed Estate, which surrounds the site on three sides. The landowner has similar objectives for the site as the Trust. He recognises the importance of the site for its wildlife and habitats.

Adjacent landowners

The owners of the house known as Hafod y Llyn also own 20 acres of woodland contiguous with the reserve woodlands and included within the stockproof fence of the reserves woodland enclosure. Effectively, all their land is managed in the same way as the reserve land. The owners are very sympathetic to the Trust's aims for the nature reserve and are happy for the current arrangement for excluding sheep from their area of woodland to continue.

Local residents

There are very few local residents as the site is isolated and distant from the local villages, both of which are 2 miles away. The few local residents adjacent to the site in scattered farms and houses are kept informed by the site manager about the reserve. The residents of Trawsgoed House take an active interest in the reserve and its wildlife, and are effectively the custodians of the breeding roost of lesser horse-shoe bats in the stable block.

The local community

There are centres of local population within 2 miles of the reserve at Llanilar and Cantre'r Gwaelod. The existence of the main reserve has not been actively publicised, and many people are unaware of its existence. A few local people walk there and almost all are very sympathetic to the aims of management.

Contractors

Much of the work on the reserve is carried out by local contract labour. This has included rhododendron control, boundary work, tree felling and safety work, scrub control, lichen survey, engineering work, etc. Contractors will continue to be used, and local contractors are always given preference unless the skills required for the task are not available locally.

3.4.2 Access

This is not an important access site. The roads to the site are very narrow and unsuitable for anything other than small cars. It is not possible to travel to the site by public transport. The site is designated as open access, i.e. there are no access restrictions, and most of the site is accessible on foot to people with average mobility and fitness. The site is visited by less than 50 people each year: these are mainly local people who visit throughout the year but most often during the summer months. People tend to visit for very short periods, usually of about half an hour; most of these are dog walkers. People visit mainly for the experience of walking through particularly attractive oak woodland or simply to exercise their dogs.

There are reserve signs at the entrance points, with a reserve map showing the pathways and tracks, basic information about the site, and health and safety information.

There are occasional visits by scientists and specialists: these are recorded in CMS.

Given the difficulty of providing access to the site, and the presence of so many similar sites with well-developed access provisions in the immediate area, the Trust has no plans to develop this as an access site.

Fig. CS1.3 The winter canopy

3.5 Bibliography

Campbell, S. and Bowen, D. Q. (1989). Geological Conservation Review – Quaternary of Wales. W. A. Wimbledon Editor in Chief. NCC, Peterborough.
Day, P. (1985). Broad-leaved Woodlands in the North Wales Region. 3 Vols. NCC, Peterborough.
Garrett, S. and Richardson, C. (1989). Gwynedd Inventory of Ancient Woodlands (provisional). NCC, Peterborough.
Hodgetts, N. G. (1992). Guidelines for Selection of Biological SSSIs: Non-Vascular Plants. JNCC, Peterborough.
Linnard, W. (1982). *Welsh Woods and Forests: History and Utilization*. National Museum of Wales, Cardiff.
Looney, J. H. H. and James, P. W. (1987). *Effects of acidification on lichens*. Interim report to NCC, Peterborough.
Meteorological Office Commercial Services (1991). *A report on the climate of three areas in West Gwynedd. Met. Office*, Specialist Consultancy Group, Bracknell.
Ratcliffe, D. A. (1968). *An ecological account of Atlantic bryophytes in the British Isles*. New Phytologist 67 365–439.
Ratcliffe, D. A. (ed.) (1977). *A Nature Conservation Review*. Cambridge University Press, Cambridge.
Roberts, R. A. (1959). Ecology of human occupation and land use in Snowdonia. Journal of Ecology, *47* 317–323.
Rodwell, J. S. et al. (1991). *British Plant Communities*. Vol 1, Woodlands and Scrub. Nature Conservancy Council, Cambridge University Press. Cambridge.
Rose, F. (1971). Unpublished report on the Cryptogamic vegetation of woodland areas in North Wales. (Reserve file)
Soil Survey of England and Wales (1983). Sheet 2 Wales. Lawes Agricultural Trust, Rothamsted Experimental Station, Harpenden.
Stewart, A., Pearman, D. A. and Preston, C. D. (1994). *Scarce Plants in Britain*. JNCC, Peterborough.

4 Nature Conservation Features

4.1 Evaluation

A full evaluation was not required for this site. This is because the Trust has extremely limited resources for management and has adopted a policy of managing sites to safeguard the main habitats and any legally protected species. There is an assumption, or hope, that all other species will be protected as a consequence of habitat protection.

The site is entirely covered by woodland, and the only protected species is a population of lesser horseshoe bats.

Legally recognised conservation features

1. Broadleaved Woodland:

   W17 – *Quercus petraea / Betula pubescens / Dicranum majus* woodland,
   W17b – Typical sub-community.
   W17c – *Anthoxanthum odoratum / Agrostis capillaris* sub-community

2. Lesser horseshoe bats *Rhinolophus hipposideros*

Table CS 1.1 Feature selection

Feature	RDB	International	European	National
Broadleaved Woodland			SAC Annex 1 habitat	SSSI feature
Lesser Horseshoe Bats	The 1996 IUCN Red List of Threatened Animals classifies this species as *Vulnerable* (VU A2c)	Appendix II of the Bonn Convention. Appendix II of the Bern Convention.	Annexes II and IV of the EC Habitats and Species Directive.	SSSI feature Schedule 5 of the Wildlife and Countryside Act 1981.

4.2 Factors (master list)

The key factors which are most likely to influence the features on this site:

- The landowner's objectives for the site
- Sheep grazing on land surrounding the site, because there is potential for trespass
- Rhododendron, mainly offsite, as a source of seed from extensive upwind flowering populations
- Past management, both woodland and agriculture
- Beech
- A tree preservation order (1950)
- Aerial pollution and climate change
- Human disturbance at the summer breeding roost
- The condition of the building at the summer roost
- The presence, offsite, of old abandoned mine workings providing hibernation sites for the bats

4.3 Objectives

4.3.1 Objective for Lesser Horseshoe Bat

Description

Horseshoe bats are easily identified by a horseshoe-shaped flap of skin called a nose-leaf which surrounds the nostrils. This amplifies the ultrasonic calls that the bat emits when searching for food. Lesser horseshoe bats can be distinguished from the greater horseshoe bat by size. These are quite small bats: the body length is around 3.7 cm, with a wingspan of 24 cm. Individuals weigh between 4 and 9 g. Unlike the greater, where the fur has a reddish hue, the lesser has long fluffy fur, pale greyish-brown on the back with a paler grey front.

The lesser horseshoe bat was originally a cave-roosting bat, although most summer maternity colonies now use buildings, particularly large, old houses and farm buildings. Most still hibernate in underground sites such as caves and old mines. Females forage within 2–3 km of the maternity roost, feeding on insects taken in flight in mixed woodland, hedgerows and tree lines.

In Britain, the lesser horseshoe bat is now found only in south-west England and Wales. It was formerly present in south-east England and the Midlands. Current estimates suggest a UK population of 14,000, divided equally between Wales and England. About 230 summer (or all-year) roosts and about 480 hibernation roosts are known. Of the latter, only 20% are used by more than ten bats. The lesser horseshoe bat is widespread throughout central and southern Europe, but has undergone severe decline in the northern part of its range. (UK Biodiversity Action Plan)

At the reserve, the lesser horseshoe bats depend on the mosaic of semi-natural habitats and woodland, at and surrounding Trawsgoed, for feeding. The valley has a number of known summer breeding and hibernation roosts, and their specific dependence on the reserve is not known. The two largest known lesser horseshoe bat breeding roosts are very close to the reserve, and both have in excess of 300 bats. One is in the upper part of a stable block, immediately adjacent to the reserve, and the other is in a private residence within half a mile of the site. In addition, there are three known underground hibernation roosts in mine workings very close to the reserve.

4.3.1.1 Vision for Lesser Horseshoe Bats:

The woodlands at Trawsgoed contribute towards maintaining a sustainable, robust and viable population of lesser horseshoe bats in the valley. There are a variety of both hibernation roosts and summer breeding roosts which are available to the bats. All the currently used roosts remain suitable for, and available to, the bats. The maternity roost at Trawsgoed contains at least 250 bats which raise sufficient young to ensure the survival of this population. The size and range of the population is not restricted or threatened, directly or indirectly, by any human activity. The population within the valley interacts with populations in adjacent areas, thus maintaining a diverse gene pool for the species.

4.3.1.2 Performance Indicators

Factors:

There are no obvious factors that can be monitored. There are several factors with implications for management, and these are dealt with in the rationale.

Attributes:

Number of adults in the maternity roost
The only attribute that can be monitored, without any risk of disturbance, is the number of adults in the maternity roost. There may be more than 300 bats present in the stable block at any time during the breeding season. However, given that the

roost was only discovered in 1997 and there is no information available prior to this date and the capacity of the roost is unknown, it is not strictly possible to set useful limits for monitoring. Ideally, the population should be kept under surveillance for a longer period to establish both capacity and variations in use. However, managers need a warning system and, consequently, provisional limits will be set and a monitoring project established.

Attribute: Number of adults in maternity roost
Upper Limit: Not required
Lower Limit: 250 bats present at any time during the breeding season

4.3.1.3 Status and Rationale

The bat population has gradually increased since its discovery in 1997. However, the limit that has been established is provisional. It is therefore not possible at this time to comment on the status of the bat population other than to assume that it is at least recovering. The implications for management are to continue as before: the roost is clearly successful. There is consequently no reason for any changes to management.

The following factors are important:

Casual disturbance by people at the breeding roost

The house and the workshop above the stable in the farmyard complex at Trawsgoed which is used by lesser horseshoe bats for breeding was let on a 50 year lease in 1997. The tenants are extremely sympathetic to the bats but need to use the room in which they roost. A false ceiling was erected in March 1998 to prevent the bats from being disturbed by activities in the room below. This seems to have resolved the situation without apparent detriment to the bats. It is difficult to establish meaningful monitoring or surveillance, but, providing a good relationship is maintained with the tenant, the risk to the bats is extremely low.

Bats' access to the breeding roost

The bats gain access to the roof space in the workshop by flying through an open window fitted with a shutter. There are no other access routes, but the shutter is ideal: it provides fly-in access without apparently affecting temperatures in the roof space. The shutter must be open from at least 1 April until all the bats have left in the autumn.

Condition of the building

The building housing the bats is in good structural condition, and no maintenance / remedial work is required in the foreseeable future. Management input will be continued liaison with the owner and tenant of the property to ensure that the reserve manager maintains a good working relationship and is aware of any potential activities or changes which may have implications for the bats.

The woodland habitat

The most significant factor is the woodland, which provides some of the main feeding habitat for the bats. It is assumed that if the woodland is maintained at Favourable Conservation Status this will be adequate to meet the requirements of the bats.

Offsite factors

There are additional offsite factors, but, unfortunately, these are not in the control of the site manager. The most significant factor is potential loss of, or disturbance to, hibernation roosts. Several are known in the valley: they are all in abandoned mine workings, some of which are regularly visited by underground explorers. All have been designated as SSSIs because of their importance to lesser horseshoe bats. This should increase their protection against development. Some mine entrances may also be fitted with lockable grilles to reduce the incidence of casual disturbance.

4.3.2 Objective for Upland Acid Oakwood

Description of the feature

The woodlands are mainly even-aged stands of mature oak, with little variation in tree age apart from occasional older trees on roadsides, track-sides and in the vicinity of Trawsgoed House. There is a concentration of older trees, with some veterans, in the Cadair Rhys Goch section of the woodlands. These were possibly retained near the house as a landscape feature. The tree canopy is largely complete, but there are some exposed, west-facing areas where a severe gale in October 2002 created a landscape of fallen large trees. Beech is a relatively frequent canopy species, but, with the exception of birch, holly and rowan, other species are very rare. The habitat is almost entirely upland acid oak woodland (NVC W17), containing two separate sub-communities: the dominant is W17b, the typical sub-community. W17c, the *Anthoxanthum odoratum / Agrostis capillaris* sub-community, occurs where soils are deeper and slightly less acidic.

4.3.2.1 Vision

The entire site is covered by a high forest, broadleaf woodland. The woodland is naturally regenerating, with plenty of seedlings and saplings particularly in the canopy gaps. There is a changing or dynamic pattern of canopy gaps created naturally by wind throw or as trees die. The woodland has a canopy and shrub layer that includes locally native trees of all ages, with an abundance of standing and fallen dead wood to provide habitat for invertebrates, fungi and other woodland species. Thc field and ground layers will be a patchwork of the characteristic vegetation communities, developed in response to local soil conditions. These will include areas dominated by heather, or bilberry, or a mixture of the two, areas dominated by tussocks of wavy hair-grass or purple moor-grass, and others dominated by brown bent grass and sweet vernal grass with abundant bluebells. There will also be quite heavily grazed areas of more grassy vegetation. Steep rock faces and boulder sides will be adorned with mosses, liverworts and filmy ferns.

The lichen flora will vary naturally depending upon the chemical properties of the rock and tree trunks within the woodland. Trees with lungwort and associated species will be fairly common, especially on the well-lit woodland margins.

The woodland does not contain any rhododendron or any other invasive alien species with the exception of occasional sycamore. There will be periodic light grazing by sheep. This will help to maintain the ground and field layer vegetation but will not prevent tree regeneration.

Please note: Chapter 3 contains an 'inspirational' version of this objective.

4.3.2.2 Performance Indicators

Factors and Limits

Grazing by sheep

Until completion of the woodland boundary fencing in 2002, sheep grazed freely throughout the woodlands, which were open to adjacent fields. There was very little unprotected recent natural tree regeneration within the woodland. Sheep grazing suppressed the development of structure in the field layer and also prevented the desirable build up of leaf litter. The woodland vegetation was predominantly of well-developed bryophytes, with little else. Following grazing control, birch, holly and, to a lesser extent, oak have begun to regenerate naturally.

It is not possible or desirable to completely exclude sheep: there will always be some low-level grazing. In the short term, grazing will be excluded to ensure a period of natural regeneration. An upper limit of 1 sheep per hectare is acceptable as this will have no significant impact on the woodland.

Factor: Sheep
Upper limit: No more than 1 sheep per hectare in any woodland enclosure
Lower limit: No less than 0.25 sheep per hectare

Alien species – Rhododendron *Rhododendron ponticum*

Rhododendron is well established in the surrounding area and in the past had infested significant areas of the site. This was cleared in 1995 and the re-growth was sprayed with herbicide in 1996. However, there is always potential for further infestation. The clear aim of management is to eradicate the species from the site. The problem is that it is very difficult, and probably impossible, to prove that there is no rhododendron in an area. The seedlings are extremely small for the first few years and even when rhododendron reaches the sapling stage it is very difficult to locate, particularly when growing in dense woodland. Once the plants begin to flower they can be seen easily, because the pink flowers stand out against the green of the woodland. Monitoring is then very simple: unskilled individuals can wander through the wood searching for flowers, and when they are found all the rhododendron, flowering and non-flowering, can be controlled. If this process is maintained, eventually the species will be eradicated.

Factor: Rhododendron
Upper limit: No flowering Rhododendron
Lower limit: Not required

Beech

Beech occurs throughout the site. Many of the trees appear to be younger than the adjacent oak trees. Due to a prolonged period of heavy grazing, beech has not successfully regenerated, but, following the removal of sheep in 2002, this situation may change. Beech is shade tolerant and able to regenerate under a dense oak canopy. It can also produce, albeit very occasionally, a heavy crop of viable mast. There is, therefore, potential for beech to become the dominant canopy species, and the consequential loss of species diversity would not be acceptable. To ensure that this does not happen, the amount of beech in the canopy will be restricted to a manageable level. Although it would be desirable to eradicate beech completely, there would be considerable practical difficulties in achieving this.

Factor: Beech
Upper limit: 5% in the canopy
Lower limit: None

Attributes and Limits

Extent and distribution
The current extent of the woodland is 67 ha. This should not decline.

Attribute: Extent
Upper Limit: Not required
Lower Limit: 67 Ha

Canopy gap creation rate

Natural regeneration in oak woodlands is only possible when there is enough light for the seedlings and saplings. Woodland that contains a dynamic, changing pattern of gaps will also, over time, deliver a structurally dverse canopy which will provide opportunities for a wide range of associated species. There is, therefore, a need to ensure that there are sufficient gaps in the canopy to provide opportunities for regeneration and that the gaps fill with the required canopy species. Three different attributes will be monitored to provide the evidence that this is actually happening: First, the overall canopy cover is defined; that is, how much of the canopy should be open at any given time. The rate at which gaps should be created is specified. Finally, the presence of viable saplings in the gaps will be monitored.

The presence of saplings will indicate that conditions are right for regeneration long before the fact that a gap is filling can be measured.

Attribute: Canopy cover
Upper Limit: Tree canopy 90% of woodland area.
Lower Limit: Tree canopy 75% of woodland area.

The canopy gap creation rate within the woodland area should be between 0.25 and 0.5% of canopy cover per annum, measured over a minimum 20 year period. (A gap is any area equal or greater than 1.5 times the height of the tallest adjacent tree, or any area of between 20 and 50 m distance across, not including areas of bare rock, etc.)

Attribute: Canopy gap creation rate
Upper Limit: 0.5% per annum measured over a minimum 20 year period
Lower Limit: 0.25% per annum measured over a minimum 20 year period

Attribute: Natural regeneration within gaps
Upper Limit: Not required
Lower Limit: 2 viable saplings per 0.1 ha of gap

(Where viable saplings are taken to be healthy/ vigorous native tree saplings reaching a minimum height of 1.5 m, consisting of species that will replenish the canopy.)

Canopy species composition

For many obvious reasons, and in particular a need to optimise opportunities for the widest possible range of locally native species, it is essential that the canopy comprises locally native species. There is no tolerance, in the longer term, of beech or any other exotic species.

Attribute: Canopy species composition
Upper Limit: 100% locally native species
Lower Limit: 95% locally native species

Dead wood

Dead wood is included as an attribute because it is an extremely useful surrogate, i.e. the presence of dead wood will indicate potential for the presence of a wide range of typical woodland species, including beetles, fungi, epiphytic lichens and hole-nesting birds (Peterken 1993). Dead wood is also a good measure of the ecological structure of a woodland: the presence of too little dead wood will be indicative of a dysfunctional or inappropriately managed woodland. A reasonable and attainable level would be about 30 cubic meters per hectare. Dead wood should consist of a mixture of fallen trees (minimum 1 of girth >40 cm dbh per hectare), fallen branches, dead branches on live trees, rot columns in living trees and standing dead trees (minimum 1 of girth >40 cm dbh per hectare).

Attribute: Dead wood
Upper Limit: Not required
Lower Limit: 30 cubic metres per hectare

Fig. CS1.4 Dead wood at Trawsgoed

4.3.2.3 Woodland – Status and Rationale

Gap creation rate within the existing wooded area has been low for many years. However, the gale of October 2002 created many new gaps. The consequence is that the canopy is much more open and the cover is now between 90 and 75%. This will not be confirmed until aerial photographs are taken. The sudden increase in gap creation rate may not be representative of the general trend in the woods: it is far too early to come to any useful conclusion. For now, with the exception of beech control (see below), there will be no artificial gap creation. However, this option will be kept open for the future in case natural processes and beech removal fail to produce sufficient gaps. The new gaps, all containing fallen dead and living trees, will certainly make a significant contribution to the structure of the woodland. In many of the gaps, there is natural re-growth of damaged trees and saplings appear to be surviving. There is now limited natural regeneration throughout the site as a consequence of sheep exclusion in 2002. The overall age structure is varied but comprises a mosaic of even-aged areas which are a reflection of past felling areas. There are no areas with a natural mixed age structure, and there are very few large, old and derelict veteran trees. The volume of dead wood remains below the specified limits, but since the gale there has been a significant increase. With the exception of the trees which blocked the roads, all other wind-throw trees have been left in place.

Rhododendron is under control: it is now virtually absent throughout the reserve. Unfortunately, the threat of further infestation remains because of the proximity up-wind of extensive areas of flowering bushes. This factor remains a threat. Beech is currently estimated at 10%: this is well above the acceptable limits.

Sheep are under control, and this can be maintained.

Management can do little to mitigate the effects of atmospheric pollution, which will tend to acidify the already extremely acidic substrates and soils that have no buffering. The effect on long-term tree health is not understood. Stressed trees may become more liable to normal pathogens. Mycorhizae are suppressed by acidification, and lack of them will impair the trees' ability to extract nutrients from the soil. However, there is currently no evidence that the woodland is being adversely affected by atmospheric pollution or acidification.

Current conservation status

The conclusion is that, although there has been an obvious and very significant improvement, the overall condition remains unfavourable. The indication, based on an assessment of the factors, is that recovery will continue. The status is, therefore, *recovering*.

The implication of this status is that the general approach to management will be maintained. Efforts will be directed towards reducing the potential for further invasion by rhododendron and ensuring that the rhododendron monitoring project is maintained. The main management activities will be liaison and participation in a partnership with the National Park, the National Trust and local land owners to control rhododendron over the wider area.

A programme to accelerate the removal of beech will be developed. A Tree Preservation Order was placed on most of the site and some surrounding woodland areas in 1950. Overall, this has a positive impact, but given that controlling beech will involve felling quite large trees permission will be required. The selective removal of beech will be managed to provide canopy gaps suitable for regeneration.

There will be no further removal of dead wood, even if footpaths have to be re-routed.

The stock-proof fences will be maintained in good condition, and any trespassing sheep will be removed as soon as possible.

The Countryside Council for Wales will be contacted for advice on atmospheric pollution and climate change.

Note: The most significant factors and the implications for management have been dealt with in the preceding assessment of status. The following two additional factors are included because they help to account for the current condition of the woodland.

Owners'/occupiers' objectives

The entire site is owned by a private owner whose objectives are similar to those of the Trust. There are no conflicts of interest.

Past woodland management

The entire woodland area has been managed in the past as productive high forest, producing timber for construction, ship building, firewood, tan bark, etc. There is no apparent evidence of coppice management in the present stand of trees. The consequence of past management is that the stand of trees throughout the woodlands is even aged and relatively uniform. Over-mature trees are rare, and tree canopy cover is variable. There is very little dead wood, either on the ground or within living trees, throughout the woodlands, except in the most inaccessible locations or in areas affected by the wind-throw episode in October 2002.

Fig. CS1.5 Canopy gaps created during a gale in 2002

5 Action Plan – Extracts

Monitoring and management projects

Bats:

Monitoring projects:

Attribute: Breeding roost numbers of adults
RA03/1 Collect data, mammals, monitor – Monitor lesser horseshoe breeding roost numbers

Management projects:

Maintain condition of the breeding roost
ML00/8 Liaise, owners/occupiers – site owner
ML00/3 Liaise, owners/occupiers- site occupier

Woodland:

Monitoring projects:

Factor: Sheep
RA04/1 Collect data, mammals, count/estimate/measure/census – Woodland sheep grazing by observation (Note: this project is described below)

Factor: Rhododendron
RF14/2 Collect data, trees/shrubs, count/estimate/measure/census – Sweep site for rhododendron

Factor: Beech
RF03/4 Collect data, vegetation, monitor – Monitor beech

Attribute: Extent
RF13/5 Collect data, trees/shrubs, monitor – Monitor extent of woodland area

Attribute: Canopy cover
RF13/6 Collect data, trees/shrubs, monitor – Monitor extent of woodland canopy

Attribute: Canopy gap creation rate
RF13/8 Collect data, trees/shrubs, monitor – monitor gap creation rate (Note: this project is described below)

Attribute: Natural regeneration within gaps
RF13/7 Collect data, trees/shrubs, monitor – monitor tree regeneration

Attribute: Canopy species composition
RF13/11 Collect data, trees/shrubs, monitor – Monitor tree canopy composition

Attribute: Dead wood
RF13/10 Collect data, trees/shrubs, monitor – Estimate volume of dead wood

Management projects:

Control rhododendron:
ML30/2 Liaise, neighbours – rhododendron control on adjacent land
ML40/1 Liaise local national authorities – rhododendron control on adjacent land
ML50/1 Liaise local community/groups – National Trust rhododendron control on adjacent land
MS00/01 Manage species, tree/shrub – rhododendron control on site (Note: this project is described below)

Control beech:
MS00/2 Manage species, tree/shrub – control beech
ML40/02 Liaise local national authorities

Woodland management/gap creation
MH02/1 Manage habitat, woodland/scrub, by thinning/group felling – silvicultural management/canopy gap creation

Control grazing by sheep:
RA04/1 Collect data, mammals, count/estimate/measure/census – woodland sheep grazing by observation
ML00/3 Liaise, owners/occupiers – Liaise with graziers
ME01/1 Boundary structures – inspect/maintain boundary structures – TPO consent

Examples of projects with project descriptions:

RA04/1-Collect data, mammals, count/estimate/measure/census – Woodland sheep grazing by observation (Note: this project description is included in the action plan)

1. Feature: Woodland
2. Purpose: To maintain an ongoing awareness of levels of sheep trespass in the woodland areas and to ensure removal of sheep once numbers exceed eight in any enclosure. (Note: light sheep grazing in a wood may be beneficial in removing competition for oak seedlings, but regular, heavy sheep grazing will cause damage to natural tree regeneration.)
3. General Background: The woodlands have been heavily grazed by sheep for a very long period before the reserve was established. Between 2000 and 2002 the woodlands were fenced to exclude stock.
4. Methodology: Walk through all woodland areas regularly, at least once a month, to check for sheep trespass. Cover all areas visually and record the number of sheep seen. If sheep are inside an enclosure, attempt to locate how they got in and arrange for the repair of any damage to fences/walls etc. If there are more than 8 sheep in any enclosure, take action to evict them as soon as possible.
5. Specific Risk Assessment: See SHI/SHA for Trawsgoed.
6. Reporting Requirement: Annual summary will be maintained in CMS.

RF13/8 Collect data, trees/shrubs, monitor – monitor gap creation rate

1. Feature: Woodland habitat
2. Attribute: Canopy gap creation rate
3. General background/bibliography

The use of aerial photography to monitor gap creation rate has been developed on this site.

4. Methodology: The method is very simple. It involves the identification of new gaps in the woodland canopy from aerial photographs. This is followed by a visit to each gap where the actual extent is measured. When the total area of gaps within a defined sample area is known, this can be compared with previous results, and the rate of gap creation can be calculated.

   (a) Equipment:

   - A site map
   - The aerial photographs
   - Computer with GIS

   (b) Location of sample collection: The sample areas are marked on the project map.
   (c) Fixed point markers: None
   (d) Sampling technique: The woodland area is divided into a series of roughly equal-sized blocks (the actual shape is not important). These divisions will enable the comparison between successive photographs and, most importantly, they can be located on the ground.

Initially, all the canopy gaps are numbered and delineated on the aerial photograph by carefully drawing a fine line around the inside of the edge of the canopy. (A gap is any open area equal to, or greater than, 1.5 times the height of the tallest adjacent tree, or any area of between 20 and 50 m distance across, but not including areas of bare rock. Open areas which appear smaller than the crown spread of the largest immediately adjacent tree are ignored.) Occasionally, a best estimate will have to be made due to the presence of shadows at the glade edges. The GIS workspace is saved to provide the information for future comparison.

On subsequent occasions this procedure is repeated, and the new gaps or extensions to older gaps are delineated.

A print of the digitised aerial photograph is taken on a field visit and used to locate, on the ground, all the new gaps within the sample areas. Ensure that each gap fits the definition given above, and ignore any that do not. Estimate the area of each by pacing or measuring the edge of the gap. The edge of the gap is defined by the edge of the canopy (imagine a perpendicular projection of the canopy on to the ground, i.e. the edge of the shadow when the sun is vertically overhead). Within each sample area, work out the **total** area of the new canopy gap.

The next stage is to calculate the gap creation rate. **G** is the area of new gaps since the last photograph. **A** is the size of the sample area. **T** is the number of years between successive photographs

The Gap creation rate expressed as a percentage of the total area $= (G/A)/T \times 100$.

For example, new gaps (G) is 1 ha, the sample area (A) is 20 ha and the time between photographs (T) is 10 years. The gap creation rate is $(1/20)/10 \times 100 = 0.0$ $5/10 \times 100 = 0.5\%$ per annum.

(e) Unit of measurement: The number of canopy gaps, the area of individual gaps in hectares, the rate of gap creation for sample areas as an annual percentage.

(f) Sampling period: Any time when the canopy is in leaf (May to October).

(g) Frequency of sampling during sampling period: Once every 12 years unless there are significant events, for example, severe wind-throw.

Risk assessment:

This is a very low-risk, indoor-based project. The general site risk assessment will be applied during the field work component.

MS00/1 Manage species, tree/shrub – rhododendron control on site

1. Feature: Woodland
2. Purpose: To arrest the spread of rhododendron on the site as quickly as possible, with the target of complete eradication from the site and adjacent land to prevent the threat it poses to the woodland.
3. General Background:

   Rhododendron ponticum is a shrub that was first introduced to Britain in 1763. It is now naturalised throughout Britain in a variety of habitats. Because the inherent strategy of the plant is to form an impenetrable thicket with a dense canopy, thus displacing all other competition, it has become a major ecological and economic problem in native woodlands and in commercial forest plantations. In some areas of Britain it has become a major threat to the survival of native woodlands and woodland species.

   Cross (1975) has written a masterful account of the biology and ecology of the species.

   Rhododendron ponticum generally prefers an acid soil, tolerates a wide range of temperatures, and occurs at altitudes of up to 530 m in the British Isles. Bushes can reach up to 7 or 8 m in height in woodland and up to 5 m in open habitats. Once established, rhododendron is a remarkably hardy plant. It responds to cutting or burning by producing a proliferation of shoots from dormant basal buds, which grow rapidly. Its branches, which are often procumbent, root into the soil when they touch it and produce an independent clone. Its waxy leaf cuticle provides it with remarkable resistance to herbicides, and its foliage is avoided by livestock because it is highly poisonous. The plant is extremely shade tolerant and there is evidence that it can chemically suppress the growth of surrounding plants (allelopathy). It flowers profusely under various (and very low) light conditions, and produces viable seed every year. Calculations made by Cross showed that each raceme (group of flowers) can produce up to 5,000 seeds. It is not unusual for large, open-grown bushes to have up to 250 racemes, and so may produce over one million seeds every year. The seeds are tiny (20,000 per gram) and are adapted for wind dispersal. They may travel some distance in woodland, possibly up to

1 km, but flowering bushes in exposed, high-altitude locations may be carried much further by wind and turbulence.

Once rhododendron seedlings are established, they grow slowly for the first few years but speed up after about 7 years, lose apical dominance and begin to develop many stems and become bushy. According to the literature, flowering occurs about 12 years after germination, or about 7 years after cutting a mature bush, but direct experience has shown the latter may be as little as 3 years.

Shaw (1984) raises the question of where rhododendron colonisation will stop and concludes that, apart from altitude (530 m in Britain), there is no obvious limit, particularly in the acidic west of the country. Although its spread is now not thought to be exponential, rhododendron is still insidiously spreading gradually from local and obvious seed sources (Thompson et al. 1992), and, in view of the threats posed to habitats and species, the case for its control could not be stronger.

Issues affecting the control of rhododendron

Strengths (properties that make it a difficult species to control):

- It has a strong coppicing ability and regrowth is vigorous after cutting. The resulting re-growth, left unchecked, is even more difficult to cut within 4 years.
- The leaves are covered by a wax layer which is extremely effective at protecting the plant against aqueous herbicide as water simply runs off.
- It has the seeding strategy of a weed. It produces millions of very small, light seeds.
- It forms dense, impenetrable stands – an excellent survival strategy.
- It has few natural predators in this country and it appears to be disease free.
- It is a long-lived species (500 years plus in eastern Turkey).
- It is extremely shade tolerant at all stages of growth (compensation point is at 2% of full daylight), and, as an evergreen, is active in winter when light levels in woodland are higher.
- Its vascular tissue is arranged in such a way that one stem is connected to one part of the root, and any stem missed during application of a systemic herbicide will not be killed, even if the rest of the plant dies.

Weaknesses:

- Seedlings/young bushes (even quite large ones) can be easily uprooted prior to multi-stem growth. However, uprooted bushes can survive and eventually re-root unless most soil is removed from the roots and the plant is hung up to allow the roots to dry out.
- Re-invasion requires a nearby seed source.

4. Methodology

Rhododendron is cut and later the regrowth is treated with herbicide.

Small bushes or saplings can be simply uprooted and hung up to dry.

Cutting is best carried out during winter to avoid disturbance to wildlife and to avoid risk of missing the bushes that are hidden by bracken or other vegetation.

Bushes can be cut by hand or with chainsaws. The very large bushes are best tackled by clearing a path into the centre and then cutting all stems near the base. Stumps should be cut as flush with the ground as possible, and never higher than 15 cm above the ground. All stumps should be cut flat, leaving no dangerous sharp points which could be an impalement hazard.

All cut spoil must be burned. It is important to achieve a hot fire base and keep it well fed, but without letting the fire base expand. Constant attention needs to be given to turning the ends of long stems in to the centre. Spoil may be burnt when wet or dry. Fires must not be allowed closer than 15 m from the fire edge to any tree trunk, moss-covered boulders or rock outcrops. Burning must not be carried out following dry periods or in high winds. All burning must cease at least 2 hours before the site is vacated at the end of the working day.

Herbicide application: foliar spray

Spraying is normally carried out from June to the end of February. The secret of successful spray treatment is to be sure that sufficient fresh, young foliage exists to absorb a lethal dose of the herbicide. At least one full growing season should elapse between initial cutting and subsequent first spraying: treatment in the late summer of the second growing season produces the best results. Foliar spray can only be applied effectively to bushes that are less than 1.2 m tall and less than 1.0 m diameter. Bushes larger than this must first be cut and allowed to regrow. The use of a strong, red, degradable vegetable dye in the spray solution prevents double spraying and missed bushes.

All the foliage must be sprayed to the point of run-off, using a spray solution consisting of 2% by volume of Roundup Biactive (36% active ingredient, approved trade product "Roundup") in water (8 l/ha) plus 4% "Mixture B" added to the solution. This is applied by knapsack sprayer at medium volume or knapsack sprayer with a "VLV" nozzle at low volume, in accordance with the instructions on the product labels and the recommendations in Forestry Commission Field Book No. 8 (*The Use of Herbicides in the Forest*).

5. Specific Risk Assessments

Environment Agency requirements – herbicide application adjacent to water courses must be licensed.

Control of Substances Hazardous to Health Regulations (COSHH) for herbicide and herbicide additives (adjuvants).

Proper use of herbicide as per the product label.

Operatives working with herbicides must have, or be closely supervised by someone who has, current National Proficiency Training Council (NPTC) PA1 and PA6 hand held applicator certificates.

Operatives using chainsaws, or other equipment or plant, or carrying out specialised techniques (e.g. roped access), must have appropriate training and certification.

Full consideration must be given to the presence of overhead electricity cables in relation to clearance operations. Bushes under these may need special clearance for any kind of treatment. If in doubt, consult the owner of the cables (local electricity supplier).

6. Reporting Requirement

   Location of data: CMS and Site Files
   Data security: CMS routine back-up procedure, with several copies and one in a fire proof safe.

Case Study 2
Access & Recreation Section of the Management Plan for Cors Caron National Nature Reserve

Paul Culyer[1] and Rosanne Alexander

Introduction to the case study

This case study contains the full access and recreation plan for Cors Caron National Nature Reserve. It is an example of a site where there is some public interest, but where there is potential to do a great deal more to provide for visitors. The following two paragraphs of description would normally form part of the main plan, but they are included here to set the case study in context.

Cors Caron is a large raised bog, 800 ha in area, lying in the agricultural heartland of Ceredigion in Wales: it is one of the few remaining largely intact examples of this once common habitat. It stands out dramatically from the surrounding countryside as a vast sweep of golden-red lying in a dish of green hills.

Cors Caron was one of the last strongholds of the red kite when populations were at their lowest, and, although they are now much more common, these birds remain traditionally associated with this area and are one of the main attractions for visitors. Birds, such as curlew, redshank, snipe and water rail breed among the bog vegetation, which is highlighted in summer by flowering plants such as bog asphodel, bog rosemary and bog bean. The spectacular landscape supports a great variety of wildlife, ranging from water voles and polecats to moths and dragonflies.

Extract from the management plan for Cors Caron National Nature Reserve Access section of the management plan for Cors Caron NNR

Contents:

[1] Countryside Council for Wales.

M. Alexander, *Management Planning for Nature Conservation: A Theoretical Basis & Practical Guide*, DOI 10.1007/978-94-007-5116-3,

- 3.4.2.3 Visit Characteristics
- 3.4.2.4 Access to the Site
- 3.4.2.5 Access Within the Site
- 3.4.2.6 Visitor Facilities and Infrastructure
- 3.4.2.7 The Reasons Why People Visit the Site
  - 3.4.2.7.1 Wildlife Attractions
  - 3.4.2.7.2 Other Features That Attract People
  - 3.4.2.7.3 Recreational Activities
- 3.4.2.8 Current and Past Concessions
- 3.4.2.9 Stakeholder Interests
- 3.4.2.10 The Site in a Wider Context
- 3.4.2.11 Educational Use
- 3.4.3 Interpretation and Information
- 3.4.4 Educational Use

5 Access Section

- 5.1 Evaluation
  - 5.1.1 Accessibility
  - 5.1.2 Access Within the Site
  - 5.1.3 Site Safety
  - 5.1.4 Implications of Stakeholder Interests
  - 5.1.5 Carrying Capacity of the Features
  - 5.1.6 Carrying Capacity of the Site
  - 5.1.7 Summary of the Evaluation
- 5.2 Access Option
- 5.3 Access Objective
  - 5.3.1 Vision
  - 5.3.2 Performance Indicators & Monitoring
- 5.4 Status & Rationale
  - 5.4.1 Status
  - 5.4.2 Rationale
    - 5.4.2.1 Legislation
    - 5.4.2.2 Access to the Site
    - 5.4.2.3 Access Within the Site
    - 5.4.2.4 Seasonal Constraints
    - 5.4.2.5 Public Awareness
    - 5.4.2.6 Excessive Demand
    - 5.4.2.7 Visitor Infrastructure
    - 5.4.2.8 Information
    - 5.4.2.9 Interpretation
    - 5.4.2.10 Education

6 Action Plan

Note: The first three sections, ‘Plan Summary’, ‘Legislation and Policy’ and ‘Description’, usually contain all the information that is relevant to the subsequent sections in the management plan. This case study contains only the information that is relevant to access planning.

Fig. CS2.1 The edge of the bog

Access & recreation section

1 Access Plan Summary

It is the Countryside Council for Wales' policy to encourage public access in so far as it does not threaten the nature conservation features of the site. Facilities at Cors Caron will be upgraded to allow more people to enjoy the site, but access will be limited in terms of area and the overall numbers of visitors. Any activities should be quiet and unobtrusive, with access onto the fragile bog being restricted to the boardwalk.

Currently, the site does not fulfil its potential to allow visitors to experience this relatively rare habitat. It is an isolated site, away from large centres of population, and the people that make an effort to visit are largely those with a specialist interest in bird watching. The bog itself is not very accessible, and most people simply view it at a distance from the main track and from viewing points on the fringes of the site. Visitors would have a much better opportunity to engage with and appreciate the site if access to the bog was improved. This could be achieved by the construction of a new boardwalk that is suitable for disabled access. A well-designed observation shelter that blends into the landscape, as part of the boardwalk loop, would make it easier for less-active visitors to enjoy the exposed and wild nature of the site. Upgrading the railway walk would make it more accessible to wheelchair users, cyclists and people with restricted mobility, and it would also make it possible for people to reach the site without travelling by car. The riverside walk should also be improved and the permit system removed.

The construction of a new car park would provide safer access and adequate parking for visitors at peak times. It should also include toilet facilities. On arrival, visitors should find clear and welcoming information that will help them to plan their visit. It may be possible to liaise with local bus companies to persuade them to stop on request at the car park.

With the prospect of increased visitor numbers, a separate interpretation plan will be necessary. It will also be important for stakeholders to be consulted and kept informed throughout any changes.

2 Legislation & Policy

2.1 Legislation

Site designation

National Nature Reserve (NNR)

Cors Caron was declared a NNR in 1955

Site of Special Scientific Interest (SSSI) (Wildlife and Countryside Act 1981)

The site was notified an SSSI in 1984. The SSSI features are:

1. Active raised bog (also SAC feature)
2. Marshy grassland on mineral soils on the river floodplain
3. Tall herb fen dominated by *Phalaris arundinacea*
4. Semi-natural broadleaved woodland
5. Bryophyte assemblage
6. Invertebrate assemblage
7. Breeding bird assemblage
8. *Sphagnum balticum*
9. *Scapania paludicola* (liverwort)
10. *Luronium natans* (floating water plantain)
11. *Singa hamata* (an orb weaving spider)
12. *Coenophilia subrosea* (rosy marsh moth)
13. *Coenonympha tullia* (large heath butterfly)
14. Subsurface stratigraphical profile comprising lacustrine clays, overlying peat and raised bog landform assemblage
15. Teifi river channels together with associated fluvial landforms

Special Area of Conservation (SAC)

Cors Caron was declared a SAC in December 2004.
The SAC features are:

- Active raised bogs
- Bog woodland

- Degraded raised bogs still capable of natural regeneration
- Depressions on peat substrates of the Rhynchosporion
- Transition mires and quaking bogs
- Otter *Lutra lutra*

Ramsar

In 1993, it was listed as a wetland of international importance by the RAMSAR Convention.

Other relevant legislation

Management of the site will be in full compliance with:

Health and Safety at Work Act 1974

All operations carried out on the site must be in compliance with this act. All specific CCW procedures must be followed:

A site hazard identification and site hazard assessment will be completed and reviewed as specified.

Specific activity risk assessments (these can be generic or site-specific) will be completed for all activities.

Occupiers' Liability Acts of 1957 and 1984

The key sections which have implications for visitors are:

The Occupiers' Liability Act 1957 sets out the duty of care to visitors – i.e. people invited or permitted to use land, whether expressly or by implication. There is an obligation to take reasonable care that visitors will be safe doing whatever it is that they have been invited or permitted to do on a site.

The duty of care does not apply to risks that adults willingly accept on behalf of themselves or those immediately in their care.

All infrastructure, bridges, boardwalks, etc. must be maintained in a safe condition at all times. All visitors must be made aware of all natural hazards.

Disability Discrimination Act 1995

The key section of the act is: It is unlawful for a provider of services to discriminate against a disabled person by refusing to provide, or deliberately not providing, to the disabled person any service which he provides, or is prepared to provide, to members of the public.

A full DDA audit was completed 9 February 2003.

The Control of Substances Hazardous to Health Regulations 2002 (COSHH)

The Provision and Use of Work Equipment Regulations 1998 (PUWER)

The Regulations require risks to people's health and safety from equipment that they use at work to be prevented or controlled. In addition to the requirements of PUWER, lifting equipment is also subject to the requirements of the Lifting Operations and Lifting Equipment Regulations 1998.

2.2 Policy Statements – Access

The Countryside Council for Wales' access policies for all National Nature Reserves:

- CCW will declare all land in CCW ownership and, whenever possible, land in CCW's control, as 'dedicated land' under the CRoW Act.
- In all cases, CCW will consult with local communities and other stakeholders before proceeding with dedication.
- For land under CCW control, for example, lease or agreements, CCW will consult with, and seek the full agreement of, all owners and occupiers before proceeding with dedication.
- Whenever necessary, access restrictions will be applied to sites, or parts of sites, where such restrictions are essential for the protection of the conservation features.

CCW will encourage the sustainable public use of National Nature Reserves in Wales in so far as such use:

- Is consistent with CCW's duty to maintain or restore the nature conservation and geological features to Favourable Conservation Status.
- Does not expose visitors or staff, including contractors, to any significant hazards.

All legitimate and lawful activities will be permitted in so far as these activities:

- Are consistent with CCW's duty to maintain or restore the nature conservation and geological features to Favourable Conservation Status
- Do not expose visitors or staff, including contractors, to any significant hazards
- Do not diminish the enjoyment of other visitors to the site

3 Description

3.4 People – Stakeholders, Access, etc.

3.4.2 Access

The statistics on visitor numbers and characteristics are taken from a survey conducted in 1997 by the Welsh Institute of Rural Studies at the University of Wales, Aberystwyth (Scott et al. 1998, Christie et al. 1998).

3.4.2.1 Visitor Numbers

The total number of visitors in 1997 was estimated at 20,000 per year (Christie et al. 1998). This will, of course, include a large number of repeat visits, so the total of individuals will be considerably less. Since 2004, electronic data loggers have been used to obtain more accurate figures. These confirm previous estimates, suggesting that numbers have remained stable, or possibly risen slightly by perhaps 10%.

3.4.2.2 Visitor Characteristics

Visitors are evenly divided between those that come from outside Wales (46%) and those coming from within Wales (47%). Of those that visit from within Wales about half are local (within 10 miles). Only 6% are from outside the UK.

The majority of visitors (65%) are couples or families, with 20% visiting alone and 10% as part of a small group. No visits are from larger groups or as more formally arranged activities by clubs or educational establishments.

Most visitors (58%) are over 45, with just 4% under 18. 60% of visitors are male and 40% female.

The majority (70%) are either working or students, with 21% being retired. Only 2% are unemployed and none are unable to work as a result of disability or illness.

3.4.2.3 Visit Characteristics

The vast majority of visitors (71%) come to watch nature or for other specialist interests, with the opportunity to see red kite being most specifically mentioned (31%). Just 7% come to enjoy the landscape qualities of the area and 16% are dog-walkers.

Of those that are on holiday, 42% describe the opportunity to visit the reserve as a major feature of their holiday, while just 18% stumbled across it by chance. Among local people, the majority (62%) visit at least several times a month.

Almost all visits (93%) last less than three hours, with over a quarter (28%) lasting less than half an hour.

3.4.2.4 Access to the Site

The B4343 runs along the eastern side of the reserve, and from this a small lay-by car park gives access to the reserve. Poor visibility makes this car park quite difficult to use, and at peak times it can be full (Scott et al. 1998). It is a remote area, situated in the agricultural heartland of Ceredigion: an area of low population, away from the main tourist routes, that is not well served by public transport. It is very difficult to reach the reserve other than by car, with the nearest bus stops being two miles away, at either Pontrhydfendigaid or Tregaron. The narrow, winding nature of the country roads linking the site to neighbouring villages means that it is not an easy journey to make on foot or by bike.

Most visitors travel to the site by car (93%), with just 1% travelling by bike and none on foot or by public transport (Scott et al. 1998).

There is informal parking at Pont Einon to the south of the reserve, off the A485. Parking at Ystrad Meurig station yard, to the north, is used by local people. Both these sites have excellent views over the bog, but they give no access onto the reserve.

3.4.2.5 Access Within the Site

The old railway track provides a well-maintained and well-surfaced route along the eastern edge of the reserve, allowing a walk of just under 5 km. It is easily reached from the lay-by car park and is suitable for wheelchair users. Although it gives excellent views across the site, it does not allow any access out onto the bog itself.

A narrow boardwalk across the bog gives access to the riverside walk, which loops back to rejoin the railway track, covering a distance of about 7 km. Although it is well maintained, the boardwalk is very narrow, allowing only single file walking, and is not suitable for people with mobility problems. The boardwalk is open to permit-holders only, with about 100 permits being issued per year.

Fig. CS2.2 The old railway track

3.4.2.6 Visitor Facilities and Infrastructure

The small lay-by car park provides space for about eight cars but has no facilities. Of 22% of visitors wanting improvements to existing services this primarily related to the car park, while the 19% of visitors wanting new facilities mainly required toilets (Scott et al. 1998). A series of eight information panels runs north from the car park, beginning with a sign to welcome visitors and give them a sense of orientation. However, some visitors find this sign difficult to use and are unaware of the best route to take on leaving the car park (Scott et al. 1998). Other panels describe the railway line, the wildlife of the track, bog plants, vegetation in adjacent ditches and hydrological management. The monochrome panels are 1 m × 1.5 m in size. Perhaps because of the enclosed and linear nature of the trail, the information panels attract a great deal of attention and are used by 84% of visitors (Scott at al. 1998).

A booklet with coloured photographs and maps gives a good introduction to visitors. It describes how the bog was created and outlines its history and its importance. There are brief descriptions of the vegetation and wildlife as well as location maps and an outline of the access options within the reserve. There is also a bird list and a booklet giving much more detailed descriptions of the vegetation, which would be useful for visitors with specialist interests. Leaflets attract much less attention than the panels and are used by only 6% of visitors, although this is possibly as a result of the leaflet box being vandalised rather than reflecting a lack of interest by visitors (Scott et al. 1998). Of those that do use the leaflets, 100% rate them as excellent (Christie et al. 1998).

The riverside walk has numbered way points, and an accompanying booklet allows this to be used as a self-guiding trail. The way-marked walk is used by 18% of visitors. Some people have commented that they would like this part of the reserve to be more accessible (Christie et al. 1998).

There is an observation tower about one and a half kilometres to the north of the car park. This acts as a bird hide and also gives spectacular views across the bog, but it is due to be demolished for safety reasons. The hide is used by 35% of visitors, which is possibly a reflection of the high number of people with a specialist interest in birds (Scott et al. 1998).

Reserve staff give about twelve guided walks a year and approximately six slide shows.

3.4.2.7 The Reasons Why People Visit the Site

3.4.2.7.1 Wildlife Attractions

Bird watching is the prime reason for people to visit the site, and this has the potential to attract visitors throughout the year. In winter there are whooper swans, hen harriers and large numbers of wildfowl. Summer brings curlew, some breeding lapwing, redshank, snipe, reed bunting and sedge warblers. Most important of all are the red kite. Although these birds are now widely seen in mid-Wales this remains an area where people traditionally come to see them, with 31% of people specifically mentioning red kite as the reason for their visit (Scott et al. 1998).

Butterflies put on beautiful displays in summer, with common blue, small pearl-bordered fritillary and commas being frequently seen. Dragonflies are abundant and receive much attention from visitors. Although the bog is spectacular, lack of good access means that people cannot easily appreciate the detail of the vegetation. Most people are unlikely to get good views of the individual plants such as bog asphodel and sundew.

Fig. CS2.3 Beautiful Demoiselle

3.4.2.7.2 Other Features That Attract People

The reserve is an area of outstanding landscape that dominates the valley north of Tregaron. Its spectacular golden-red colouring floods the area and it clearly stands out as being very different to the surrounding countryside. Surprisingly, this stunning landscape does not seem to play a major part in attracting visitors, with just 7% citing the landscape quality as a reason for visiting the site (Scott et al. 1998). However, 35% of visitors believe that 'quietness' is an important factor in their enjoyment, and this is, of course, very closely linked to the landscape in terms of the size and remoteness of the reserve.

Fig. CS2.4 Reflection in a bog pool

3.4.2.7.3 Recreational Activities

Activities on the site are limited to quiet enjoyment, which includes walking, cycling and bird watching.

3.4.2.8 Current and Past Concessions

There are no concessions on the site.

3.4.2.9 Stakeholder Interests

There are a number of tenant farmers who farm parcels of land within the reserve. This is limited to rough grazing for sheep. The fishing rights, for salmon and trout, are privately owned. Shooting rights over the northern and central section of the reserve are retained by the former owner of the land.

3.4.2.10 The Site in a Wider Context

This is a very isolated site, and there is nothing similar in the local area. Pony trekking was once important, but has declined for practical reasons associated with the individuals providing the service, and this does not indicate any lack of demand for the facility.

3.4.2.11 Educational Use

The comprehensive visitor survey conducted during June to September 1997 recorded no educational use (Scott et al. 1998), although this clearly coincided with the holidays for most educational establishments. Reserve staff do, in fact, make an effort to provide for local schools and universities, giving about eight to ten guided walks a year for primary and high schools and one or two for universities. In addition, there may be one guided walk for a more distant university and two or three self-guiding school or college groups.

5 Access

5.1 Evaluation

Current demand is strongly biased towards people who already have an interest in natural history, and this is largely dominated by birdwatchers. Bogs are not generally perceived as attractive to the casual visitor and even the spectacular landscape has little appeal. They have an air of danger that is partly real, as a result of the difficult nature of the terrain, but people also have a false perception of bogs as barren, hazardous places more suited to the will-o'-the-wisp than to human visitors. If people were aware of the true nature of the bog they would be much more likely to want to visit it. If they could see the wonderful detail of the plant life, experience the openness and sense of peace or hear the variety of birdsong they would begin to understand why it is so special. However, there is little point in trying to use these assets to attract visitors because the inaccessibility of the site greatly limits the potential for people to enjoy these aspects.

5.1.1 Accessibility

Parking is possible for up to eight cars but, because the car park is basically a lay-by, carelessly parked cars can result in it quickly becoming full, making the site inaccessible during periods of peak demand.

5.1.2 Access Within the Site

The main part of the site is too dangerous and too fragile to be accessed by visitors without the provision of significant infrastructure. Access is mainly limited to the railway track running along the edge of the reserve and to viewing points at either end of the reserve. There is a boardwalk that gives some opportunity for visitors to get out onto the bog, but this is narrow and quite difficult to walk on, and does not take people to the areas that they would most like to see. Access to the boardwalk is for permit holders only, and this is likely to deter most visitors who believe (incorrectly) that it is necessary to have a specific reason in order to obtain a permit.

5.1.3 Site Safety

The bog is a naturally forbidding place, and the difficulty of the terrain is immediately apparent. People are unlikely to be tempted to walk on it and it would certainly be unsafe to do so.

5.1.4 Implications of Stakeholder Interests

The potential for visitors causing problems for tenant farmers is very limited because there are currently so few visitors. However, any attempt to extend access to the site is likely to be perceived by them as a problem, and they would be concerned, for example, about the possibility of dogs worrying livestock. In reality, increased visitor numbers are very unlikely to have any impact on farming activities, but it is essential to negotiate with tenants before any changes take place to ensure that they feel included in the process. Although there were some conflicts of interest concerning shooting rights, these have been fully resolved.

Local residents with properties overlooking the bog may feel that increased access is detrimental to them, but they will also have some benefits from being able to make use of the improved facilities. Again, it is essential to liaise with neighbours to ensure that any negative impact is minimised.

If more people were attracted to the reserve, there would be improved opportunities for local farmers to diversify into tourism. This could be important in an area that generally has few visitors.

5.1.5 Carrying Capacity of the Features

The bog itself is extremely fragile and has no carrying capacity without the provision of a boardwalk. The largest raised mire, which is relatively intact, is a rare example of this type of habitat, and it should remain undisturbed in order to protect it and to retain its wilderness qualities. There is a risk that people will disturb ground-nesting birds in the immediate vicinity of where they are walking, and dogs are particularly likely to cause problems. Wintering flocks of wildfowl are also vulnerable to disturbance, and this will limit any potential to open up large areas of the bog. Public access to the bog should be restricted to a boardwalk.

5.1.6 Carrying Capacity of the Site

The site is extremely large, and a boardwalk would encroach only on a very small area, leaving the rest of the bog largely undisturbed. Given the remoteness of the area, away from large centres of population or tourist routes, any improvements to access would be unlikely to attract enough people to damage the wilderness qualities of the site. Any development of car parking facilities should be such that it did not allow for excessive numbers of people.

5.1.7 Summary of the Evaluation

Currently, the site does not fulfil its potential to allow visitors to enjoy such a rare and fascinating habitat, but there would be little point in trying to attract more people because parking and access are not adequate. The existing boardwalk is not easily accessible, and people do not feel encouraged to use it. Parking and access for walking, bird-watching and wheelchair use could be improved enormously without detriment to the site. The remote location and relatively large size of the site mean that excessive numbers of visitors are unlikely to be a problem.

5.2 Access Option

Access is encouraged, but is limited in terms of the area of access and overall numbers. Activities should be quiet and unobtrusive, such as walking, bird watching and photography, with cycling and horse riding permitted on the railway track. Access to the bog should be restricted to the boardwalk.

5.3 Access Objective

To encourage the sustainable and inclusive public use of Cors Caron in so far as such use is consistent with maintaining the nature conservation features at Favourable Conservation Status and provided that visitors are not exposed to any hazards.

5.3.1 Vision

Vision 1 – simple version – describing the facilities and infrastructure

There is a wide range of information available to attract people to the site, including leaflets and a web site. Signs at the car park and on local roads make it easy to find.

The car park has disabled bays, toilet facilities, seating and shelter as well as picnic areas and secure spaces for bikes. The Ystwyth Trail, following the route of the old railway line, makes it possible for people to reach the site by bike or on foot from the nearby villages of Tregaron and Pontrhydfendigaid. Buses can stop at the reserve car park, allowing people to make use of local bus services. An information panel introduces visitors to the reserve and helps them to plan their visit. There are clear descriptions of the options available, where seating and shelter can be found, and, particularly for people using wheelchairs, an indication of the distance between turning and passing places.

From the car park, visitors can follow the old railway line, a broad, smooth track that runs the entire length of the reserve and gives good views out over the bog. Alternatively, they can use the boardwalk to gain access to the bog itself. This is a circular route of 1.5 k which provides a good, level surface that is accessible to wheelchair users. There are seats and information panels at intervals along the route. An observation shelter, looking directly over pools which may be used by breeding and wintering birds, gives distant views out over the floodplain of the River Teifi. For visitors wishing to see more of the reserve, the riverside walk, which is not suitable for disabled visitors, can be accessed from the boardwalk. This covers a distance of about 7 k and follows the banks of the river for part of its length before returning to the railway track.

Vision 2 – describes the facilities, infrastructure and the experience that visitors can expect to enjoy at the site

There is a wide variety of information available to attract people to Cors Caron, including leaflets and a website. The car park at the reserve is clearly signposted and easily accessible from the main road, and gives an immediate feeling of being welcoming and well cared for. There are disabled bays, toilet facilities, seating and shelter as well as picnic areas and secure spaces for bikes. As an alternative to travelling by car, the Ystwyth Trail, following the peaceful route of the old railway line, gives an ideal opportunity to reach the site by bike or on foot from the nearby villages of Tregaron and Pontrhydfendigaid. Buses can stop at the reserve car park, allowing people to make use of local bus services. Information panels provide a perfect introduction to people unfamiliar with the reserve and help them to plan their visit. People using wheelchairs and less-agile walkers will discover that the main routes are accessible to them, and that the boardwalk provides a good, secure surface with plenty of space to manoeuvre. There are clear descriptions of the options available, where seating and shelter can be found, and, particularly for people using wheelchairs, an indication of the distance between turning and passing places.

The old railway line provides a broad, level track that runs the entire length of the reserve. It is a partially tree-lined trail that gives superb views over the reserve, but perhaps the true highlight of any visit is to follow the boardwalk out onto the bog itself. The reserve is spectacular at any time of the year, but its appeal is unsurpassable in early summer. As visitors leave the car park on a smooth pathway the sound of birdsong drifting from the trees draws them immediately into their new surroundings. The sun sifts through the translucent green of the new leaves, while butterflies flit through the dappled light. After a short distance, the boardwalk peels away from the track and, as the shelter of the trees is left behind, the view opens up to reveal the full sweep and grandeur of the site. The bog lies in a vast bowl rimmed with hills, and the landscape stretching out appears untamed and exciting compared to the gentle greenness of the surrounding trees and fields.

A pool butts up against the side of the path, and the sunlight catches the iridescence of dragonflies' wings as they dart and meander above the water, occasionally resting on the boardwalk at the feet of passers-by. Overhead, birds soar through a sky that appears endless above such an open landscape. Occasionally, it may be possible to glimpse the spectacular sight of a hobby plunging down to snatch a dragonfly. Staring skyward may also bring the reward of seeing the magnificent, fork-tailed silhouette of a red kite. Though they may be seen frequently now in mid-Wales these once-endangered birds remain a powerful emblem of these special places that were their only stronghold. With so much to see it would be easy to miss the subtler sights and sounds: the piping of redshank or the softly melodic, bubbling call of the curlew.

Fig. CS2.5 Dragon's wings

As visitors travel further out onto the bog, the tussocky landscape is scattered with small pools: sharp and glinting fragments of reflected sky. By this point people will have realised that they have found their way into the sort of terrain that would normally be inaccessible. This is a rare experience for anyone, but for someone with restricted mobility, who may feel excluded from truly wild places, it offers an almost unimaginable freedom. Ahead is the observation shelter, a building of such soft, natural colours and flowing curves that it appears to have grown from the landscape. This, together with the regular seating along the boardwalk, gives confidence to anyone who may be wary of embarking on a walk into a nature reserve. Along the route, beautifully carved information panels highlight some of the details of the surrounding landscape and its wildlife. Inside the shelter, a wall of windows looks out across a pool and then on over the flood plain of the Afon Teifi. The stunningly open outlook contrasts with the feeling of protected seclusion inside the building.

As they follow the boardwalk beyond the shelter visitors begin to get a sense of the extraordinary structure of a raised bog as they see the land ahead of them rising up in a smooth dome. Here the hummocky lawns of sphagnum mosses spread like a densely textured tapestry. The colours threaded through it range from vibrant green to jewel-bright, ruby red. Spikes of bog asphodel splash it with yellow while the bog rosemary brings a subtler wash of pink. Silky puffballs of cotton grass appear to float above the surface making striking white highlights. In this peaceful atmosphere visitors are more aware of the snatches of bird song scattered all around. Perhaps the most uplifting of all is the soaring song of the skylark as it trickles back down to earth with a ringing purity.

After one and a half kilometres the curve of the boardwalk brings people almost back to their starting point, and for a moment it may seem strange to have returned so easily to the 'real world' after a journey that has taken them into such a different place. The more adventurous may want to extend their visit to take in the riverside walk. This leads off the main boardwalk down to the Afon Teifi and covers a distance of about 7 km, allowing people to experience a little more of the sense of remoteness. They can follow the meandering river banks accompanied by birds, such as sedge warblers, grasshopper warblers and reed bunting, while across the river the faintly rippling reeds slice the sunlight into sparkling ribbons.

In winter the reserve presents a different face. With the rest of the countryside dull and drained of colour, it fills the dish between the hills like a pool of red spilled across the landscape. While many places have been churned to mud by winter rains, the boardwalk continues to provide a secure surface for anyone who wants to venture out. Visitors may hear the quiet whistling of teal from the scattered pools or see a hen harrier gliding overhead. Herons, with broad, blunt wings, imprint their distinctive silhouettes onto the sky. Occasional flocks of birds, perhaps lapwing or fieldfare, twist and wallow, sketching stippled patterns in the air. For those with the patience to wait, there is a fleeting moment of brilliance just before dusk. In the light of the setting sun the bog flames golden-red before the sudden cold of winter twilight sends visitors heading back to the car park.

5.3.2 Performance Indicators & Monitoring

Any improvements to the parking and access facilities are likely to change the number of visitors. In this instance, limits could not be set immediately but would be determined from the results of surveillance conducted over the first 5 years of the new regime.

1. The total annual number of visitors, or a representative sample, for the whole, or part, of the site. (This can be used to measure trends.) Until the implications of any possible changes are established by surveillance, a lower limit should be

set at the annual number of visitors indicated in the last survey. The upper limit would be the level at which the quality of visitors' experience is diminished by overcrowding. This could be measured by questionnaires.
2. The number of educational groups. A lower limit should be set at the current level.
3. Level of satisfaction measured formally by, for example, structured questionnaires or visitor surveys.
4. The number of complaints or compliments.

5.4 Status & Rationale

5.4.1 Status

Although visitors are provided with good opportunities to view the site, access onto the bog itself is very limited and the status of access provisions could therefore be considered as unfavourable. The bog can be reached by means of a very narrow boardwalk that is suitable only for reasonably fit people. It allows no access for people with any sort of mobility problems. The boardwalk is accessible to permit holders only, which actively deters people from visiting and gives the impression that members of the public are not generally welcome. There is no seating or shelter on the bog, which can discourage people from setting out.

The car park is not easy to see or access from the road and it can be full at peak times, preventing potential visitors from gaining access to the reserve. The lack of information in the car park makes it difficult for visitors to plan their visit. There is very little opportunity for people to visit the reserve other than by car.

5.4.2 Rationale

5.4.2.1 Legislation

When considering possible changes to the reserve it is necessary to comply with all Health and Safety, Public Liability and Disability Discrimination legislation. This means that any new boardwalk must be suitable for disabled people. Planning permission must be obtained for any changes to the parking and toilet facilities.

5.4.2.2 Access to the Site

The construction of a new car park would provide safer access and adequate parking for visitors at peak times. It should include disabled bays and secure spaces for bikes. Level pathways should provide easy access from the car park to the reserve. It may be possible to liaise with local bus companies to persuade them to stop on request at the car park. The completion of the Ystwyth Trail will extend and upgrade the old railway track, greatly improving access by bike and on foot from nearby villages.

5.4.2.3 Access Within the Site

The most pressing need is to provide better access to the bog. This would be achieved by the construction of a new boardwalk, preferably a loop that would give visitors a chance to see the different vegetation communities on the bog. The boardwalk must be suitable for disabled access, with frequent turning and passing places for wheelchairs. It should also provide seating for people who are not able to undertake a long walk without stopping to rest.

The riverside walk should be improved and the permit system should be removed, making it open access for all visitors, with just occasional closures for management requirements.

5.4.2.4 Seasonal Constraints

The site has attractions for visitors at all times of the year and the new boardwalk would make it accessible in all weathers.

5.4.2.5 Public Awareness

Improvements in access would increase the appeal of the site to a wider range of visitors. It is important that more people are aware of what the site has to offer. Publicity could be improved by providing good information for internet users and by ensuring that leaflets are available in some of the more mainstream places likely to be visited by tourists. This may help to attract visitors with more general interests as well as the high proportion of specialists that currently use the site.

5.4.2.6 Excessive Demand

In such a remote site, excessive demand is not envisaged, but the size of the car park should be such that it does not allow for over-use of the site.

5.4.2.7 Visitor Infrastructure

Many visitors would welcome better facilities at the car park, and the provision of toilets, some limited shelter and a picnic area would greatly improve people's experience of the reserve. A well-designed observation shelter that blends into the landscape, as part of the boardwalk loop out onto the bog, would make it easier for less-active visitors to enjoy the exposed and wild nature of the site.

The tower hide must be demolished for safety reasons. This should be replaced by another hide in the same location. The new hide should be on a raised bank to give an elevated view over the bog.

5.4.2.8 Information

The car park should be clearly signposted at the site and on local roads. Leaflets should be available both at the car park and locally. On arrival, visitors should find a clear and welcoming information panel that will help them to plan their visit. It should set out the options available and, particularly for wheelchair users and less mobile visitors, it should give a clear indication of the distances between turning and resting places.

5.4.2.9 Interpretation

With the prospect of increased visitor numbers, a separate interpretation plan will be necessary. Interpretation should be relatively low-key, but it should include a well-produced booklet describing the wildlife, history and importance of the reserve, and information panels along the paths and boardwalk highlighting the main areas of interest without being intrusive.

5.4.2.10 Education

Given the remoteness of the site there is unlikely to be a high demand for educational facilities. The main focus should be directed towards local schools. Ideally, every child attending a local primary school should be given the opportunity to visit the reserve and take part in a guided walk during their final year. Other schools should be accommodated as far as possible.

Educational packs should be available, consisting of a number of separate inserts so that they could be tailored to the requirements of the curriculum and the age of the recipients. These would be suitable for primary schools, high schools and universities

6 Action Plan

The following are two examples of the many management projects required to provide the management infrastructure necessary to support public use.

Project Plan 1

Project title: Resurfacing the old railway walk

Year/s when the project is active: 2007/8

Event/s within a year: 1 event – October to March

Zones or compartments: 00

Expenditure: £12,000 **Staff Time: Reserve** manager 8 days

Project priority: 1

Justification for the project (i.e. the intended outcome):
The Countryside Council for Wales is in the process of improving public access at Cors Caron National Nature Reserve (NNR). As part of this programme of works it is necessary to improve the surface of the old railway walk.

Potential impact on other features:
None

General background/bibliography:
This track has been open to the public since 1967 following the closure of the railway.

Project methodology:
The work will be carried out by a contractor.

1 Specification of works:

1.1 Repair to potholes

1.1.1 All potholes are to be repaired along the bridge over the River Teifi to the entrance at Ystrad Meurig, a distance of approximately 800 m (see attached map).
1.1.2 The potholes should be cleared of standing water before filling. This should be done by baling or pumping.
1.1.3 The sides of all potholes should be excavated to give vertical sides before repairing.
1.1.4 Sub-base type II should be used as the fill. The contractor is welcome to suggest other suitable material.
1.1.5 Where a number of potholes occur in close proximity, then the area should be excavated as a whole and resurfaced.

2.2 Resurfacing

2.2.1 A section of the railway, approximately 1,400 m by 3 m wide, needs resurfacing. Only minor excavation should be needed to clear soil and mud from the surface. Excavated material can be deposited over the sides of the railway embankment – NOT on the verges.
2.2.2 100 mm of sub-base type II shall be laid down and compacted. The contractor is welcome to suggest other suitable material.
2.2.3 The contractor will aim to make a cambered surface to shed water.

All work to be completed and invoiced by 16th March 2008

3. Conditions of contract

3.1 Before tendering, the contractor should make an appointment with the reserve manager. He/she should satisfy him/herself as to the full extent and character of the work and site conditions affecting the contract. The tender quotation should be for all the work necessary to complete the contract to the required conditions and specification.

3.2 The contractor shall prove that they possess a fully comprehensive insurance policy to a minimum value of £2,000,000 to cover any claims arising from CCW or any third party in respect of damage or negligence on the part of the contractor. **A copy of the certificate must be sent with the tender**, unless previously supplied

3.3 Site hazards: The contractor should carry out a risk assessment to the satisfaction of CCW for the scheduled work. The work area is adjacent to a public access route within the NNR. Suitable precautions to minimise risk to public & staff should be taken by the contractor. **A copy of the risk assessment must be forwarded with the tender.**

3.4 The contractor shall carry out any mutually agreed variations or additional work after discussion with CCW staff. Both parties to the contract must confirm variations and any extra cost involved in writing.

3.5 The contractor can sub-let or assign any of the work involved in this contract, but not without the knowledge and written consent of CCW.

3.6 All requirements of the Health and Safety at Work Act 1974, and all relevant regulations and codes of practice must be adhered to. Failure to comply with this requirement will result in the job being stopped immediately and until such time as the contractor is able to comply.

3.7 All users of chainsaws must hold the relevant National Proficiency Test Council competence certificate. **Copies of certificates must be forwarded with the tender,** unless these have been provided previously.

3.8 All plant & machinery to be used for the work must have current certificates of worthiness where appropriate. **Copies of certificates must be forwarded with the tender.**

3.9 Vehicle access routes and use on site must be agreed with CCW.

3.10 Suitable extra precautions, such as warning signs and/or diversion barriers, must be taken for sections of the work on or near public access areas of the site.

Site hazards

The Railway Walk is generally open to the public, so the contractor must ensure that appropriate safeguards are in place throughout the work.

Deep ditches and river channels that are liable to flood after heavy rain.

Project Plan 2

Project title: Construction of boardwalk

Year/s when the project is active: 2007/2008

Event/s within a year: 1 event – October to March

Zones or compartments: 10

Expenditure: £38,000 **Staff Time:** Reserve Manager 6 days

Assistant Reserve Manager 8 days

Project priority: 1

Justification for the project (i.e. the intended outcome):
The Countryside Council for Wales is in the process of improving public access at Cors Caron National Nature Reserve (NNR). As part of this programme of works it is necessary to construct a boardwalk which will provide access for the visiting public. The boardwalk will be suitable for wheelchairs and compliant with the Disability Discrimination Act.

Potential impact on other features:
None

General background/bibliography:
See management plan.

Project methodology:
The work will be carried out by a contractor.

1 Specification of works:
Section 1
Remove from site and dispose of approximately 65 m of existing boardwalk.
Section 2
Construct 875 m of new boardwalk
Section 3
Design and construct a ramp to allow wheelchair access from the old railway line to the boardwalk at point X on the accompanying map.

2 Boardwalk design

The boardwalk is to be 1.2 m wide between the kick boards. The kickboards should be 75 mm high.

The boardwalk must have a maximum gradient of 1:12.

The maximum distance between passing places should be 150 m.

There should be no slope across the boardwalk.

The decking boards are to be transverse across the boardwalk and spaced a maximum of 10 mm apart.

The decking boards are to be supported by two 100 mm × 100 mm stringers on the outside edges and a 100 mm × 50 mm stringer down the centre. The decking is to be secured by two galvanised nails at each support point.

The stringers are to be supported and fixed to two 100 mm × 50 mm crossbeams fixed to 100 mm × 100 mm posts by 16 mm galvanised coach bolts. These supports should be at least every 3 m. Each support post is to be dug or driven into the ground to a depth of 1500 mm where the boardwalk is raised less than 200 mm above ground. Where the boardwalk is raised to a greater height than this the support posts should be dug or driven 2,000 mm into the ground.

The timber sections should be as follows:

Decking boards	150 mm × 40 mm	× 1,300 mm
Crossbeams	100 mm × 50 mm	× 1,500 mm
Stringers	100 mm × 100 mm	× variable length
Support post	100 mm × 100 mm	× variable length
Kick board	50 mm × 75 mm	× variable length

The tenderer should provide prices for constructing the boardwalk in:

(a) untreated sweet chestnut
(b) pressure treated Douglas fir

(both to be free of large knots and other defects)
Non-slip surfacing – Tenax Geogrid Type LBO 220 to be stapled to the boardwalk decking. CCW can provide details of one supplier of this material.

Storage and on site movement of materials
It is important to minimise the use of vehicles on site and avoid compacting or churning the surface. CCW will provide support to the contractor for transporting materials on site with a tracked ATV. Due to the presence of rare and sensitive species, the locations for the temporary storage of new and redundant materials must be agreed with the CCW Reserve manager.

Other requirements
The site is open to the public while work is in progress. The contractor will be responsible for the provision of signs and barrier tape to warn the public and deter entry to work areas. The contractor must take all reasonable steps to ensure that materials stored on site are safe.
Note: The conditions of contract follow a similar pattern to those in the previous example.

The management plan was implemented in 2008

Fig. CS2.6 The boardwalk leads across the bog

Fig. CS2.7 The new observation shelter

References

Christie et al. (1998). *The effectiveness of interpretation, its economic impact and recreational use of Welsh National Nature Reserves*. University of Wales, Aberystwyth.

Scott et al. (1998). *A survey of visitors to Welsh National Nature Reserves*. University of Wales, Aberystw

Case Study 3
The Relationship Between Species and Habitat Features

Martin Vernik[1], Jurij Gulic[1], and Mike Alexander[2]

Abstract This case study is included because it is an excellent demonstration of the relationship between species and habitat planning. Even when a habitat is not recognised as a special feature on a site, if it supports a species which is a special feature the habitat is best treated as a feature. For Natura 2000, or any other, sites where a site-specific version of Favourable Conservation Status defines the management objectives there is an obligation to protect the habitat. The definition of FCS for a species feature requires that sufficient habitat must exist to support the population in the long term. The specific requirements of the individual protected species will determine the condition of the supporting habitat.

Note: The information presented in this case study is based on a Eurosite Management Planning Workshop held in the Topla Landscape Park, Slovenia, during May 2006. The workshop participants represented Eurosite, The Institute for the Republic of Slovenia for Nature Conservation, the CMS Consortium and exeGesIS (an environmental & IT consultancy based in Wales)

The aim of the workshop was to demonstrate and test the planning structure given in Chapter 2 of this book and the CMS software on a Slovenian Natura 2000 site.

1 The Topla Landscape Park – Summary Description

Topla is an IUCN Category V Landscape Park situated in northern Slovenia, close to the border with Austria. The total area of the park is 15 km^2. It lies in Črna na Koroškem local community, the current population of the protected area is 25

[1] IRSNC Maribor Slovenia.

[2] CMS Consortium Wales.

M. Alexander, *Management Planning for Nature Conservation: A Theoretical Basis & Practical Guide*, DOI 10.1007/978-94-007-5116-3,

individuals. The park, which has been protected since 1966, is the only protected area in this part of Slovenia.

The park is dominated by the 2,125 m Mount Peca, a vast and complex mountain area in eastern Karavanke. It has typical karst features with pristine areas of diverse high-mountain habitats, particularly forests, grasslands, rocks, cliffs and shrub communities.

The Topla valley is one of the most spectacular and picturesque alpine valleys in Slovenia. It has well-preserved, man-made habitats, including managed forest, mountain meadows and pastures, with a wealth of beautiful alpine flowers and many butterflies. The diversity of semi-natural habitats is further enriched with smaller fens and marsh communities which contain endangered flora and fauna. These areas are also particularly important for species diversity, including many endemic endangered and protected species.

The valley is also well known for its interesting tectonic features (contact of Eurasian and Adriatic tectonic plates), and the lead/zinc ore deposits are of global interest because of their sedimentary origin.

In addition to outstanding wildlife and geological interest, the valley is also of considerable cultural, ethnological and historical importance. The typically dispersed settlement pattern, containing a few large, self-sufficient farms, is well preserved. The valley and its surroundings contain many relicts of a once thriving mining industry, which are testament to the history of the entire region. Abandoned sawmills and farms tell the story of the massive depopulation following the Second World War. This was mainly a consequence of the extremely demanding and inhospitable environmental conditions.

The area is also a Natura 2000 SPA and pSAC:

The SPA features are:

Ptarmigan *Lagopus mutus helveticus*
Black Woodpecker *Dryocopus martius*
Capercaillie *Tetrao urogallus*
Hazel grouse *Bonasa bonasia*
Tengmalm's owl *Aegolius funereus*
Pygmy owl *Glaucidium passerinum*
Golden eagle *Aquila chrysaetos*
Black grouse *Tetrao tetrix*
Peregrine *Falco peregrinus*
Three-toed woodpecker *Picoides tridactylus*

The pSAC habitat features are:

Calcareous rocky slopes with chasmophytic vegetation
Siliceous rocky slopes with chasmophytic vegetation
Hydrophilous tall herb fringe communities of plains and of the montane to Alpine levels
Alpine and sub alpine calcareous grasslands
Siliceous alpine and boreal grasslands
Alpine and boreal heaths

Illyrian *Fagus sylvatica* forests (*Aremonio-Fagion*)
Scrub with *Pinus mugo* and *Rhododendron hirsutum*
(*Mugo-Rhododendretum hirsuti*)

2 The Case Study

This case study demonstrates the relationship between two SPA woodland bird species, capercaillie *Tetrao urogallus* and the three-toed woodpecker *Picoides tridactylus*, and two forest habitats. One is a pSAC habitat: Illyrian *Fagus sylvatica* forests (*Aremonio-Fagion*). The second is an unscheduled, commercially managed mountain spruce forest.

The study comprises four summarised objectives which have been extracted from the management plan *(additional notes which are not contained in the original plan are shown in italics)*. The objectives are all written in the present tense, that is, they are a description of the condition that is required for each feature: this does not represent the current condition (see Chapter 15 for a full explanation).

2.1 SPA Feature – Capercaillie Tetrao urogallus

Description

This forest-dwelling grouse is an extremely large, turkey-like bird with broad wings and tail. Apart from dark brown wing coverts, the plumage is an overall dark slate-grey with white barring on the tail.

Objective

The Topla Park is an important breeding site for a robust, resilient and viable population of capercaillie. The distribution of the population (shown on the attached map) is maintained or increasing. There are at least 4 to 5 lekking areas, with a distribution which is maintained or increasing (see attached map). There is sufficient suitable habitat for this species, including safe nesting sites and a secure breeding environment. The impact of predators is insignificant. The size and range of the population are not restricted or threatened, directly or indirectly, by human activity. Lekking and nesting birds are not disturbed by human activities during the breeding season.

Performance indicators

Factors
The most **significant** factors are:

The number of suitable lekking areas:

There are currently 4–5 lekks that have been used for many years. This factor is linked to the condition of the forest habitat. It is such an important factor, and one that can be monitored, that it will be used as a performance indicator. The distribution of the lekks within the park should also be recorded to ensure that the distribution is maintained.

Factor: Number of lekking areas
Upper limit: not required
Lower limit: 4

Food:

The adult diet is predominantly vegetation: rowan berries *Sorbus aucuparia,* young beech *Fagus*, willow *Salix* and larch buds *Larix*. The young feed mainly on insects and spiders, and ants are also considered an important food source in this area. The number of anthills is consequently a useful indicator of an adequate food supply. Competition from other herbivores, for example deer, for food is a potential factor, but there is no evidence that it is a problem. It is not possible, at this stage, to use food supply as performance indicator because there are insufficient data to develop limits. Two surveillance projects should be established:
A project to measure the quantity of suitable vegetation available to the birds

Factor: Available food – vegetation
Upper limit: not required
Lower limit: To be established

A project to record (sample) the number of ant nests within the capercaillie breeding areas

Factor: Number of ant nests (sample)
Upper limit: not required
Lower limit: To be established

Note: *Once there has been sufficient surveillance to establish the relationships between food and the population of birds, the initial surveillance projects can be converted to monitoring projects. This is a very common situation encountered when management plans are prepared for the first time for a protected area. It can sometimes be very frustrating when the essential performance indicators cannot be quantified because basic information is not available. We should remind ourselves that planning is a developmental and iterative process: with time and careful preparation the plan will fulfil its functions.*

Anthropogenic factors:

- The development of new forest roads and forest operations at inappropriate times of the year, for example, during the lekking season, could be a problem.
- Inappropriate fencing, for example, to protect young trees from grazing, must be avoided as the capercaillie can fly into these fences.
- Human disturbance, for example, skidoos, mountain bikes, the collection of forest fruits and hunting, could have an impact on the population. There is no current evidence of any impact.

All these activities can be controlled within the park.

Note: *It is not possible to establish limits for any of these activities. However, it is important that surveillance projects are established and the activities recorded. If there is any change or decline in the capercaillie population this information can be interrogated.*

The forest habitat as a factor:

The Topla capercaillie require:

- A forest with a diverse age and physical structure, with Norway spruce present in the canopy.
- A dynamic, shifting pattern of small, irregularly-shaped, open glades. These may occur naturally, but, if this does not happen, intervention management will be considered. (The forest has a long history of management and, as a consequence, some areas are very even-aged.)
- Veteran trees, mainly larch, Norway spruce and Scots pine, near the lekking areas. These veteran trees are used by the capercaillie as display trees.
- Un-fragmented continuous areas of habitat to ensure that the birds do not become isolated.

Note: *To avoid duplication, these factors will be dealt with when the objectives for the forest are prepared. (These are factors which can affect the capercaillie, but they are also attributes of the forest. It also follows that capercaillie are a factor which influences the forest condition or, at least, the objective for the forest.)*

Attributes

Number of birds attending lekks:

The only useful and measurable attribute is the number of birds using the lekking areas in the spring. This is reasonably easy to measure, and there should be a direct relationship between the number of birds observed in this way and population trends. There are currently insufficient data to establish limits and a monitoring project. A surveillance project should be initiated, and, in time, this can be converted to a monitoring project.

Attribute: Number of birds attending the lekks
Upper limit: not required
Lower limit: To be established

2.2 SPA Feature – Three-Toed Woodpecker *Picoides tridactylus*

Description

It is a medium to large woodpecker, twice the size of a lesser-spotted woodpecker. With the exception of the males, which have a yellow crown, the plumage is entirely black and white. Birds are easily identified because of the clearly defined bars on their flanks. Their preferred habitats in Slovenia are the spruce forests at middle latitudes in the mountains.

Objective

The Topla Park is an important breeding site for a robust, resilient and viable population of three-toed woodpeckers, with at least 15 breeding pairs recorded each year. The distribution of the population (shown on the attached map) is maintained or increasing. There is sufficient suitable habitat for this species, including safe nesting sites and a secure breeding environment. The size and range of the population are not restricted or threatened, directly or indirectly, by any human activity.

Performance indicators

Note: *The objective, with performance indicators, factors and limits, follows exactly the same pattern as demonstrated in the previous objective for capercaillie. To avoid repetition, only the 'forest condition' as a factor and an attribute for the population of three-toed woodpeckers are included.*

Factors

The forest habitat

The Topla three-toed woodpeckers require:

- A forest habitat with more than 50% spruce trees in the canopy.
- Sufficient trees with nest holes, which are surrounded by a high density of spruce trees with a diameter of 20–30 cms dbh.
- Sufficient standing dead conifers.
- Sufficient deciduous veteran trees.
- Un-fragmented areas of forest of at least 300 ha.

Attributes

The only attribute that can be easily measured is the size of the breeding population.

Attribute: Total size of breeding population
Upper limit: not required
Lower limit: 15 pairs

2.3 pSAC Feature – Illyrian Fagus sylvatica forests (Aremonio-Fagion)

Description

This is a beech forest distributed within the Dinarides (Dinaric Alps) and the associated ranges and hills. There are outliers and irradiations in the south-eastern Alps and in the mid-Pannonic hills. In these areas the Illyrian beech forest is in contact with, or interspersed among, medio-European beech forests. Species diversity is greater than in the Central European beech woods (see species list in Appendix 1).

The beech forest is a pSAC feature, and, consequently, there is an obligation, regardless of the associated populations of SPA birds, to maintain the forest at Favourable Conservation Status.

The woodland is, at best, very marginal breeding habitat for the capercaillie and the three-toed woodpecker. However, it is an essential component of the wider forest mosaic which provides for these species. Given that the beech forest is so important, its condition should not be significantly modified to meet the requirements of the birds. There are a few opportunities where very limited compromise is possible, for example, the tolerance of low levels of Norway spruce in the canopy.

Objective

The beech forest:

- Occupies at least the area shown on the site map.
- Contains a canopy dominated by locally native species which will include a very small Norway spruce component.
- Is in a condition which is suitable for the regeneration of the canopy species.
- Has a diverse age structure which includes viable saplings and veteran trees, mainly deciduous species, but occasional larch, Norway spruce and Scots pine are tolerated.
- Contains a dynamic, shifting pattern of naturally-occurring, small, irregularly-shaped, open gaps in the canopy.
- Contains sufficient standing and fallen dead wood.
- Supports the full range of associated species.
- Contains some large, un-fragmented, continuous areas of forest.

Note: *The objective is presented as a series of bullet points in order to emphasise the individual attributes.*

Performance indicators

Factors

Alien species – Norway spruce
Norway spruce is not a native species. However, entirely as a consequence of the presence of capercaillie, there will be tolerance of a low level of spruce. There should be no more than 10% spruce in the canopy. Presently, as a consequence of past silvicultural management, there is approximately 30%, and this has been the case for around 100 years. Limits for Norway spruce are essential:

Factor: Norway spruce
Upper limit: 10% of the canopy
Lower limit: 5% of the canopy

Grazing and browsing

Animals, including roe deer, red deer and chamois, are thought to be a problem. The impact of these animals, focusing on the suppression of natural regeneration, should be monitored. Currently, there is a national surveillance project looking at the effect of browsing on sapling survival: this may have implications for Topla. Regardless of the national scheme, a local surveillance project should be established, and, eventually, this could lead to the formulation of a monitoring project. Attempts to measure the factor, the grazing animals, will be prohibitively expensive. In this case, the obvious alternative is to measure the attribute which will change as a consequence of the factor, i.e. regeneration.

Human activities

The forest is managed in accordance with the Slovenian National Forest Plan and a local Forest Management Plan. The Forest Service, which has considerable powers, can control and prescribe all the management work within this forest. There will be a requirement for some negotiation to resolve any potential conflict with the Natura 2000 plan. Some of this feature falls within a protected zone which means that there can be no timber extraction because the trees protect farms lower down the slopes from avalanches. It is not possible to establish limits for activities. However, surveillance or compliance monitoring should be introduced to ensure that all management operations and other human activities are recorded.

Attributes

Forest cover

The forest covers at least 35% of the total area of the pSAC. This was the extent at designation and is the minimum area permitted. The distribution of the forest is shown on the attached map.

Attribute: Area covered by the beech forest
Upper limit: (This will be determined by the lower limit for other pSAC habitats.)
Lower limit: 35% of the pSAC

Canopy species

This is a semi-natural woodland, and the canopy must be dominated by locally native species. (A full list of species is given in Appendix 1, below.)

Attribute: Canopy species
Upper limit: Not required
Lower limit: 90% of the canopy will comprise locally native species

Regeneration

Natural regeneration is essential. This can be suppressed by a number of different factors. Grazing has been recognised as a potentially significant factor. Three to four years following a good mast year for beech there should be sufficient saplings to ensure a reserve cohort of potential recruits for the canopy. A method for quantifying and measuring the saplings should be developed.

Attribute: Number of viable saplings
Upper limit: Not required
Lower limit: To be determined

Old veteran trees

Old veteran trees, mainly deciduous species (with limited tolerance for larch, Norway spruce and Scots pine), are an essential component of this forest. They provide opportunities for a wide range of fungi, lower plants and invertebrates. In this particular situation, they are also extremely important for the bird populations. Currently, the number of veterans is probably below an acceptable level. This is because of the long history of silvicultural management.

Attribute: Number of veteran trees
Upper limit: Not required
Lower limit: To be determined

Dead wood

The presence of dead wood, standing trees, dead limbs on live trees, and fallen trees and branches are an essential component of the woodland. It is an extremely useful surrogate, i.e. the presence of dead wood will indicate potential for the presence of a wide range of typical woodland species, including beetles, fungi, epiphytic lichens and hole-nesting birds. Dead wood is also a good measure of the ecological structure of a woodland: the presence of too little, or too much, dead wood will be indicative of a dysfunctional or inappropriately managed woodland. This type of woodland should contain between 15 and 20 cubic metres per hectare of dead wood (this value requires confirmation).

Attribute: Volume of dead wood per hectare
Upper limit: Not required
Lower limit: Provisional 15 cubic meters per hectare (to be confirmed)

Canopy gaps

The forest should contain a dynamic, shifting pattern of naturally-occurring, small, irregularly-shaped, open gaps in the canopy. This will ensure long-term structural diversity in the forest. The gaps can also be important for the capercaillie.

Attribute: Canopy gap creation rate
Upper limit: 0.5% of the canopy per annum
Lower limit: 0.25% of the canopy per annum

(These are provisional values for the gap creation rate. They should deliver a forest with a canopy cover of between 75 and 85%.)

Areas of un-fragmented, continuous forest cover

As a consequence of the natural distribution of this woodland community and the impact of generations of forest management, the woodland is fragmented, with few large areas of continuous cover. Capercaillie require large areas of un-fragmented forest, and the three-toed woodpecker requires areas of at least

300 ha. (Although the beech forest is not strictly suitable for the birds it is interspersed with areas of spruce forest which will also contribute to providing suitable habitat for this species.) The beech forest will form a component of the large, un-fragmented areas: there is little or no potential for large areas comprising only beech forest.

Attribute: Areas of un-fragmented continuous forest cover
Upper limit: Not required
Lower limit: To be determined (the number of areas of at least 300 ha of continuous cover)

2.4 Feature – Mountain Spruce Forest

Description

This woodland type is mainly a modified version of the original beech forest. For many generations, forest management has been one of the most important sources of income in the Topla valley. Foresters and farmers have gradually introduced alien species, mainly Norway spruce, to provide a commercial crop. The consequence is that the coniferous areas dominate the landscape.

The mountain spruce forest is not recognised as an important conservation feature. However, it became clear when considering the specific requirements of both capercaillie and three-toed woodpecker, along with the other SAC species, that the native beech forest cannot provide the full requirements of these species. Both are, in fact, reliant on the managed spruce forest, which together with the native beech forest forms a large-scale landscape mosaic. For this reason, a full objective with performance indicators has been prepared for the spruce forest. In other words, it will be treated as a habitat feature. The alternative would be to treat the individual woodland attributes as factors associated with each bird objective. For the sake of clarity, this example has focused on two SPA bird species: in reality, there are many more. The consequence of not treating the woodland as a feature would be a hopelessly repetitive and potentially confusing plan. In addition, the species will be protected through controlling forest management operations, and so it is essential that all these operations are brought together in a single location.

Objective

The managed spruce forest:

- Occupies at least the area shown on the site map.
- Has a diverse age structure.
- The areas where three-toed woodpecker nest contain a canopy which comprises at least 50% spruce. In the remainder of the forest, the canopy comprises at least 50% coniferous species, which include Norway spruce, Scots pine and larch.

- Over the entire area, the canopy comprises at least 25% native deciduous trees, which include beech, ash, rowan, willow and sycamore. These are maintained through natural regeneration with limited intervention management if this fails.
- Contains sufficient dead wood, both fallen and standing dead trees and limbs, to provide for the full range of associated species. There are sufficient dead conifers for the three-toed woodpecker.
- Contains sufficient trees with nest holes for the three-toed woodpecker. These will be surrounded by a high density of old spruce with a diameter of 20–30 cm dbh.
- Contains sufficient deciduous and coniferous veteran trees.
- Is a component of un-fragmented areas of forest that are at least 300 ha in extent.
- At higher altitudes, adjacent to the *Pinus mugo* communities, there is a belt, 200 m deep, that has an open canopy of about 20% with a dominance of Norway spruce with larch.
- Has a soft edge between the spruce/larch community and the *Pinus mugo* communities.
- Contains a dynamic, shifting pattern of small, irregularly-shaped, open canopy gaps. These may occur naturally, or they may be created during forest management operations.

Note: *The performance indicators, factors and attributes with limits follow exactly the same pattern as demonstrated in the preceding objective for the beech forest. Consequently, there is no purpose in repeating them in this section.*

3 Discussion

This case study demonstrates the relationship between species and habitat. The objectives for both areas of forest are influenced by the requirements of the birds. The beech woodland is a feature of the pSAC, and although there can be some compromise to meet the needs of the birds this cannot place the forest at risk. For example, there can be no more than 10% of conifers in the canopy. However, because both the SPA bird species are dependent on the managed spruce forest the objective for these areas is entirely guided by the requirements of the birds. This is not an uncommon situation as it reflects the reliance of so many important and threatened species on semi-natural or cultural landscapes, in this case a commercially managed forest. As an interesting aside: conservation managers readily recognise the importance of plagioclimatic or semi-natural vegetation, for example, grasslands and heath, regardless of the presence of important species. Therefore, there is no contradiction in extending the list of recognised features on a site to include managed forest.

Whenever a management plan is prepared for a site which contains protected species, the species objectives, with particular emphasis on their habitat requirements, should be completed *before* the habitat features are tackled.

Appendix 1:

Plant species of the Illyrian *Fagus sylvatica* forests:

Fagus sylvatica, F. moesiaca, Acer obtusatum, Ostrya carpinifolia, Abies alba, Quercus cerris, Sorbus graeca, Tilia tomentosa, Anemone trifolia, Aremonia agrimonioides, Calamintha grandiflora, Cardamine trifolia, C. waldsteinii, Corylus colurna, Cotoneaster tomentosa, Cyclamen purpurascens, Dentaria eneaphyllos, Dentaria enneaphyllos, Dentaria trifolia, Doronicum austriacum, Epimedium alpinum, Euphorbia carniolica, Hacquetia epipactis, Helleborus niger ssp. *niger, H. odorus, Knautia drymeia, Lamiukm orvala, Lamium orvala, Lonicera nigra, Omphalodes verna, Pancicia serbica, Primula vulgaris, R. hypoglossum, Ruscus* spp. *Saxifraga lasiophylla, Scopolia carniolica, Scrophularia scopolii, Sesleria autumnalis, Vicia oroboides.*

EC (2003). *Interpretation Manual of European Union Habitats*. European Commission DG Environment.

Case Study 4
Marsh Fritillaries at Rhos Llawrcwrt National Nature Reserve – An Example of Adaptable Planning

David Wheeler [1]

Abstract This case study is based on the management of a marsh fritillary butterfly *Eurodryas aurinia* population at Rhos Llawrcwrt National Nature Reserve from 1992 to 2006. It demonstrates the adaptable approach to planning and management as illustrated by the development of a conservation objective. The most significant changes in the objective are a consequence of developments in management planning and our understanding of what an objective should be. The objectives are also adapted to reflect improvements in science, and particularly a better understanding of the survival strategies and requirements of the species.

1 Site Description and History

Rhos Llawrcwrt National Nature Reserve (NNR) is an area of marshy grassland in mid Ceredigion, west Wales. The site supports an internationally important population of the marsh fritillary butterfly *Eurodryas aurinia*, which is one of the largest in the UK. The site is located approximately 9 km east of the coast and 1 km south west of the village of Talgarreg.

---

[1] The Countryside Council for Wales.

M. Alexander, *Management Planning for Nature Conservation: A Theoretical Basis & Practical Guide*, DOI 10.1007/978-94-007-5116-3,

First mentioned in 1214 in a charter granted to the Cistercian monks of Whitland, Llawrcwrt has been part of the local agricultural economy for at least nine centuries and probably much longer. The site is typical of many wet pastures in the county, having a long history of extensive grazing by cattle, horses and sheep, but also some cultivation, including planting of crops such as potatoes and black oats. Agricultural practice, often ad hoc in nature, has been fundamentally important in determining the nature of flora and faunal communities that exist today. Influenced by geology, geomorphology and other physical factors, this management has led to the development of an open wetland landscape relatively free of scrub and woodland. Llawrcwrt Farm first came to the attention of local naturalists and the Nature Conservancy in the early 1970s when it was recognised as an area of 'relatively unmodified' marshy grassland with associated diverse ranges of plants and invertebrates. Following further survey, part of the farm was notified as a Site of Special Scientific Interest (SSSI) in 1979, and in 1983 Grade 1 status in the Nature Conservation Review was confirmed. At this time, it was clear that the marsh fritillary population was the largest in the county and one of the largest in the UK. Extensions to the SSSI followed, and in 1985 the Nature Conservancy Council purchased most of the designated area, which was declared an NNR in 1986. Since declaration, the SSSI and the NNR have been further extended. The SSSI qualifying features are:

Marshy grassland
Neutral grassland
Marsh fritillary *Eurodryas aurinia*
Slender green feather moss *Hamatocaulis vernicosus*

Rhos Llawrcwrt is a Special Area of Conservation designated under the EU Habitats Directive for its populations of marsh fritillary and slender green feather moss *Hamatocaulis vernicosus.*

The NNR now covers a total of nearly 66 ha and is divided into two sections separated by c 200 m of improved pasture. The western block covers 54 ha and includes approximately 24 ha of rhos pasture and 26 ha of agriculturally improved grassland. The eastern block, known as Cors y Clettwr, covers approximately 12 ha, most of which is rhos pasture.

An inspection of any large-scale maps of south and west Wales will reveal that the word 'rhos' is a common component of many place names and has long been associated with the 'bogs' in valley bottoms that were so characteristic of the region. As an ecological term, rhos pasture is used for a specific mixture of vegetation communities. In south west England this mixture of communities is known as culm grassland. The following are extracts taken from the English Nature Wildlife Enhancement Scheme report *Management Guidelines for Culm Grassland* (1991):

> Culm Grassland is not easy to describe in terms of better-known wildlife habitats (which is why it has been given a name of its own), but in essence is a complex of wet acidic grassland, wet heath, fen and mire communities. Most sites contain several of these elements, each grading into one another to form a close-knit mosaic of distinctive appearance.

> In strict phytosociological terms, culm grassland (rhos pasture) is composed of a variety of mire vegetation communities. However it is aptly named 'Grassland' in lay terms because purple moor grass *Molinia caerulea* is characteristically constant at a high percentage cover and sites are traditionally grazed by domestic livestock.

The references to 'mosaic' in the English Nature definition of Culm grassland and the NVC survey are particularly significant. The NVC map for the western block of Rhos Llawrcwrt illustrates the fine scale of the mosaic at this site.

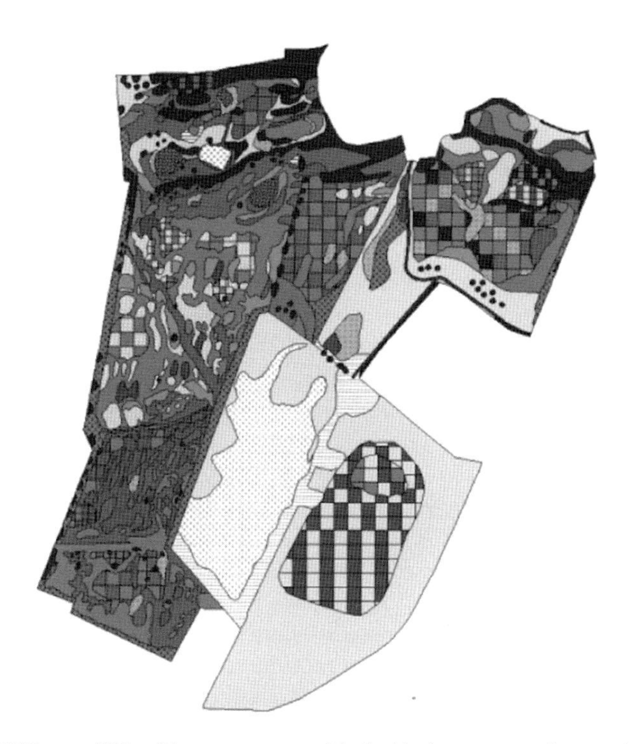

Fig. CS4.1 NVC map of Rhos Llawrcwrt western block: this demonstrates the complex vegetation mosaic.

2 Marsh fritillary

2.1 Ecology and underlying principles of management

The marsh fritillary butterfly is one of the nine British fritillary species. It is a reddish brown insect with cream, yellow and dark markings and has spiny larvae that are dark in colour. It is primarily a wetland species and has a very restricted range throughout Europe that in Wales is associated with marshy grassland habitat containing the larval food-plant devil's-bit scabious *Succisa pratensis*. Individual site populations are prone to wide fluctuations in size; there are tales from Ireland in the nineteenth century of massive 'outbreaks' of caterpillars causing villagers to barricade their homes and rake up huge piles for burning (Thomas and Lewington 1991). The butterfly occurs in groups of populations known as metapopulations, where each population is no more than c 5 km from another population. Occasional interchange of adults between individual populations reduces the occurrence of localised extinctions. Up to the 1930s it is very likely that the Rhos Llawrcwrt fritillaries would have been part of a large metapopulation.

In west Wales, adult emergence usually starts in the 3rd week in May and is finished by the 3rd week in June. The butterflies are very weak fliers and even mature hedges can act as barriers to movement. They are particularly susceptible to poor weather during the flight period. Wind and rain can reduce the chance of successful pairing and egg laying. The existence of natural shelter in the breeding habitat can be very important and, in years when the weather is poor and the population is low, may be a critical factor in successful breeding.

Adults are rarely on the wing after the 1st week in July. Eggs are laid in rows on the underside of leaves of the food-plant and hatch after 3 weeks. The caterpillars immediately spin a dense silken web within which they feed until all leaves are devoured. They then move en masse until the next plant is located and repeat the process again. This continues until late August when the caterpillars spin a dense silk web deep in the vegetation, often in the base of grass tussocks. They spend the winter months in this web and reappear on the first warm days of spring, often as early as late February. As air temperatures are low at this time of year and the caterpillars must raise their body temperature to feed, they cluster together above the vegetation, sometimes on top of webs, to form a black mass of bodies that absorbs the heat of the sun. Feeding in webs continues until eventually they start to wander away from the webs and feed as individuals until pupation, which will normally start from mid-April. Optimum habitat structure may be described as a mix of both short sward vegetation and tussocky vegetation to provide areas for both basking and shelter. The desired sward height at the end of the grazing season is usually quoted as being between 8 and 25 cm. This structure can only be maintained by light summer grazing with cattle or ponies. Stocking rates need to be in the order of 0.2–0.4 lu/ha/annum (livestock units per hectare per annum), although this will need to vary from year to year to take account of vegetation growth.

Apart from the usual range of predators, including ground beetles and spiders, the marsh fritillary is parasitised by tiny parasitic wasps. In Britain there are two species, but in Wales the wasp *Apantiles bignelli* is the main parasite. This wasp injects eggs into the caterpillars; wasp grubs emerge from these eggs and feed on the butterfly caterpillars until they are ready to spin their own cocoons and finally emerge as adult wasps. Up to 70 wasps may emerge from a single caterpillar and each generation of butterfly caterpillars may be host to up to 3 generations of wasps. The second generation of wasp grubs over-winters in the butterfly caterpillars. The wasp can have a devastating effect on fritillary populations, killing up to 75% of caterpillars in some years.

The relationship between the parasite and fluctuations in the butterfly population is not fully understood, but a reasonable model is based on a multi-year cycle where the wasp population expands to a point that causes a crash in the butterfly population which in turn causes a crash in the parasite population. This host-parasite cycle is affected by many other parameters, including the weather. Cool, sunny springs favour the butterfly, the caterpillars of which warm up quickly by communal basking. Wasp pupae develop much more slowly in lower temperatures, and wasps emerge too late to parasitise the butterfly caterpillars before they pupate. A cycle of fixed length between peaks in butterfly population is therefore possible but unlikely. As is the case with most host-parasite relationships, the parasite alone is unlikely ever to be the cause of long-term decline or extinction in the host.

For a site manager, the challenge is to distinguish responses in the butterfly population to 'natural factors', such as the parasite or climate, from factors over which the site manager is able to exert some control, such as habitat structure and quality.

2.2 *Status of marsh fritillary in 1983*

It was not until 1983 that any data on the status of the butterfly at Rhos Llawrcwrt was collected. Adrian Fowles (CCW entomologist) surveyed all known marsh fritillary sites in Ceredigion, using direct counts for all adults at small sites and counts along a series of parallel transects over suitable habitat at larger sites (Fowles 1983). Fowles recorded 413 adults at Rhos Llawrcwrt, by far the highest count in the survey that year, representing c 75% of the total count for all sites. Extrapolation of the count data indicated that a conservative estimate of total marsh fritillaries flying on the site on the count day would be in the order of 700. Assuming that was a peak figure for the entire emergence period, it was suggested that in excess of 2000 adults emerged at Rhos Llawrcwrt in 1983 (Fowles 1983). However, the Ceredigion-wide survey also confirmed that Rhos Llawrcwrt was an isolated site and not part of a larger metapopulation. It was therefore very vulnerable.

2.3 *Establishment of permanent transects in 1984*

In 1984, Fowles established a Pollard Walk to collect data on adult butterflies, following the methodology described in the National Butterfly Monitoring Scheme (Pollard 1977). This was based on a transect route in the western block of the NNR, approximately 1.7 km long, that was sampled throughout the flight period. From 1985, this route was walked once each week during the flight period, but only with climatic parameters in which the butterflies would be active. All adult butterflies within a defined distance from the recorder were counted. The total number of butterflies counted for all weeks was than recorded as an annual adult index. Fowles also set out a series of 53 100 m^2 transects to collect data on larval webs. These web transects were located so as to representatively sample all potential breeding habitat available in the western block at the time. The transects were recorded during early September, and the total number of larval webs observed was recorded as an annual larval web index.

3 The Development of a Management Objective:

3.1 Late 1980s: The First Management Plan

The first NNR management plan was drafted during the late 1980s. A considerable amount of information on the site was brought together in this exercise. The description was a large and detailed document, but the section on evaluation and objectives was never completed. Objectives were written for the rhos pasture and other habitats. No specific objectives were written for the marsh fritillary, but management requirements were expressed as prescriptions under the rhos pasture objective.

Objective: 'To maintain and encourage the rhos pasture'. (This is typical of the period. It carries very little meaning, and there is no explanation of what is to be maintained or what 'encourage' actually means.)

Action plan (prescription):

- Maintain grazing at low levels throughout the year with no removal during flowering time in order that the Molinia sward is maintained.
- Monitor grazing regime by regular counts and mapping.
- Monitor vegetation through established quadrats and fixed-point photography.
- Monitor invertebrates communities and key rhos spp.
- Maintain butterfly transects and larval web counts of *E. aurinia*
- No burning
- No alterations to hydrology

The management and monitoring projects were identified but not described.

Please note that the term 'monitoring' is used inappropriately: in reality, these are surveillance projects (see Chapter 5).

3.2 1991–1992: The Development of First Fritillary Objective

In 1991 I was appointed reserve manager, and set about familiarising myself with the site and the existing plans. I made an assessment of the condition of the rhos pasture on the western block as marsh fritillary habitat and concluded that it was generally in poor condition. Although the average sward height was 17 cm in September 1991, there were some areas of rank sward above 30 cm in height, and there was more than 5 cm of dense plant litter over much of the pasture. Of particular concern were a number of large patches of purple moor grass *Molinia caerulea* that showed vigorous growth around the patch edge and appeared to be expanding. However, adult index data showed that the butterfly population had recovered from a crash following the high count in 1983, and the index of 1204 in 1991 was the highest it had been since 1984. The first larval web data since 1984 was recorded in 1991. In 1984, the web index was 95, which suggested a total of c 3870 webs on the site. In 1991, the web index was 154, confirming the high population.

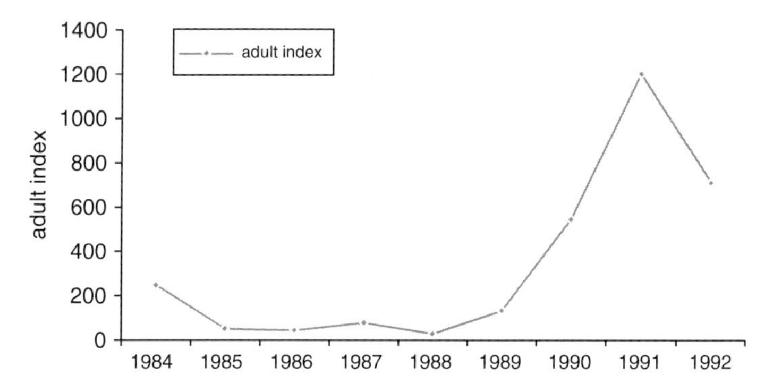

Fig. CS4.2 Marsh Fritillary adult index 1984–1992

I decided that larval web data should be collected annually. The web index would be a useful complement to the adult data. Much of the data would be collected by David Woolley, who was the immediate NNR neighbour and farmer at Llawrcwrt Farm. In this, I was extremely fortunate as David had an intimate knowledge of the site dating back to the early 1980s and was an enthusiastic amateur entomologist. David was also the main grazier for the NNR and was sympathetic to the needs of the conservation management programme.

A period of 'recovery management' was required. The NNR had been grazed lightly by cattle under grazing licence during the period June to September. Stocking rates for the western block for 1991 were only 0.14 lu/ha/annum. I determined to increase the stocking levels, but the condition of the pasture made this an unattractive prospect for graziers, and no extra cattle were available. I opted to use winter grazing by ponies to break up the litter layer, and in the winters of 1991 and 1992 a herd of up to 20 ponies spent c 4 weeks on the western block during December/January. This was only possible because there was suitable fall-back land on the NNR in the form of agriculturally improved pasture. The ponies spent the majority of the winter months here, and without the improved pasture there was insufficient incentive to attract the pony grazier to stock the NNR.

My first revision of the management plan started in 1992 and was written in the Conservation Management System (CMS) (see Case Study 5). An objective was produced for each of the SSSI features. Guidance at the time did not require that the objective described a required condition for the feature and so the use of a standard phrase 'maintain and enhance' merely recognised that the status of each feature should be maintained and implied that, if possible, it should improve. Attached to this objective were a number of surveillance projects. Sward height data and data on the status of the larval food-plant had also been collected from the larval web transects. This data had the potential to inform management and so it was logical to attach this to the objective statement.

1992 marsh fritillary objective: 'To maintain and enhance the marsh fritillary population at Rhos Llawrcwrt.'

Prescription 1. Collect annual data on the status of the butterfly and its habitat.

Projects:

- Surveillance of annual status of marsh fritillary adults (index calculated from data from transect)
- Surveillance of annual status of marsh fritillary larval webs (index calculated from data from 53 transects)
- Surveillance of sward height (index calculated from data from 53 transects)
- Surveillance of marsh fritillary larval food-plant (index calculated from data from 53 transects)

Prescription 2. Manage the vegetation by a controlled grazing programme.

Projects:

- Grazing programme and records – management of stock and grazing levels to maintain the appropriate vegetation structure.

Prescription 3. Provide shelter for egg-laying adults

Projects:

- Manage shelterbelts and hedges – establishment and maintenance of a system of shelterbelts and hedges around all field boundaries.

Objectives for the marshy grassland and neutral grassland made no reference to marsh fritillary, and there was, therefore, a very poor connection in the plan between the species and the management of its habitat.

3.3 1994–95: First Objective Revision

In the early 1990s there was relatively little in the literature on habitat management for marsh fritillary, and, following the 1983 survey, I was very aware of the potential fragility of the isolated population at Rhos Llawrcwrt. I had initially been concerned that increasing stocking rates might have an adverse effect on the butterfly population, particularly as, after the first winter of pony grazing, the 1992 adult index fell to 711 from the 1991 value of 1204. However, the larval index increased from 154 in 1991 to 236 in 1992. Clearly, the index data needed careful interpretation, but the population did not appear to be suffering from the change in grazing management. In 1993 and 1994, both indexes remained high. Superficial observation and some simple surveillance of vegetation structure told me that the average sward height was decreasing, the litter layer was breaking up and the larger patches of Molinia were no longer increasing in area. All this data gave me the confidence to continue with the winter grazing.

By 1994, the concepts of 'favourable condition', 'attributes' and 'limits' had been introduced to management planning. My next revision of the plan needed to incorporate these into the objectives. Favourable condition for marsh fritillary needed to take account of the cycle of the population in response to the parasitic wasp and other natural factors.

Population size was the obvious attribute to utilise, and I was fortunate in already having data on this attribute in the form of adult and web indexes. The reliability of adult index data was more sensitive to conditions at the time of collection: poor weather for a couple of weeks during the flight period might prevent adults from flying and suppress the annual index to a point where it was an unreliable indicator of population size. The accuracy of any adult count also relied heavily on the competence and skills of the observer. Conversely, the data collection window for larval webs was relatively long and so optimum conditions could be selected. Webs were relatively easy to locate and they did not fly away before they could be identified. I, therefore, opted to use the larval web index as an attribute of population size. The collection of adult index data would continue as a surveillance project. Based on the precautionary principle, I decided to set a lower limit for the attribute of 50 webs. If in any year the index fell below 50, I would critically examine the management and quality of habitat to identify potential problems. The larval web surveillance project therefore became a monitoring project. Planning guidance introduced the concept of limits that were used to express a range for any of the factors that may affect a feature. For the first time, the plan could include some quantified requirements for the habitat. I selected abundance of larval food-plant and average sward height as factors for which I could set operational limits. There were already surveillance projects for these two factors, and I could use historic data to support the identification of limits. There was still no reference to the habitat requirements of the butterfly in the vegetation feature objectives.

1995 marsh fritillary objective: 'Maintain the marsh fritillary population in favourable condition where the annual larval web index is greater than 50.'

Current condition: Unfavourable

Factor – larval food-plant: Devil's-bit scabious *Succisa pratensis* is the sole food-plant of the marsh fritillary on Rhos Llawrcwrt NNR. Given the wide fluctuations in butterfly population size, it is fundamentally important that a large scabious population is maintained in a healthy state across the western block, which contains the core breeding areas of the butterfly. Appropriate grazing is essential to maintain the scabious in a healthy condition, where it grows in open sward conditions, unshaded by other plants. This maintains the plant in a prostrate form with multiple rosettes of leaves growing from a short stem, which are suited to the feeding requirements of the larvae. In 1991, a peak in the size of the butterfly population occurred. In September 1991, surveillance data was collected from a series of 53 transects which were representative of butterfly breeding habitat across the western block. Despite a very large larval population, the food-plant was not exhausted by late September. Transect data gave a percentage index of 37, representing the abundance of the food-plant. It is therefore reasonable to set an operational limit of 30 for the percentage index.

Factor: The % index for the abundance of *Succisa pratensis* on 53 transects representative of marsh fritillary habitat on the western block:
Upper limit: not required
Lower limit: not less than 30.

Factor – average sward height of the rhos pasture in the western block: The sward in rhos pasture breeding habitat must not be allowed to become tall and rank. Rank swards inhibit the growth of the larval food-plant and prevent the basking activity of the larvae, which is crucial during cool springs to promote feeding activity. Conversely, very short, over-grazed sward will provide little shelter in poor weather for larvae and adults. Over grazing may also adversely affect the status of scabious. The identification of limits for sward height is based on historical data collected from a series of 53 transects which are representative of butterfly breeding habitat across the western block, evidence from other marsh fritillary sites, and discussion with colleagues and other site managers.

Factor: The average sward height on 53 transects representative of marsh fritillary habitat on the western block will be as follows:
Upper limit: 25 cm
Lower limit: 8 cm

The management prescriptions described in the 1992 objective remained unchanged.

The revision of the management plan in spring 1995 was immediately followed by a spectacular crash in the larval web and adult indexes.

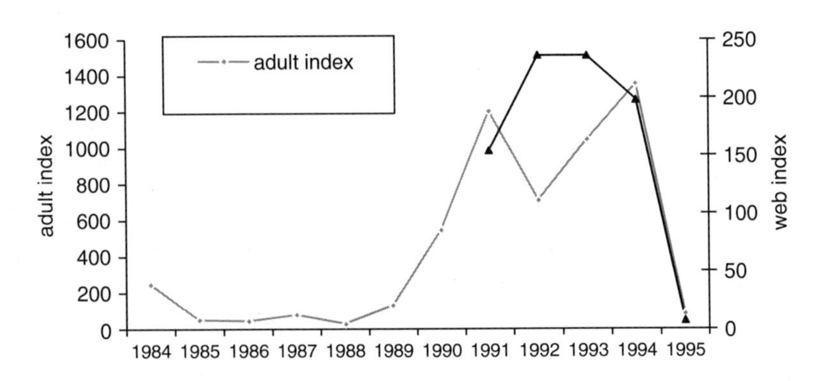

Fig. CS4.3 Marsh Fritillary adult and larval web indexes 1984–1995

The adult index indicated population levels similar to the mid 1980s. The larval web index was 7, well below the lower limit, triggering a response from me to attempt to establish what had caused the crash.

- Based on casual observation and subjective assessment, parasite cocoons on the butterfly larvae had not been obviously abundant in September 1994 at the time of web data collection.
- I had access to reasonably local meteorological data and I checked the records for the 1994 adult flight period. As far as I was able to judge, conditions were generally suitable for most of the flight period.
- Grazing levels on the western block had been gradually rising since 1991, with a subsequent fall in the average height of the sward: in September 1994 average height was 11 cm. The previously dense litter layer was far less in evidence.
- The larval food-plant surveillance data indicated that abundance had decreased slightly but not, in my view, enough to cause the butterfly population to crash.

I concluded that the habitat was in generally good condition and, although I had no evidence, I could only assume that the most likely cause of the crash was the parasitic wasp. As confident that I could be that management was still appropriate, I continued with the grazing programme, stocking at between 0.5 and 0.6 lu/ha/annum until 1997.

By summer 1997, the various patches of Molinia had contracted in size and the grazing programme could enter a maintenance phase, with reduced stocking levels and no winter grazing by ponies after 1998. In 1999, there was finally evidence of a rise in the butterfly population with the larval web index reaching 55.

I had been concerned for a while that the meteorological data I had access to was rather inadequate. I needed to review critical periods in the annual butterfly life cycle to determine whether weather conditions may have had a significant influence on breeding success. Finally, in early 2000, funding became available to install a fully automatic weather station on site. This measured all the relevant parameters, including air temperature, wind direction and speed, and rainfall.

In the summer of 2000, very large numbers of adults emerged and, for the first time, I got a hint of the 'plagues' described in historical references. Clouds of marsh fritillaries were everywhere on the NNR, with many individuals being observed in the surrounding improved fields. Thousands of larvae ate their way through the available devil's-bit scabious, and some then wandered off into surrounding improved pasture in search of more. Neighbouring farmers were in no danger of needing rakes, but, clearly, this was a year when the Rhos Llawrcwrt population could have contributed to adjacent populations – had they existed!

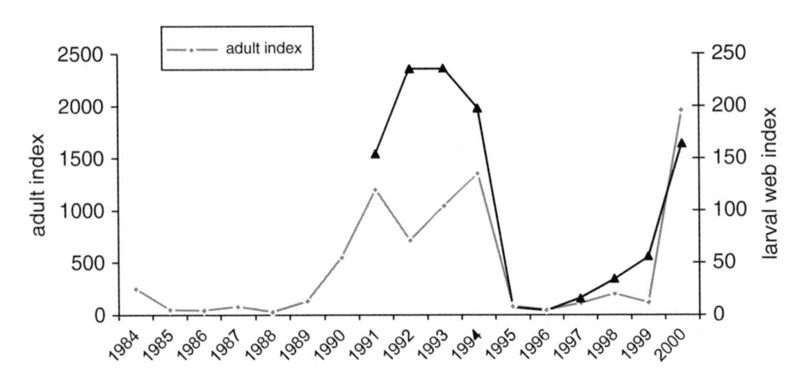

Fig. CS4.4 Marsh fritillary adult and larval web indexes 1984–2000

The adult index was 1957, the highest yet recorded. The larval web index in the autumn was 164. Interestingly, the web index had been higher in the previous peak years of 1992–1994, although the adult index was lower throughout that period. Data from the new meteorological station confirmed that the weather through the 2000 flight period had been mainly warm and sunny with little wind, and so there was no reason to suppose that egg-laying had been suppressed because of poor conditions. This demonstrated that indexes needed careful interpretation when comparing annual data and that their value is in providing broad indications of population trends.

3.4 2000: Second Objective Revision

I started a further revision of the management plan in 2000. Planning guidance now focused on 'Favourable Conservation Status'. I had by now realised the shortcomings of previous objectives in respect of the butterfly population cycle. I wanted a revised objective to take account of the variable population size over time, but it still had to reflect the viable status of the butterfly on the site. I now had larval web data which spanned a full cycle of the butterfly population from the peak in 1991–1994, through the years of very low web numbers in 1995–1997, and then the gradual rise to the next peak in 2000. During the previous 10 year cycle, the index fluctuated between 4 and 236. On the basis of pragmatism, I selected a 'mid' point representing a lower limit to be exceeded for at least half of the cycle. Factors and management prescriptions remained unchanged from the 1995 objective. A revised layout was adopted.

2000 marsh fritillary objective: To maintain the marsh fritillary population at Rhos Llawrcwrt in favourable condition where:

Performance indicators:
Attribute – Larval web index:
Upper limit: not required

Lower limit:	The larval web index will exceed 100 for 5 years in any 10 year period.
Factor – Status of food-plant:	The % index for the abundance of *Succisa pratensis* on 53 transects representative of marsh fritillary habitat on the western block
Upper limit:	not required
Lower limit:	not less than 30
Factor – Average sward height:	The average sward height on 53 transects representative of marsh fritillary habitat on the western block will be as follows:
Upper limit:	25 cm
Lower limit:	8 cm

Status of Feature: Favourable Maintained

Between 2000 and 2005, the fritillary population crashed and, again, I had to re-examine management practices.

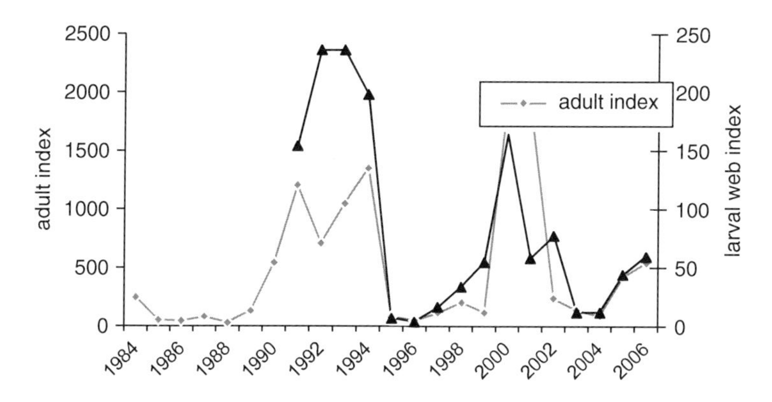

Fig. CS4.5 Marsh fritillary adult and larval web indexes 1984–2006

Remaining confident that management was appropriate on the western block, I was now anxious to pursue the reversion of agriculturally improved fields in the NNR that had the potential to revert to a rhos pasture type vegetation. These fields had been used to manage the grazing programme up to 2000, taking stock coming off the 'unimproved rhos pasture' when required. Ongoing research into appropriate reversion techniques was indicating that at least some improved land had the potential to revert to rhos pasture and marsh fritillary habitat. A programme of management was based on biomass removal to reduce nutrient levels. The long-term aim was to revert all improved pasture and thereby significantly extend the amount of marsh fritillary habitat on the NNR. The winter grazing of the improved pasture by ponies in the 1990s had created a certain amount of poaching, leading to the spread of soft rush *Juncus effusus* in some fields. This proved difficult to control and required intensive cutting programmes followed by grazing of the aftermath.

3.5 2005: Third Objective Revision

By 2005, further research on marsh fritillary had been completed. One of the most significant outcomes was an understanding of the extent of the habitat that a population requires to ensure long-term viability. The management planning guidance was also revised, and the objective structure was changed to make it more accessible through the inclusion of a 'vision'.

The 2005 revision of the management plan, in accordance with new guidance notes (Fowles 2004), took account of the entire NNR/SAC. Previously, the objective had focused on the western block because it contained the majority of the butterfly population. The revision continued to build on past objectives and experience.

2005 marsh fritillary objective:

Vision for the marsh fritillary population

There will be a very large butterfly population at Rhos Llawrcwrt which will be viable in the long term. Because the marsh fritillary is parasitised by a wasp, the number of butterflies in the population will vary over a cycle of several years, but, during the peak years, a visitor taking a walk through the site on a sunny day in June will see several hundred adult butterflies. In these years, the caterpillars, feeding communally in silken webs on their food-plant, devil's bit scabious, will be found in their thousands throughout large areas of Llawrcwrt and Cors y Clettwr.

Rosettes of the food-plant will be both very numerous and widespread throughout the cattle-grazed rhos pasture, growing amongst a short turf of grasses, sedges and flowering herbs, with scattered tussocks of purple moor grass and rushes providing shelter for the caterpillars in wet weather. This colourful wet grassland mosaic will extend throughout Llawrcwrt, Cors y Clettwr and the fields which were drained and reseeded for agriculture in the 1980s but have reverted back to rhos. Dense mixed hedges of hawthorn, hazel, mountain ash and other locally native species grow around the boundaries and between fields and offer vital shelter to the breeding adult butterflies during poor weather in what is otherwise a very exposed landscape with little shelter.

There are a number of smaller breeding populations of marsh fritillary on rhos pasture sites within 5 km of the National Nature Reserve. Butterflies from Rhos Llawrcwrt will occasionally visit and breed on these sites, and butterflies from the smaller populations will visit Rhos Llawrcwrt. This exchange of butterflies will help to keep all populations in a healthy condition.

Performance indicators

Factors and limits

Guidance notes from Adrian Fowles, CCW Senior Invertebrate Ecologist (Fowles 2004):

> Research on the marsh fritillary suggests that the species requires a minimum area of suitable habitat on a metapopulation site to be viable in the long term.
>
> On rhos pasture habitats, marsh fritillary larvae will only feed on devil's-bit scabious; the abundance of this food-plant is directly related to the survival of larvae. Sward height also affects survival.

Very short swards are unsuitable as they provide no refuge for larvae in cold wet weather. Tall rank swards are also unsuitable because they do not support vigorous populations of the larval food-plant, devil's-bit scabious, and they do not provide basking opportunities for larvae. All larvae must bask in the sun to raise body temperature to allow feeding activity.

Sward structure and status of the larval food-plant in defining types of habitat quality:

Definition of Good Condition marsh fritillary habitat: Grassland with Molinia abundant where, for at least 80% of sampling points, the vegetation height is within the range of 10–20 cm. cm (when measured using a Borman's disc) and Succisa pratensis is present within a 1 m radius. Scrub (>0.5 m tall) covers no more than 10% of area.

Definition of Suitable marshy grassland: Stands of grassland where Succisa pratensis is present at lower frequencies but still widely distributed (>5% of sampling points) throughout the habitat patch and in which scrub (>0.5 m tall) covers no more than 25% of area. Alternatively, Succisa may be present at high density in close-cropped swards.

Note: Available habitat is defined as the total of good condition and suitable habitat.

There is very limited habitat in the landscape surrounding Rhos Llawrcwrt and the nearest occupied marsh fritillary site is approximately 7 km distant. There are a few former rhos pasture sites on neighbouring farms which have potential for reversion, but all these sites are small. CCW owns land adjacent to the SAC which is currently improved pasture but which has some potential for reversion to rhos pasture. However, reversion management will take many years and the focus for management must be on maximising marsh fritillary habitat on the SAC.

The definitions for 'good' and 'suitable' habitat may be used at Rhos Llawrcwrt with a single modification. Sward data collected from the parts of the site which have supported a strong population of the butterfly historically indicate that a sward with a height range of between 12 and 25 cm is more appropriate. The definition of good condition habitat becomes:

Grassland where for at least 80% of sample points the vegetation is within the range 12–25 cm and Succisa is present within a 1 m radius. Scrub >0.5 m tall cover no more than 5% of the area.

Factor: **Extent and distribution of marsh fritillary habitat**
Upper limit: not required
Lower limit: Within the SAC boundary:
26 ha of available habitat (the likely distribution is c 9 ha in the core compartments, c 12 ha in the low density compartments and c 5 ha in the compartments for reversion), including 10 ha of good condition habitat.

Factor: **Quality of marsh fritillary habitat**
Upper limit: not required
Lower limit: 50% of the marsh fritillary habitat within compartments 5 and 6 is described as 'good condition habitat with dense *Succisa*'

And

50% of the marsh fritillary habitat within compartment 14 is described as good condition

And

Scrub covers less than 10% of compartment 15

Attributes and limits

Population size: This is the only attribute that may be used as a performance indicator for this species. Other attributes of quality, such as productivity and sex ratio, are difficult and time consuming to measure. In relation to determination of population size, it is most appropriate to consider the larval stage. Eggs are difficult to find and, for this reason alone, abundance cannot be systematically evaluated. Data collected on adult butterflies cannot always be relied on to indicate population size because of observer difficulties related to the mobility of the butterflies. The abundance of larval webs will therefore be considered as the sole attribute for this species.

Abundance of larval webs: Larvae feed communally in webs, which are easily observed and reasonably static. Historically, a large amount of sample data on webs has been collected from a series of 53 transects in the current 'core' area for the butterfly on site – compartments 8,5,6,7 and 10. Data has been analysed annually and an 'index' calculated which represents the total number of webs on all transects observed in any one year. In order to allow comparison with historical data, the performance indicators will include reference to the annual index collected on the 53 transects.

In addition, the performance indicators must take account of web abundance elsewhere on site, namely compartments 2, 3, 4, 12 and 14, which have historically been occupied at low density, and compartments 16 and 18, which are currently semi-improved grassland but have potential for reversion.

Because of the host/parasite relationship with the wasp, the marsh fritillary population fluctuates significantly over time. Historical data from the site indicates that there is approximately a 10 year cycle between population peaks. If the butterfly population were not being affected adversely by other factors, it should be relatively high for half of that period.

The following site-specific performance indicator has been devised:

Attribute: **Larval webs**
Upper limit: not required
Lower limit: over any ten year period:

The web index count will be greater than 50 for at least 5 of the years.
And
Annually:

- Larval webs continue to be present in all management compartments
- A minimum total of 50 webs are present in compartments 12 & 14
- A minimum total of 50 webs are present in compartments 16, 18, 22 and 24

Annual data on adult abundance will be collected as a surveillance project. Although adult data is not as reliable as larval data, it may still be used to confirm that any annual cohort of larvae are successfully producing adult butterflies and that there is a direct relationship between trends in the abundance of larvae and adults.

Status of marsh fritillary: unfavourable recovering

The 2005 revision of Favourable Conservation Status meant that, for the first time since 1992, the status of the feature was deemed to be unfavourable. This is because there is insufficient 'good' and 'suitable' habitat on the NNR at present. However, there is potential to increase the area of both categories of habitat through reversion of the improved pastures on the NNR and achieve Favourable Conservation Status in the future.

4 Summary

Management of the marsh fritillary population must take account of changes in the butterfly population, changes in factors affecting the population and new research informing understanding of the ecology of the butterfly. An adaptable approach to management is therefore essential. Management objectives must also be adaptable in response to better understanding of ecological systems.

The site manager will need to balance short-term benefit against long-term gain. For example, in retrospect, I would like to have avoided the winter poaching of the improved fields that resulted in the spread of *Juncus*. Science is an essential management tool. It supports important judgements on management. However, the site manager must interpret scientific data very carefully. Setting of stocking levels was critical to the management of the habitat on the NNR. The graph below shows stocking rates and vegetation height and demonstrates that the relationship between the two is not straightforward. At low stocking levels, stock do not graze an entire enclosure evenly but preferentially select certain areas. The response of the vegetation to grazing is not the same every year, mainly because growth rates are influenced by climate conditions. A large element of informed judgement is often required.

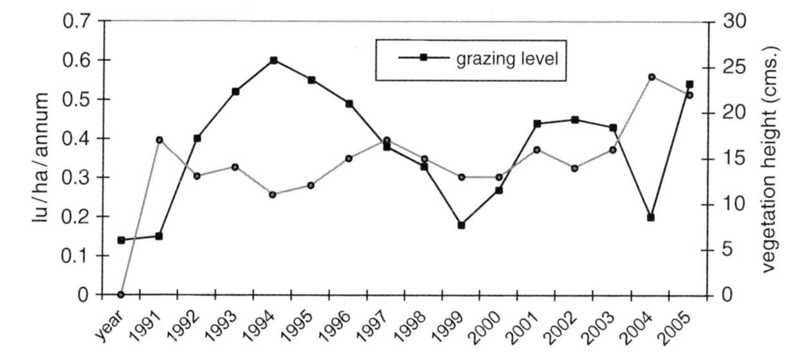

Fig. CS4.6 Grazing Level and Vegetation Height 1992–2006

Index data from 2005 to 2011 shows that the butterfly population continued at relatively low levels with the larval web index varying between 17 and 60. Poor weather during the adult flight period, which affects successful pairing, is a possible explanation for the lack of further peaks in population.

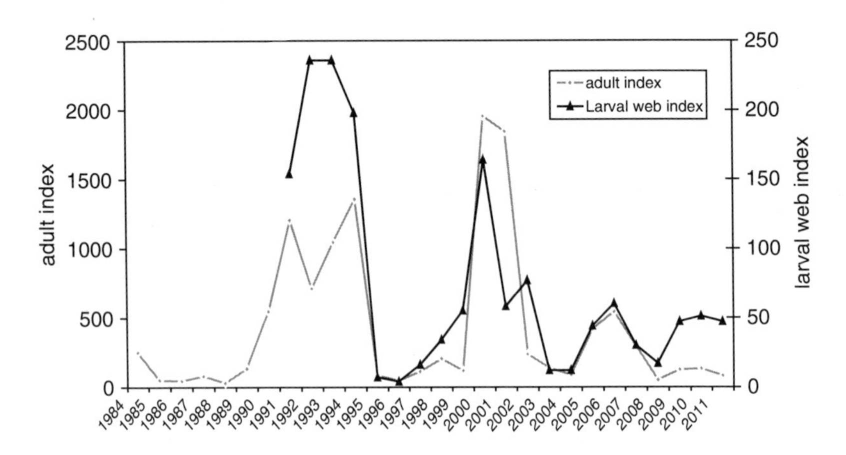

Fig. CS4.7 Marsh fritillary adult and larval web indexes 1984–2011

From 2006, management effort, in the form of biomass removal to reduce nutrient status, has been focussed on reversion of 24 ha of improved pastures, with considerable success achieved in one area. This field was a rye grass and clover sward in 1990 and in 2011 had more than 100 vascular plant species associated with marshy grassland including an established population of *Succisa pratensis*. Although the sward structure is still insufficiently developed, marsh fritillary adults were prospecting for egg-laying sites here in 2010 and 2011.

Because breeding activity outside of the core areas is most likely to be initiated during peak breeding years, for example, 1991–94 and 2000 at this site, I am now interested in investigating options for using such peaks as a means of encouraging breeding in reverted habitat, particularly through larval migration. A problem with maintenance management is that the actions of large grazing animals suppress the food-plant by eating it and the butterfly by treading/lying on the larvae. In this context, and perhaps unusually, 'maintenance management' for marsh fritillary produces sub-optimal breeding conditions. It may be possible to encourage localised increases in breeding success by temporarily excluding cattle from parts of the core breeding area. If these un-grazed areas are appropriately located, some larval migration to the reversion pastures may occur. This management option is not without risk and clearly cattle would have to be reintroduced before dense tussocks of Juncus and Molinia started to dominate areas.

Given that the existing population is still one of the largest in the UK, it would be very easy to be complacent and be content with the population being confined to the current areas of unimproved rhos pasture on the western block. This would be nothing short of foolhardy. There is an absolute requirement to maximise breeding habitat for the butterfly and ensure that it is populated at this site. Who knows what the future will bring and what it will mean for this butterfly? The precautionary principle must be applied and every effort made to increase the long-term viability of the population at Rhos Llawrcwrt.

References

Bulman, C. R. (2001). *Conservation biology of the marsh fritillary butterfly Euphydryas aurinia.* Ph.D., University of Leeds.

Fowles A. P. (1983). *Report of the Marsh Fritillary Survey of Ceredigion*, Unpublished report, Countryside Council for Wales, Bangor.

Fowles, A. P. (1984). *Population Studies of the Marsh Fritillary Colony at Rhos Llawrcwrt NNR, Dyfed*, Unpublished report, Countryside Council for Wales, Bangor.

Fowles, A. P. (1991). *Aspects of Monitoring at Rhos Llawrcwrt NNR*, Unpublished report, Countryside Council for Wales, Bangor.

Fowles, A. P. (2004). *Conservation objective for marsh fritillaries on marshy grassland.* Unpublished report, Countryside Council for Wales, Bangor.

Fowles, A. P. (2005). *Habitat quality mapping for marsh fritillary populations.* CCW Staff Science Report No. 05/5/1. Unpublished report, Countryside Council for Wales, Bangor.

Pollard, E. (1977). *A method for assessing changes in the abundance of butterflies.* Biol. Conserv, 12, 115–134

Robertson, J. and Wheeler, D. (2002). *Reserve focus: Rhos Llawrcwrt NNR, Ceredigion.* British Wildlife 13 (3), 171–176.

Thomas, J. and Lewington, R. (1991). *The Butterflies of Britain and Ireland.* Dorling Kindersley, London.

Wilkinson, K. (2005). *Rhos Llawrcwrt SAC/NNR marsh fritillary SAC monitoring 2005.* Unpublished report, Countryside Council for Wales, Bangor.

Woolley, D. (1991 – 2006). *Rhos Llawrcwrt NNR marsh fritillary studies.* Unpublished report, Countryside Council for Wales, Bangor.

Case Study 5
Computers and Management Planning

James Perrins[1] and David Mitchel[1]

Abstract This case study is included to demonstrate the essential role that computers can play in the preparation of plans and, more significantly, the management of planning systems. The study focuses on the development and function of the standard UK system, the Conservation Management System (CMS).

1 Introduction

Many people end up in environmental management because of a love of the outdoors and the natural environment, so spending time behind a computer might seem something of an anathema. However, computers and the Internet are now becoming such an integral part of everybody's existence that they are impossible to ignore. It was Bill Clinton, in a speech at Berkeley University in 2002, who said:

> When I took the office of president in 1993, there were only 50 sites on the World Wide Web – five zero. When I left office, there were 350 million and rising. There are probably around 500 million now.

This just emphasises the rate of growth, but what underlies it is vastly increased public access to information. As environmentalists we find ourselves having to fight for resources, and an inevitable consequence of this is the need to publicise our activities to gain wider public support. The Internet is an incredibly powerful tool to make the wider public aware of what we are doing and why we are doing it. Too often, conflicts come about through misunderstanding caused primarily by a lack of communication. Thoughtful use of Information Technology can help to prevent this. Computers are, of course, not only useful for advertising and communicating what we are doing: they are essential to the day-to-day, hands-on running of a site, and probably everything in between.

---

[1] exeGesIS SDM Ltd. Talgarth, Powys, UK.2

M. Alexander, *Management Planning for Nature Conservation: A Theoretical Basis & Practical Guide*, DOI 10.1007/978-94-007-5116-3,

2 History of the Use of Computers for Management Planning

In order to understand the development of the use of computers in management planning in Britain it is necessary to go back to at least 1968 when the Nature Conservancy (NC) were discussing a system established by the Smithsonian Institution in Washington for recording natural phenomena. One of the key people in those early days was George Peterken, who wrote,

> …we were explaining the existing characteristics of these places in terms of their past, particularly past human activities. It seemed likely to me that we would need to do this on NNRs, so I developed the idea in association with South Wales region and Aston Rowant NNR. (NCC 1986)

These discussions led to the development of the Event Record System, which was intended to provide a standardised means of recording all significant events on the NNRs. By the late 1970s the event records were entered on a simple computer database. This first attempt at capturing site management information on computers in the UK involved mainframe computers; there was nothing else available at the time. The programme was written in a system called Prime Information and was intended to capture historic records. It had little, if anything, to do with planning, but it had some potential benefits in terms of centralising and making searchable a list of all notable events that had occurred on sites. The main problem was getting the information onto these central systems. Because the system used mainframe computers all the input terminals were based in central offices. Out-posted site staff had to complete the record cards laboriously by hand, and these were then sent to the offices where the information was entered in the computer. As there was little or no direct benefit to the individuals concerned with site management, inevitably, it became one of those jobs that never got completed.

The Event Record System was reviewed in 1986 (NCC). The review recognised the value of a standard, systematic approach to site recording but found many failings. The most significant were: recording was not linked to management plans, the classification of events was difficult, and recording was inconsistent and incomplete. The most important outcome of the review was the introduction in 1986 of the Project Recording System (PRS). PRS was, as its name implied, a simple recording system, only slightly more sophisticated than the original Event Record System. The project classification system (see project codes in Chapter 17) was introduced, and all the work and unplanned events on a site were recorded each year. The information was entered on paper forms and sent to central offices for entry in the database. The output or reports were a series of simple, printed standard lists. Unfortunately, they had little relevance to site management. The main failing was that, once again, the site managers were not aware of any direct advantage to their sites or to themselves and, as a consequence, were very reluctant users. There was an important lesson learned here. When people are expected to use a system it is essential that they gain significant and direct benefits for themselves. This is the only means of ensuring that they are sufficiently motivated to use the system.

During 1987, Mike Alexander developed the use of PRS as a paper-based system and working with a colleague, Tim Reed, attempted to implement the system on all the NNRs in Britain. They made one major and very significant change: PRS became PPRS, the Project Planning and Recording System. The link between recording and planning, recognised by Peterken in 1986, was established by simply preparing batches of the recording forms, using a photocopier, and, instead of using these to record information, they were used to describe or plan the intended project.

In late 1988, Mike Alexander and James Perrins, initially working in a private capacity, developed PPRS on a database management system called Advanced Revelation. Functionally the system was very much driven from the ground up, i.e. features were developed that the site managers wanted in the system. The information required for organisational purposes was captured incidentally as a component of the information that site managers collected for their own purposes. This was really the start of dynamic planning systems in the UK, and it can be claimed that all subsequent developments have simply been ongoing refinements of the planning process. Over the subsequent years, this system has been developed to keep pace with both user expectation and technological advances. It now uses Microsoft SQLServer as the underlying database and is now named Conservation Management System International (CMS*i*).

In addition to developing PPRS and the management planning process, this small team, Alexander, Perrins and Reed, also obtained substantial funding from British Petroleum and the Directorate General Environment, Nuclear Safety and Civil Protection (DG XI), Commission of the European Communities. The funding was used to finance development and training, and to purchase PCs for RSPB, The National Trust, The Wildlife Trusts, The Wildfowl and Wetland Trust and the Nature Conservancy Council. These were the first computers provided by any of these organisations for nature reserve management and probably the first used for professional conservation management in the UK. The computers were intended mainly to trial CMS, but they also demonstrated the value of computers as a general tool for conservation management.

3 CMS Consortium

The initial development of CMS was undertaken by James Perrins and Mike Alexander. It was, in part, financed by the Nature Conservancy Council (NCC). As CMS was developed, more and more organisations within the UK became interested in using the software and having an input in, or control over, its future development. In order to accommodate this interest and commitment, a partnership comprising the newly established UK country agencies and all the major nature conservation NGOs was established. James Perrins and Mike Alexander handed ownership of CMS to the partnership.

The Countryside Council for Wales (CCW) assumed a lead role in the partnership and between 1993 and 2004 provided an office and additional facilities.

Initially, the partnership employed staff to support and develop CMS. However, in 2004 it was felt that it was more cost effective to out-source this service to a commercial company. The partnership was re-established as a consortium and entered into an arrangement with a commercial company, exeGesIS Spatial Data Management.

2010 saw a major change in the development of the Consortium with three of the largest Dutch nature conservation organisations joining the Consortium, namely Staatsbosbeheer (www.staatsbosbeheer.nl), Natuurmonumenten (www.natuurmonumenten.nl) and de12Landschappen (http://www.de12landschappen.nl/). This brought fresh impetus both philosophically and financially to CMS and a new version (CMS*i*) was created complete with free in built GIS, role based customisation and editing controls, simple tools to translate the programme into any language and a strengthening of the corporate use of the programme whilst not compromising its use for site managers.

The CMS Consortium is now a not-for-profit limited company which supports, promotes and formally guides the development of the CMS software. In addition, the Consortium is committed to developing best practice in management planning and promoting management planning as an essential component of nature conservation and countryside management. Currently (2012), the Consortium comprises 10 members representing both the government and non-government sectors in the UK and the Netherlands. The work of the Consortium is directed by a small management board representing the members and users.

4 CMS Users

CMS is used by most of the major conservation and land management organisations in the United Kingdom and now the Netherlands. There are presently several hundred CMS software licenses issued to over 250 organisations. Although CMS has a long history in the UK, with its easy tools for language translation, it has also been used in a number of other countries including:

- The Netherlands
- France
- India
- Malaysia
- Costa Rica
- Estonia
- Slovenia
- Bulgaria
- Croatia
- Ukraine
- Canada

5 CMS Functionality

The aim of CMS*i* is to liberate a management plan from being a reference document, that may or may not be used, and turn it into a dynamic system which is central to site management. CMS*i* provides the tools that staff need to drive their work plans and to ensure that a management plan is an integral component of the day-to-day management of a site. More than this, by linking recording with planning the process becomes a powerful reporting tool. It then provides very significant benefits at an organisational level. Once plans from different sites can be brought together, reports which represent the organisational perspective and requirements can be produced with very little effort.

Learning from experience, or adaptive management, is the key to CMS*i*. By storing and linking the management plan with recorded work in a database, the manager can, at any time, review what has happened on a project over a defined period. Hard facts, not anecdotes, are at hand to allow them to review the effects of site management, learn from that experience and modify their actions if necessary. In other words, planning is a dynamic process which can always be kept up to date.

6 The Planning Process in CMS*i*

6.1 Information Management

Most plans begin with an information section, which contains details of relevant legislation and policy, followed by a general description. A very common problem within organisations is that the style and structure of every management plan can be different, often so different that it becomes difficult to compare plans from site to site. When CMS*i* is used, organisations can select a standard structure or, alternatively, develop their own individual approach to organising the information in a plan. They are then able to apply their chosen structure to all their sites. The benefits are significant: it is easier to share plans, or planning experience, between sites, and reporting at an organisational level becomes much easier and more effective. When following an established data structure, planners can concentrate on the information without being overwhelmed by the process. CMS*i* provides a simple template which is ready and easy to complete. The system is dynamic: changes can be made to the information when necessary. Whenever a printed version is required, a full and up-to-date version is always available.

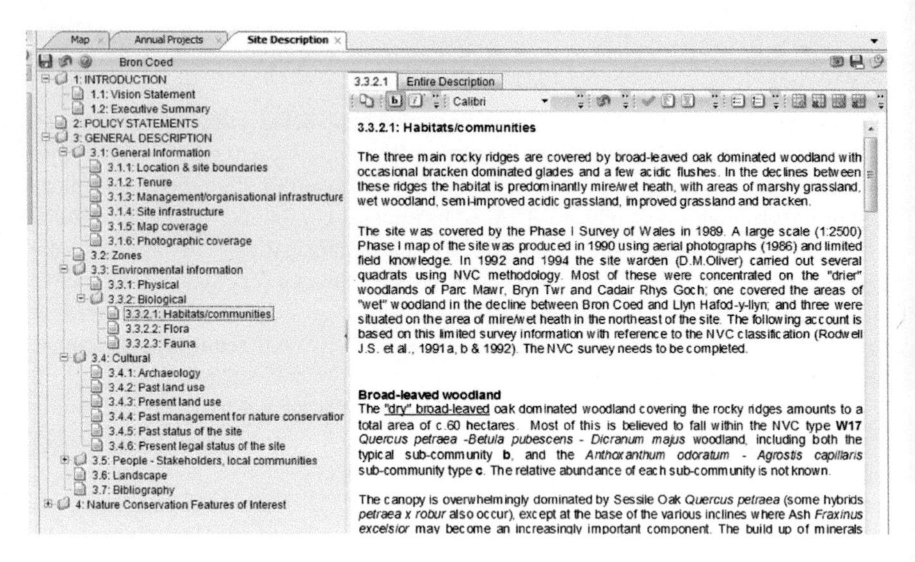

Fig. CS5.1 CMS*i* Site Description screen

6.2 *Features, Objectives and Action Plan*

CMS*i* contains all the information required to plan the features, including the associated objectives and performance indicators. This information is linked to the individual project plans and project records which, taken together, comprise the action plan. These components of CMS*i* are displayed and accessible through the planning tree, which provides a visual guide to aid both data entry and access. At any point on the tree, the user can click on a title to bring up full descriptions of the feature, objective, project plan or annual project report. The following example also shows how recording is integrated into the planning process and how data can be stored and presented in the system.

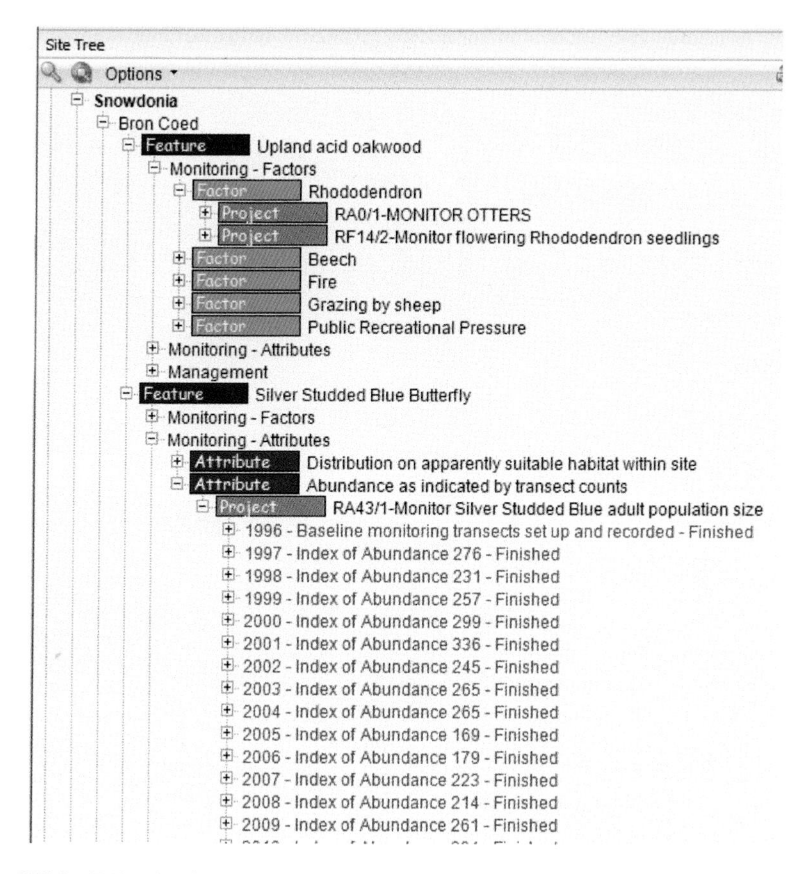

Fig. CS5.2 CMS*i* Planning Tree

The new ability to integrate mapping with planning is invaluable. So much site work is spatially relevant, so the ability to plan where work should happen whilst being able to see other GIS datasets is very important. Similarly, by confirming that the work did actually occur in that location or by digitising where it did happen, an historical record is built up spatially of where work occurred, which is much more efficient than trying to describe it in words.

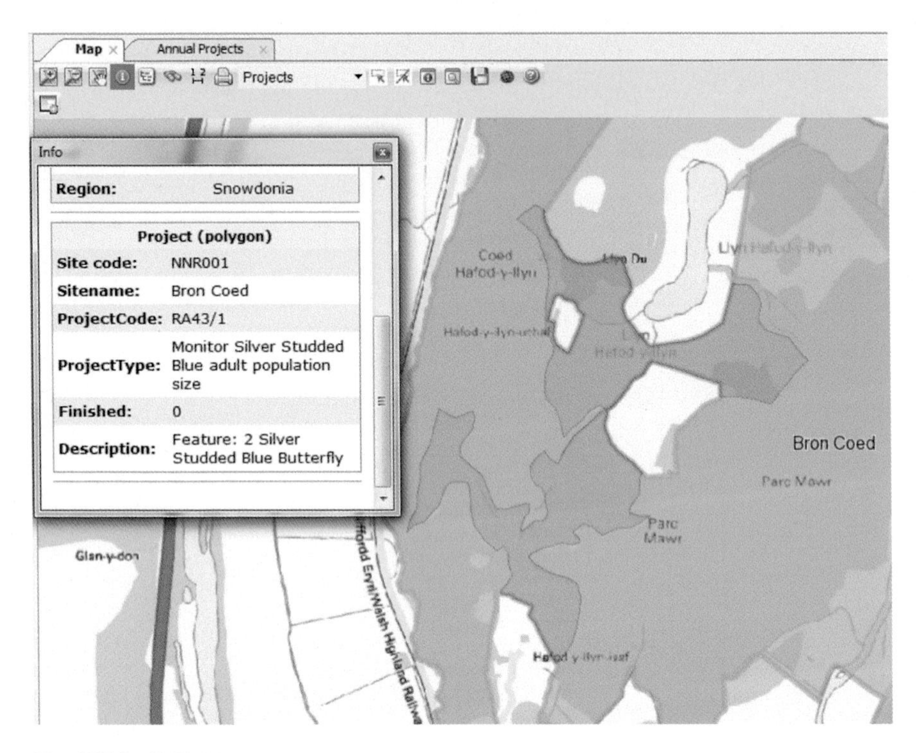

Fig. CS5.3 CMS*i* Map

6.3 Recording in CMSi

The integration of planning and recording is one of the key benefits of using CMS*i*. This can only be achieved if the amount of work required to maintain records is kept to a minimum. Hard-pressed staff cannot devote endless hours to recording. There are two quite distinct types of information required by site managers and the planning process:

The first are the simple records of day-to-day activities. These can be entered in the CMS*i* project diary, which provides a convenient and easily accessed facility for the entry of the sort of information that a reserve manager would traditionally hold in their personal notebooks. The difference is that these entries are fed directly into a system where they are securely stored and easily accessible. Users can also enter work records directly from the map, which has the added bonus of filtering all projects to those planned to occur at the point on the map where the user right clicks the map. This is one example of improving workflows in CMS*i*.

There is also an essential need to maintain a succinct and accurate summary account of progress on all management projects. If a manager has maintained a record of activities in the CMS*i* diary, albeit as a sequence of simple notes, this can be used to generate the summary account. This information is central to the adaptive management process. In addition, all site managers will be called on to prepare a variety of standard or special annual and other organisational reports. Information obtained from the diary is used to produce the summary reports for each project. These, in turn, are used to meet the data requirement for all other reports.

The ability for remotely based staff, volunteers or partner organisations to record work completed directly into the database via a web browser is a strong feature of the system. The user here logs on to the remote work recording tool and is presented with a list of sites and projects filtered to those allocated to them. They can then record their work done, print off work plans and read the methodologies for the projects they have to do.

6.4 CMSi as a Reporting Tool

CMS*i* has evolved as a tool for site management. There are numerous functions to aid site managers in answering questions about their sites. For example, a project tagging system allows any individual project to be linked to a particular funding source. This could be an external donor, a corporate target or a wider strategic plan. This allows these projects to be selected when reports are required. A manager can rapidly answer questions like, "What work have I done on externally funded projects over the last year?" or "What Biodiversity Action Plan work have I done over the last month?" The ability to rapidly produce this type of report for headquarters and external use can save site managers a significant amount of time and effort.

Another example of a function which is frequently required by site managers is the ability to prepare a wide variety of work programmes. This can be particularly important when a manager is responsible for a multitude of sites. It can be all too easy for key projects on smaller sites to be forgotten. However, by maintaining all the plans on CMS managers are always reminded of what they and their staff have to do.

Staff Work Plan

01/02/2007 To 28/02/2007

CMS

ESTATE WORKER

Site: Bron Coed

Project		Start Date	End Date	Anytime	Priority	Time (Days)	Allocated	A M J J A S O N D J F M
ME01 / 1	inspect / maintain boundary structures	01/04/2006	31/03/2007	No	1	3.00	No	
ME22 / 1	PP Inspect for and remove risks from trees on access routes	01/10/2006	31/03/2007	No	1	1.00	No	
ML30 / 2	RHODODENDRON on adjacent land	01/04/2006	31/03/2007	No	1	0.00	No	
MS00 / 1	RHODODENDRON CONTROL	01/06/2006	31/03/2007	No	1	4.00	Yes	
ME02 / 1	inspect / maintain / replace bridge	01/04/2006	31/03/2007	Yes	1	0.00	No	

Fig. CS5.4 A Work Plan report printed from CMS*i*

CMS*i* can be run at many different levels. Some organisations run CMS*i* corporately as well as locally. The most efficient way is to store the data on a centrally hosted Terminal or Citrix server, allowing users to log on remotely from different locations and efficiently use the same database. CMS*i* can also be run locally, if internet connections are not good enough, with the local plans merged later into a corporate read only data-volume: tools are provided to help this process. Once the corporate dataset is created, an organisation can run reports drawing information from all aspects of planning and management from all or any of their sites. Questions like, "How many of our statutory site features are at Favourable Conservation Status?" or "How much staff time have we spent on invasive alien control on our sites?" can be readily answered.

6.5 Sharing Data

We began this chapter by discussing the benefits of sharing information and the role the Internet can play in this. Up until now, however, we have been discussing ways to capture and manage data on a local or organisational scale.

In order to share data more widely, we need to move data onto the Internet. CMS*i* is not web-based, and it is tempting to assume that the reason for this is historic, in that it began life before the option of web-based systems became a reality. Whilst this is partly true, it does not present the whole picture. There is often a requirement to use CMS*i*, and similar systems, at fairly remote locations, where Internet access is poor, expensive or even non-existent. Until web access truly becomes available worldwide, at reasonable speeds and a sensible price, this is likely to remain the case. As a consequence, the only option available to systems that need to be used beyond conventional office environments is a stand-alone desktop component. However, this need not prevent the sharing of management planning experience over the Internet.

All too often, the same planning problem is encountered by different managers scattered around the country, who are not in communication with each other. Each will independently seek a solution, usually repeating, for no good reason, the work of others. The less experienced will possibly fail in their initial attempts. In addition, some organisations have an obligation to keep stakeholders, neighbours or their members informed about management on their sites. Finding a practical solution to these problems has been a long-term goal of the CMS Consortium.

Many organisations put their management plans on the Internet, but this is only a partial solution. The CMS Consortium is developing a system which will allow organisations to put all their CMS*i* management plans, in a fully searchable format, onto their intranet or Internet. This will allow anyone access to run a wide range of queries which interrogate the management plan data. It will, for example, allow community groups to learn from the approach taken by a large professional conservation organisation. Local people will be able to discover why work is being carried out on reserves in their area

The essential need to share management information has been recognised by other UK organisations and in particular:

- **Conservationevidence.com.** This website aims to share experience of management practice, describing successes and failures for others to learn from. The database has a worldwide series of case studies. It is project rather than site focused.
- **The Centre for Evidence Based Conservation at the University of Birmingham. (http://www.cebc.bham.ac.uk).** This centre conducts systematic reviews on the effectiveness of conservation management and policy.

7 Other Software Tools Available and Currently in Use

In addition to management planning software, there is an extremely wide range of software that can help the planning and management of sites. Many site managers might consider email as the single most beneficial computer based tool in use today. It provides an almost incredible ability for people to share experiences, regardless of nationality, location, time zones, etc. Similarly, a wide range of standard office software (word processors, spreadsheets, image handling software, etc.) is used far more widely by site managers than any of the software discussed in this study.

However, we will focus on the more specialist packages which are of particular relevance to site managers. In addition to planning systems, most site managers will have two specific requirements: a system for managing species records and a mapping or Geographic Information System (GIS).

7.1 Species Recording Software Used in the UK

Recorder

This program has been funded by the Joint Nature Conservation Committee (JNCC). It has a range of National Biodiversity Network (NBN) standards and concepts built into the program and links into the NBN species dictionary project. It is designed for small-scale biological recorders and large Local Record Centres or national schemes and societies. It incorporates mapping, allowing recorders to click on a map where they made an observation and thus eliminating one of the major sources of error in biological recording – the generation of a grid reference. For more information see (http://jncc.defra.gov.uk/page-4592).

MapMate

MapMate is another successful recording package with similar functionality to Recorder. Initially designed for moth recording, it is now widely used for all types of biological recording. For more information see (http://www.mapmate.co.uk.)

AditSite

This is a long running programme which, in common with MapMate, uses the NBN species dictionary. For more information see (http://www.aditsite.co.uk/intro.htm)

Indicia

A new web based tool aimed at collating species data via the internet funded by the OPAL network. See http://code.google.com/p/indicia/.

In reality, the choice of programme is not so important. The bigger issue is that the various packages must be able to talk to each other and to the National Biodiversity Network Gateway (http://www.searchnbn.net). This is a UK-wide Internet resource for viewing biological data which has been pooled from a wide variety of sources, including conservation organisations, local record centres and national schemes and societies. The NBN Gateway is rapidly becoming a very important resource for site managers.

7.2 Geographical Information Systems (GIS)

There is a range of information that can be conveyed most efficiently by using maps. The precise location of a species or habitat on a site is obviously essential information required by all site managers. There is a need to locate planned work and to record where it actually took place. Some sites contain sensitive areas where certain activities are excluded. Maps can be extremely useful when managing public access: there is often a need to provide a variety of access maps. Public liability and health and safety issues (for example, the need to provide locations of abandoned mine workings, cliff edges or deep water) often generate a need for maps.

The two most popular GIS systems are MapInfo (www.mapinfo.com) and ArcView (http://www.esri.com.), whilst OpenSource programmes like QuantumGIS and MapWindow are also becoming popular.

7.3 Integrating the Software Tools

Although there have been some attempts to develop a single programme that meets the wide diversity of management requirements on a site these have generally failed. They are extremely expensive to develop and maintain, often are totally bespoke to that organisation and are very vulnerable. The creation of links between the various software tools is an alternative and practical approach to unifying all the requirements of site management. CMS is an example of this approach. CMS can link with *Recorder* to generate site species lists or species lists related to particular projects. In this scenario, data are not entered in CMS but the reports are created in CMS without having to open *Recorder*.

The free internal mapping in CMS*i* allows the integration of a wide variety of spatial data directly in CMS*i* by the ability to load that data into the mapping. Hence, as an example, managers of sand dune sites can see the location of dune slacks and other habitat or monitoring datasets when planning habitat work on that site.

Whilst CMS*i* contains free internal mapping, some advanced GIS functions, like advanced editing, spatial queries, exporting to other GIS file formats, etc., are only possible through full GIS. With the addition of the linking software (MapLink), CMS*i* can link with MapInfo or ArcView. The map object can be highlighted in CMS*i,* and, when the MapLink button is clicked, MapInfo or ArcGIS opens centred on this object. Advanced editing can now be undertaken on the map object held directly in the CMS*i* database. Spatial queries can also be run and the results passed back into a CMS*i* filter or into the CMS*i* Sites Tree. Likewise, the CMS*i* layers can be displayed in full GIS outside of CMS*i* entirely, making the CMS*i* data available to a wider user group.

New tools in CMS*i* cater for deeper integration with other databases. The ability to export and import xml files from the database allows this data to be exchanged with other databases if they can accept data imports from this file format. In the Netherlands, for example, financial data is exchanged between the corporate finance system and CMS*i*.

8 Summary

CMS*i*:

- Enables the implementation of management planning as an adaptive process. Management plans become dynamic, working documents that are readily updated.
- Is a project recording system as well as a planning system, so it allows efficient reporting on work completed and the outcomes of that work.
- Can be used for a single site or as a powerful corporate tool for resource planning, monitoring and information sharing across any number of sites.

Computers encourage the use of standard approaches and layouts, making a management plan more readily understood by others. Once data are held in electronic format, the sharing of knowledge and approaches, both between managers and more widely with members of the public, becomes a much easier exercise.

Linking with mapping and GIS systems enables the management plan to be integrated with data from a wide range of other sources, potentially in completely unrelated formats.

Love them or hate them, computers are not going away, and our obligations to explain and justify the actions that we take are going to increase. We need to embrace technology and make use of what it can offer, but, at the same time, we must make sure that it is working for us rather than forcing us to become slaves to the machine.

Reference

NCC (1986). *A Review of event record system and proposed project recoding System.* Unpublished report, Nature Conservancy Council, Peterborough, UK.

Glossary

Action plan A plan of action for a specific period of time containing several individual projects that describe specific actions. The information contained in the individual projects is aggregated to produce a wide variety of work and resource plans.

Adaptive Management A cyclical, adaptable management process which allows site management to: respond to natural dynamic processes; accommodate the legitimate interests of others; adapt to the ever-changing political and socio-economic climate; and, in the long term, succeed, despite uncertain and variable resources.

Anthropocentric The idea that humans are the central or most significant species; also called instrumental values.

Anthropogenic Something of human origin; the consequence of a human action or intervention.

Attribute An attribute is a characteristic of a feature that can be monitored to provide evidence about the condition of the feature.

Audit A critical examination of the performance of the plan, or a part of the plan, so as to measure the quality of the plan and its implementation, carried out by the management organisation (internal audit) or by an independent authority not directly associated with the site (external audit), usually at the invitation of the management organisation.

Evaluation Evaluation is simply the means of identifying, or confirming, which of the features on a site should become the focus for the remainder of the planning process.

Factor A factor is anything that has the potential to influence or change a feature, or to affect the way in which a feature is managed. These influences may exist, or have existed, at any

M. Alexander, *Management Planning for Nature Conservation: A Theoretical Basis & Practical Guide*, DOI 10.1007/978-94-007-5116-3,

time in the past, present or future. Factors can be natural or anthropogenic in origin, and they can be internal (on-site) or external (off-site).

Favourable Conservation Status (FCS) FCS is the desired status of a habitat or species, at any geographical scale from its entire geographical range to a defined area within a site. Although the concept of FCS originates in international and European treaties and directives, it is a concept that can be used for any wildlife management plan anywhere.

Feature Nature conservation features can be a habitat, a community or a population. Other features of interest can include geological, geomorphological, archaeological and historical features.

IUCN Protected Area An area of land and/or sea especially dedicated to the protection and maintenance of biological diversity, and of natural and associated cultural resources, and managed through legal or other effective means. (Protected areas are categorised according to their primary management objective.)

The IUCN Protected Area Categories System IUCN protected area management categories classify protected areas according to their management objectives.

Category: Ia Strict Nature Reserve

Category: Ib Wilderness Area

Category: II National Park

Category: III Natural Monument or Feature

Category: IV Habitat/Species Management Area

Category: V Protected Landscape/ Seascape

Category: VI Protected area with sustainable use of natural resources

Management Management is about taking control to achieve a desired outcome. ‘Control’ does not necessarily imply taking an action. It can, for example, mean ‘enabling’ a process.

Monitoring Surveillance undertaken to ensure that formulated standards are being maintained

NCR criteria The UK Nature Conservation Review (NCR) criteria are recognised as the standard or conventional approach to identifying important nature conservation sites, and are also used as a basis for identifying biological site features. They are: size, diversity, naturalness, rarity, fragility, typicalness, recorded history, position in an ecological/geographical unit, potential value, intrinsic appeal.

Objective The description of something that we want to achieve.

Precautionary Principle Where there are threats of serious or irreversible damage, lack of full scientific certainty shall not be used as a reason for postponing cost-effective measures to prevent environmental degradation.

Project A project is a clearly defined and planned unit of work.

Rationale The rationale is the process of identifying, in outline, the most appropriate management for the various site features.

Recording Making a permanent and accessible record of significant activities (including management), events and anything else that has relevance to the site.

Site A site is the area covered by a management plan. It can vary in size from less than a hectare to a large National Park covering many square kilometres. The term is used synonymously with area.

SMART Objectives **S**pecific, **M**easurable, **A**chievable, **R**elevant, **T**ime-based

Specified Limits Specified limits define the degree to which the value of a performance indicator is allowed to fluctuate without creating any cause for concern.

Stakeholder A stakeholder is any individual, group, or community living within the influence of the site or likely to be affected by a management decision or action, and any individual, group or community likely to influence the management of the site.

Surveillance Making repeated standardised surveys in order that change can be detected.

Survey Making a single observation to measure and record something.

Wilderness IUCN definitions: Large area of unmodified or slightly modified land, and/or sea, retaining its natural character and influence, without permanent or significant habitation, which is protected and managed so as to preserve its natural condition. And:Ecosystems where, since the industrial revolution (1750), human impact (a) has been no greater than that of any other native species, and (b) has not affected the ecosystem's structure. Climate change is excluded from this definition.

Zones Sites may be divided into zones to meet a wide variety of purposes, for example, to describe management actions or to guide or control a number of activities.

Index

M. Alexander, *Management Planning for Nature Conservation: A Theoretical Basis & Practical Guide*, DOI 10.1007/978-94-007-5116-3,